Y0-AAA-481

**Progress in Mathematics**
**Vol. 11**

**Edited by**
**J. Coates and**
**S. Helgason**

Birkhäuser
Boston · Basel · Stuttgart

# 18th Scandinavian Congress of Mathematicians

## Proceedings, 1980

**Erik Balslev, editor**

**1981**

**Birkhäuser**
**Boston · Basel · Stuttgart**

Editor:

Erik Balslev
Matematisk Institut
Universitetsparken
Ny Munkegade
8000 Aarhus C
Denmark

Library of Congress Cataloging in Publication Data
Skandinaviske matematikerkongres (18th : 1980 : Århus, Denmark)
18th Scandinavian Congress of Mathematicians proceedings, 1980.
(Progress in Mathematics ; 11)
Bibliography
1. Mathematics--Congresses. I. Balslev, Erik. II. Title. III. Series: Progress in mathematics (Boston, Mass.) ; 11.
QA1.S5 1980 510 81-1615
ISBN 3-7643-3040-6 AACR2

CIP — Kurztitelaufnahme der Deutschen Bibliothek
Scandinavian Congress of Mathematicians (18, 1980, Århus):
Proceedings/18. (Eighteenth) Scandinavian Congress of Mathematicians: 1980
Edited by Erik Balslev
Boston:Basel:Stuttgart:Birkhäuser, 1981.
(Progress in Mathematics:11)
ISBN 3-7643-3040-6
NE: Balslev, Erik (Hrsg.): GT


ISBN 3-7643-3040-6
Printed in USA

CONTENTS

# PREFACE

The 18th Scandinavian Congress of Mathematicians took place in Aarhus during the week 18th - 22nd of August 1980.

The program consisted of 10 survey lectures (75 minutes) and 3 parallel series of 9 more specialized lectures (45 minutes). The papers in the present volume cover most of the lectures given at the congress. The material contained in the survey lectures of S.Helgason and R.Phillips and the specialized lectures of T.Lindström, A.duPlessis and V.Thomée has been published elsewhere. Abstracts of the two latter talks are included in a collection of abstracts published in connection with the congress (Matematisk Instituts Various Publications Series No. 33 , Aarhus 1980).

Any article with several authors is listed under the name of the author who presented the work at the congress.

Financial support of the congress by the following agencies is gratefully acknowledged:

Statens Naturvidenskabelige Forskningsråd,
Aarhus Universitets Forskningsfond,
Dansk Matematisk Forening,
Livs- og Pensionsforsikringsselskaberne,
De forenede Telefonselskaber, Jydsk Telefonselskab.

Erik Balslev

# SURVEY LECTURES

# BMO - 10 years' development

Lennart Carleson

## Introduction.

1. The space BMO - functions of bounded mean oscillation - was introduced in 1961 by John and Nirenberg [20]. The definition can be given in the following way. Let $f(x)$ be a measurable function on $R^n$. For every cube $Q$ with faces parallel to the coordinate planes let $m_Q$ be the median of $f$, i.e. the largest number so that $m(\{x \mid x \in Q,\ f(x) \leq m_Q\}) \leq \frac{1}{2} mQ = \frac{1}{2}|Q|$ say. Then $f \in BMO$ if there are numbers $\alpha < \frac{1}{2}$ and $\lambda < \infty$ so that for all $Q$

$$(1.1) \qquad m(\{x \mid x \in Q,\ |f(x) - m_Q| > \lambda) < \alpha |Q| .$$

John and Nirenberg used this concept in studies of problems concerning partial differential equations. The reason why the class was useful is that the definition is self-improving. If we assume (1.1), not only does $f \in L^p_{loc}$ for all $p$ but we even have

$$(1.2) \qquad m(\{x \mid x \in Q,\ |f - m_Q| > \lambda) \leq \text{Const. } e^{-\frac{\lambda}{\varepsilon}}$$

for some $\varepsilon > 0$. For later use, let $\varepsilon(f)$ be $\inf \varepsilon$.

Let $E_Q(f)$ be the mean value of f over Q. It is now clear that we can use $E_Q(f)$ instead of $m_Q$ in (1.1) and (1.2) and define the following norm

$$\|f\|_*^2 = \sup_Q \frac{1}{|Q|} \int_Q \left| f(x) - E_Q \right|^2 dx.$$

In this way, BMO becomes a (non-separable) Banach-space of functions modulo constants and BMO$\supset L^\infty$/consts.

The intensive development of the theory started with Fefferman's fundamental paper [12](1971) and the purpose of this lecture is to summarize what I consider to be the most interesting aspects of this 10-year-development. The selection is obviously personal and this should not be considered as a complete survey. Let me mention that interesting results have been obtained by Swedish young mathematicians - Dahlberg, Strömberg and S. Janson in particular. I should have liked to report more on these results but time does not permit this. I have had the great advantage to discuss this exposition with P. Jones and am very grateful for his comments and assistance.

## I. Representation theorems for BMO

2. Fefferman's result [12] can in the case $n = 1$ be formulated as follows:

<u>Theorem 1</u>. $f(x) \in \mathrm{BMO}(R^1)$ *if and only if*

$$f(x) = \varphi_0(x) + \int \frac{\varphi_1(t)}{x-t}\, dt, \qquad \varphi_i \in L^\infty.$$

The singular integral is the Hilbert transform $H\varphi_1$. It can be interpreted in many ways; we can e.g. first take $\varphi_1 \in C_0^\infty$ and then consider closures in $L^2$.

An immediate consequence is the characterization of the dual of the Hardy-space $H^1$. By the classical M. Riesz theorem we know that for $1 < p < \infty$, $Hf \in L^p$ iff $f \in L^p$. The Hardy space of analytic functions in $y > 0$, belonging to $L^p$ on $y = 0$ is therefore the same as $L^p$ and has $H^q$ as dual space. However for $p = 1$ this is false and the dual of $H^1$ was not known before Fefferman's paper.

<u>Theorem 2</u>. $(H^1)^* = \mathrm{BMO}$.

Theorem 1 can be generalized to $R^n$. We then need several singular transforms. The natural candidates are the Riesz-transforms

$$R_j\varphi = \varphi * \frac{x_j}{|x|^{n+1}}$$

and we have

<u>Theorem 1'</u>. $f(x) \in \mathrm{BMO}(R^n)$ *if and only if*

$$f(x) = \varphi_0(x) + \sum_1^n R_j(\varphi_j), \qquad \varphi_j \in L^\infty.$$

If $H^1(R^n)$ is defined as the space for which $f$, $R_j f \in L^1$, $j = 1,\ldots,n$, then Theorem 2 holds for $R^n$.

Although $R_j$ are natural basic singular operators - mainly because of the Stein-Weiss theory [29] - they are not the only candidates when $n \geq 2$. An interesting question which is still open, is to describe which singular operators $S_1,\ldots,S_m$ can be used in Theorem 1'. This is intimately connected with another open problem: to give a construction of $\varphi_j$ in Theorem 1'. I shall return to this later.

Fefferman's proof is based on the following *extension-property* of $f \in BMO$. Let (C) denote the class of positive measures in $R^n \times (y > 0)$ such that for every cube Q in $y \geq 0$ with one face on $y = 0$

$$\text{(C)} \qquad \mu(Q) \leq \text{Const.}(\text{side }(Q))^n.$$

Then we have:

Theorem 3. $f \in BMO(R^n)$ *iff* $y|\nabla u|^2 dxdy \in (C)$ *where* $u(x;y)$ *is the harmonic extension of* $u$ *to* $y > 0$.

This result should be compared to the following recent extension theorem of Varopoulos [31], [32].

Theorem 4. $f \in BMO(R^n)$ *iff there exists* $F(x;y)$, $y > 0$ *so that* $F(x;y) \in C^\infty(y > 0)$ *and* $|\nabla F| dxdy \in (C)$ *and* $\lim_{y\to 0} F(x;y) = f(x)$ *a.e.*

It should be remarked that the harmonic extension in Theorem 3 does not in general work.

3. In 1971 another fundamental paper [2] in this area appeared. In this Burkholder-Gundy-Silverstein proved that if $g(x) \in L^1(R^1)$ and $u(x;y)$ is the harmonic extension to $y > 0$ then $g(x) \in H^1$ if and only if $\sup_{|x-\xi|<y} |u(\xi,y)| \in L^1$. The "only if" part is classical Hardy-Littlewood-theory but the "if" part was a completely new result. Later Fefferman-Stein [13] proved that the result holds in $R^n$ and that we need only take $\xi = x$. In the result we may even replace the Poisson-kernel P in $u(x,y) = P_y * g$ by a rather arbitrary positive rapidly decreasing function h.

From the BMO-side this result follows from the following representation theorem ([4] and later papers, in particular by Uchiyama [30]).

Theorem 5. $f(x) \in BMO(R^n)$ *if and only if*

$$f(x) = \int h_{y(t)}(x-t)\varphi(t)dt, \quad \varphi \in L^\infty.$$

This opens the road to a probabilistic theory of BMO (see e.g. Garsia [16]) for martingales. Let me only discuss the simplest case.

Denote by $\omega$ dyadic cubes in $R^n$ and by $S_m$ functions constant on dyadic cubes $\omega$ of side $2^{-m}$. We say that

$f \in BMO(D)$ (D = dyadic) if

$$\sup_{\omega} \frac{1}{|\omega|} \int_{\omega} |f(x) - E_{\omega}(f)|^2 dx < \infty.$$

If $g \in L^1_{loc}$ we can define

$$g_m(x) = \frac{1}{|\omega|} \int_{\omega} g(\xi) d\xi, \quad x \in \omega.$$

Then $g_m \in S_m$ and in relation to our discussion above a natural definition is that $g \in H^1(D)$ if $\sup_m |g_m| \in L^1$. We now have

Theorem 6. $H^1(D)^* = BMO(D)$.

We should note that $BMO(D) \supset_{\text{strictly}} BMO(R^n)$.

However there is a close relation between the two cases. The following result was recently proved by Garnett-Jones [15].

Theorem 7. $BMO(R^n) = \lim_{M\to\infty} \int_{|\xi_j|<M} BMO(D+\xi) \frac{d\xi}{(2M)^n}$

*(where* $(D+\xi)$ *stands for a translation of the dyadic system by the vector* $\xi$*).*

This result gives analogous results for $H^1$ (see Davis [10]).

4. If $g(x) \in L^p(T)$, $p > 1$, it is well-known that $g(x)$ can be expanded in a series of Haar-functions $h_{\omega}$,

$$g(x) = \sum_{\omega} c(\omega)h_{\omega},$$

where the Haar functions are indexed by the dyadic intervals $\omega$ that support them. Every series $\sum \gamma(\omega)h_{\omega}$ with $|\gamma(\omega)| \leq |c(\omega)|$ also belongs to $L^p$ and $h_{\omega}$ is called an unconditional basis [28].

If we take $f(x) \in BMO(D)$ it follows immediately that

$$\sum_{\omega \subset \sigma} |c(\omega)|^2 \leq \text{Const.}\ |\sigma|. \tag{4.1}$$

Conversely (4.1) characterizes BMO(D).

The corresponding result for BMO(T) is the following [5]. For a function $\varphi(x) \in \text{Lip}\ 1$ with support in $(-\delta, 1+\delta)$ let $\varphi_{\omega}$ be $\varphi$ translated and scaled to the dyadic interval $\omega$ and normalized with $L^2$-norm 1.

Theorem 8. *There is a function* $\varphi$ *so that* $f \in BMO$ *iff*

$$f = \sum c(\omega)\varphi_{\omega},$$

*with convergence in* $L^2$, *where* $c(\omega)$ *satisfies* (4.1).

It was recently proved by P. Wojtaszczyk [33] that the same result holds for the (orthogonal) Franklin system. We obtain Maurey's result [26], that $H^1$ has an unconditional basis. This paper by Maurey is the fundamental one in this area.

The functions $\varphi_{\omega}$ in Theorem 8 have mean value zero. We now instead start from $\psi$ on $(-\delta, 1+\delta)$ which is *positive* and define $\psi_\omega$ similarly but with normalization $\int \psi_\omega = 1$. Then we have the following result [14].

Theorem 9. $f \in \mathrm{BMO}$ *if and only if* $f$ *can be represented*

$$f(x) = \sum a(\omega)\psi_\omega,$$

*where*

$$\sum_{\omega \subset \sigma} |a(\omega)| \leq C|\sigma|.$$

Let me end this exposition by mentioning the following illuminating representation theorem by Coifman and Rochberg [8]:

Theorem 10. $f(x) \in \mathrm{BMO}(R^n)$ *if and only if*

$$f(x) = \alpha \log F^*(x) - \beta \log G^*(x) + \varphi(x)$$

*where* $F, G \geq 1$ *and* $\in L^1(\mathrm{loc})$ *and* $F^*(0), G^*(0) < \infty$ *and* $\varphi(x) \in L^\infty$.

## II. Applications

### A. *Complex analysis*

5. The BMO technique has been of fundamental importance for understanding $H^\infty$ for the upper half plane and

in particular for the solution of $\bar{\partial}u = h$ with $u \in L^{\infty}$. The most spectacular is the following result of T. Wolff.

Theorem 11. *If* $h$ *satisfies*

$$\left.\begin{array}{r} |h|^2 y\,dx\,dy \\ |\partial h|\,dx\,dy \end{array}\right\} \in (C)$$

*then* $\bar{\partial}u = h$ *has a bounded solution.*

From this result one easily gets a proof of the Corona theorem. It should be observed that the result is proved only in the plane.

P. Jones [21], [23] has made a systematic study of the $\bar{\partial}$-problem. This has among other things enabled him to give a constructive proof of Theorem 1 using $H^{\infty}$-interpolation.

In connection with conformal mappings let us note the following two results.

Theorem 12. *If* $\varphi(z)$ *is univalent in* $|z| < 1$ *and* $\neq 0$ *then* $\log \varphi(e^{i\theta}) \in \mathrm{BMO}(T)$.

Theorem 13. *If* $\varphi$ *maps* $|z| < 1$ *onto a Lipschitz-domain, then* $\log|\varphi'(e^{i\theta})| \in \mathrm{BMO}(T)$.

Theorem 12 is due to A. Baernstein [1] and Theorem 13 follows immediately from Theorem 1 and will be used later.

Let me finish this part with the following recent result of Koosis [25].

Theorem 14. *Given* $w(\theta) > 0$, $w \in L^1(T)$, *Then there is a set* E *of positive measure such that*

$$(5.1) \qquad \int_E |Hf|^2 d\theta \le \text{Const.} \int f^2(\theta) w(\theta) d\theta$$

*iff* $w^{-1} \in L^1$.

A similar result holds for any dimension and any singular transform and any p, $1 < p < \infty$, instead of 2 or the maximal function $f^*$ (for p the condition reads: $\int \omega^{\frac{1}{p-1}} d\theta < \infty$).

B. *Weighted inequalities*

6. The inequality (5.1) is an example of $L^2$-inequalities with different weights. The same problem with the same weight was solved in [18] by Helson-Szegö.

Theorem 15. *The inequality*

$$(6.1) \qquad \int (H\varphi)^2 \omega dx \le \text{Const.} \int \varphi(x)^2 \omega(x) dx$$

*holds for all* $\varphi \in C_0^\infty(R)$ *if and only if*

$$\omega(x) = \exp\{b_0(x) + \frac{\pi}{2} Hb_1(x)\}$$

*where* $b_0 \in L^\infty$ *and* $|b_1(x)| < 1$.

This result shows that inequalities of this type are intimately connected with $\log \omega$ as a function in BMO. This relationship has now been completely clarified through the work of Muckenhaupt, Hunt, Wheeden and Fefferman-Coifman [19], [6]. It turns out that the weights satisfying

$$\int \left\{ \begin{matrix} f^*(x)^p \\ Sf(x)^p \end{matrix} \right\} \omega(x)\,dx \leq C \int f(x)^p \omega(x)\,dx \tag{6.2}$$

are the same for the maximal function $f^*$ and any singular operator $S$.

If we take a cube $Q$ and choose $f(x) = \omega(x)$, $x \in Q$, $f = 0$ otherwise we find for $f^*$ the necessary condition

$$\left(\frac{1}{|Q|}\int_Q \omega(x)\,dx\right)\left(\frac{1}{|Q|}\int_Q \omega(x)^{\frac{1}{1-p}}\right)^{p-1} \leq C \tag{$A_p$}$$

<u>Theorem 16</u>. $(A_p)$ *is necessary and sufficient for every inequality* (6.2).

The condition $(A_p)$ needs some clarification. First, large and small values of $\omega(x)$ do not interact. If $m_Q$ is the median, $(A_p)$ means

$$\begin{cases} \dfrac{1}{|Q|}\displaystyle\int_Q \omega(x)\,dx \leq Cm_Q \\ \dfrac{1}{|Q|}\displaystyle\int_Q \omega(x)^{1-p}dx \leq Cm_Q^{1-p} \end{cases}$$

so large and small values are regulated separately and differently. If $\varphi(x) = \log \omega(x)$ we immediately see that $\varphi \in \mathrm{BMO}$. As a matter of fact [14],

$$\inf p = 1 + \varepsilon(\varphi)$$

if $e^{\pm\varphi} \in (A_p)$ and $\varepsilon(\varphi)$ was defined in the introduction.

For $n = 1$ we know via Theorem 15 when $\varepsilon(\varphi) = 1$. The corresponding problem for $p \neq 2$ and $n \geq 2$ is (and will probably remain) unsolved but Garnett-Jones [14] have proved that

$$\inf \sup_{j \geq 1} \|\varphi_j\| \approx \varepsilon(\varphi)$$

where $\varphi_j$ are taken from Theorem 1'.

The limiting cases for $(A_p)$ are:

$(A_1)$ $\qquad \omega^*(x) \leq C\omega(x)$

$(A_\infty)$ $\qquad \dfrac{\omega(E)}{\omega(Q)} \leq C\left(\dfrac{|E|}{|Q|}\right)^{\delta}$ $\quad$ for some $\delta > 0$

where $\omega(E) = \int_E \omega dx$. Observe that if $p < q$ then $(A_p) \subset (A_q) \subset (A_\infty)$ and actually (Muckenhaupt, see [6])

$$(A_\infty) = \cup (A_p).$$

The condition $(A_1)$ is most easily understood. We have the following result (Jones [23]).

Theorem 17. $(A_p) = (A_1) \cdot (A_1)^{1-p}$.

## C. *Quasi-conformal mappings*

7. In the proofs of the sufficiency of the $(A_p)$-condition inverse Hölder inequalities are very essential and $(A_\infty)$ implies

$$\left(\frac{1}{|Q|}\int_Q \omega(x)^{1+\delta}dx\right) \leq C\left(\frac{1}{|Q|}\int_Q \omega dx\right)^{1+\delta}$$

for some $\delta > 0$. It is well known (Gehring [17]) that these inequalities are fundamental also in quasi-conformal mappings. The connection is expressed in the following theorem (Riemann [27])

Theorem 18. *An* ACL-*homeomorphism* $h: R^n \to R^n$, $n \geq 2$ *is quasi-conformal if and only if* $\{\varphi \circ h\}$, $\varphi \in$ BMO, *is again the space* BMO.

On the line it was noted by Jones that the quasi-conformal ACL-homeomorphisms are those which satisfy the $(A_\infty)$ condition for $\log h'$.

The connection with quasi-conformal mappings goes further. We can speak about BMO$(\Omega)$ for an open set $\Omega \subset R^n$ by restricting ourselves to $Q \subset \Omega$. Jones [22] has proved

Theorem 19. BMO$(\Omega)$ = BMO$|_\Omega$ *for a Jordan domain* $\Omega \subset R^2$, *if and only if* $\partial\Omega$ *is a quasi-circle.*

He has also described the general case as follows. If $f \in BMO(R^n)$ and Q and Q' are adjacent dyadic cubes with side (Q) = 2 side (Q') then

$$(7.1) \qquad |E_Q(f) - E_{Q'}(f)| \leq C\|f\|_*.$$

A similar result holds for $BMO(\Omega)$ if Q and Q' have distance to the boundary $\leq$ sides. If $Q_1$ and $Q_2$ are two dyadic cubes in $\Omega$ (7.1) gives a restriction on $|E_{Q_1}(f) - E_{Q_2}(f)|$ in terms of the length of a chain connecting the two cubes. There is a similar - more stringent-condition for $R^n$. The necessary and sufficient condition for extensions for an arbitrary $\Omega \subset R^n$ is that these two conditions are comparable.

D. *General domains*

8. The maximal function definitions of $H^1$ discussed in Section 3 generalizes to domains where surface area of the boundary and normals have appropriate definitions. A general type of such domains are Lipschitz domains $\Omega$ where the boundary is locally defined by a Lipschitz function. We now have two choices of $H^1$: we can either assume that the maximal function is summable with respect to harmonic measure $\omega$ or with respect to surface area $\sigma$ on $\partial\Omega$. For the first choice, Dahlberg [9] has shown that Theorem 2 remains valid if we define BMO on $\partial\Omega$ by inequalities

$$\int_Q |f-c_Q|\,d\omega \leqq \text{Const. } \sigma(Q).$$

For the second choice, Fabes, Kenig and Neri [11] have shown the same result with $\sigma$ and $\omega$ permuted.

In 1977 Calderòn [3] proved that the Calderòn-Zygmund theory remained valid for singular integrals defined on continuously differentiable curves in the plane. This solved a long open problem. A precise formulation can be given as follows.

Let $\Gamma$ be a continuously differentiable curve which extends to infinity and separates the plane into two domains $\Omega^+$ and $\Omega^-$. To avoid any difficulties at $\infty$, let us assume that $\Gamma$ coincides with straight lines sufficiently far out. Calderòn's result implies that if $f \in L^q(\Gamma)$, then there exist $f^+$ and $f^-$ analytic in $\Omega^+$ resp. $\Omega^-$ so that on $\Gamma$ $f = f^+ + f^-$.

If $F^+$ and $F^-$ map the upper and lower half-planes, conformally onto $\Omega^+$ resp. $\Omega^-$, then by Theorem 13, $\log |F^{\pm}{}'| \in \text{BMO}$ and actually have $\varepsilon = 0$.

By looking at the images under $(F^{\pm})^{-1}$ of one point on $\Gamma$ we get a 1-1 map $x \leftrightarrow h(x)$ of $R$ and $\log h' \in \text{BMO}$ and $\varepsilon(h') = 0$. Some rather standard considerations show that the following theorem by Coifman-Meyer [7] implies the Calderòn theorem.

<u>Theorem 20</u>. *There is a universal constant* $\eta > 0$ *such that if* $\|\log h'\|_* < \eta$ *then every* $f \in BMO(R)$ *can be written*

$$f(x) = f^+(x) + f^- \circ h(x)$$

*when* $f^\pm \in BMO(R)$ *and have analytic extensions to the upper, resp. lower half planes.*

For the proof, we wish to show that the operator $U_h = f \circ h$ has a commutator with the Hilbert transform H which satisfies (in $L(BMO, BMO)$)

$$\|H - U_h H U_{h^{-1}}\| \leq C \|\log h'\|_{BMO}$$

and this is done by making a deformation $0 \leq t \leq 1$, $h_t' = (h')^t$ to the identity. It is interesting that this part of the approach resembles the fundamental idea in Calderõn's proof and the problem whether the result holds for arbitrary Lipschitz curves (without the restriction on the oscillation of the normal, represented by $\eta$ in Theorem 20) remains open.

REFERENCES

[1] BAERNSTEIN, A., Univalence and bounded mean oscillation, Michigan Math. J. 23(1976), 217-223.

[2] BURKHOLDER, D.L., GUNDY, R., and SILVERSTEIN, M., A maximal function characterization of the class $H^p$, Trans. Amer. Math. Soc. 157(1971), 137-153.

[3] CALDERON, A.P., On the Cauchy integral on Lipschitz curves and related operators, Proc. Nat. Acad. Sci. U.S.A. 74(1977), 1324-1327.

[4] CARLESON, L., Two remarks on $H^1$ and BMO, Adv. in Math. 22(1976), 269-277.

[5] --- , An explicit unconditional basis in $H^1$, Institut Mittag-Leffler Report No. 2, 1980.

[6] COIFMAN, R.R. and FEFFERMAN, C., Weighted norm inequalities for maximal functions and singular integrals, Studia Math. 51(1974), 241-250.

[7] COIFMAN, R.R. and MEYER, Y., Le théorème de Calderòn par les méthodes de variable réelle, Comptes Rendus Acad. Sci. S. A 289(1979), 425-428.

[8] COIFMAN, R.R. and ROCHBERG, R., Another characterization of BMO, Proc. Amer. Math. Soc. 79(1980), 249-254.

[9] DAHLBERG, B., A note on $H^1$ and BMO, A tribute to Åke Pleijel, 23-30.

[10] DAVIS, B., Rearrangement inequalities for Hardy spaces, to appear in Trans. Amer. Math. Soc.

[11] FABES, E., KENIG, C, and NERI, U., Carleson measures, $H^1$-duality and weighted BMO in non-smooth domains (preprint).

[12] FEFFERMAN, C., Characterizations of bounded mean oscillation, Bull. Amer. Math. Soc. 77(1971), 587-588.

[13] --- , and STEIN, E.M., $H^p$ spaces of several variables, Acta Math. 129(1972), 137-193.

[14] GARNETT, J. and JONES, P.W., The distance in BMO to $L^\infty$, Annals of Math 108(1978), 373-393.

[15] --- , BMO and dyadic BMO (preprint).

[16] GARSIA, A., Martingale inequalities, W.A. Harris, 1973.

[17] GEHRING, F.W., The $L^p$ integrability of the partial derivatives of a quasiconformal mapping, Acta Math. 130(1973), 265-277.

[18] HELSON, H. and SZEGÖ, G., A problem in prediction theory, Ann. Math. Pure Appl. 51(1960), 107-138.

[19] HUNT, R.A., MUCKENHAUPT, B., and WHEEDEN, R.L., Weighted norm inequalities for the conjugate function and Hilbert transform, Trans. Amer. Math. Soc. 176(1973), 227-251.

[20] JOHN, F. and NIRENBERG, L., On functions of bounded mean oscillation, Comm. Pure Appl. Math. 14(1961), 415-426.

[21] JONES, P.W., Carleson measures and the Fefferman-Stein decomposition of BMO($\mathbb{R}$), Annals of Math. 111(1980), 197-208.

[22] --- , Extension theorems for BMO, Indiana Math. J. 29(1980), 41-66.

[23] --- , Factorization of $A_p$ weights, Annals of Math. 111(1980), 511-530.

[24] --- , $L^\infty$ estimates for the $\bar{\partial}$ problem (preprint).

[25] KOOSIS, P., Moyennes quadratiques pondérées de fonctions périodiques et de leurs conjuguées harmoniques, to appear in Comptes Rendus Acad. Sci. Paris.

[26] MAUREY, B., Isomorphismes entre espaces $H_1$, to appear in Acta Mathematica 1980.

[27] REIMANN, H.M., Functions of bounded mean oscillation and quasiconformal mappings, Comm. Math. Helv. 49(1974), 260-276.

[28] SINGER, I., Bases in Banach spaces, Springer, 1970.

[29] STEIN, E.M. and WEISS, G., On the theory of harmonic functions of several variables I, Acta Math. 103(1960), 25-62.

[30] UCHIYAMA, A., A remark on Carleson's characterization of BMO, Proc. Amer. Math. Soc. 79(1980), 35-41.

[31] VAROPOULOS, N.Th., BMO functions and the $\overline{\partial}$ equation, Pacific J. Math. 71(1977), 221-273.

[32] --- , A remark on BMO and bounded harmonic functions, Pacific J. Math. 73(1977), 257-259.

[33] WOJTASZCZYK, P., The Franklin system is an unconditional basis in $H_1$ (preprint).

# FINE TOPOLOGY AND FINELY HOLOMORPHIC FUNCTIONS

Bent Fuglede

The fine topology has its origin in the concept of thin sets (ensembles effilés) as introduced by Brelot in 1939 [2]. His motive was to unify the notions of an irregular boundary point for the Dirichlet problem and an unstable point for the Keldych problem. In doing so, he also found it advantageous to replace the famous Wiener criterion by more manageable criteria related directly to subharmonic functions. Accordingly, Brelot called a set $E$ in $\mathbb{R}^n$ thin at a point $x \in \mathbb{R}^n$ if either $x$ is not a limit point of $E$, or else if

$$u(x) > \limsup_{\substack{y \to x \\ y \in E \setminus \{x\}}} u(y)$$

for some subharmonic function $u$ defined in a neighbourhood of $x$. (Since $u$ is upper semicontinuous, we always have $\geq$ here.)

Immediately after, H. Cartan observed, in a letter to Brelot, that an equivalent definition of thinness (in the case $x$ not in $E$) would be to say that $E$ is thin at $x$ if and only if the complement of $E$ is a neighbourhood of $x$ in the weakest topology on $\mathbb{R}^n$ making all subharmonic functions continuous. In dimension $n > 1$ this new topology is stronger than the Euclidean topology, and Cartan called it the fine topology.

These equivalent notions of thin sets and fine topology quickly proved a natural and important element in the study of more refined properties in potential theory. For quite some time, however, there was a tendency to consider the fine topology primarily as an artifice - a means of expressing results more intuitively than by direct reference to thin sets. (For instance, an irregular boundary point for a domain is nothing but a finely isolated point of the complement.)

This attitude was understandable in view of certain pathologies of the fine topology. Thus - although Hausdorff, completely regular, and Baire - the fine topology is not normal, and it does not satisfy the axioms of countability, nor has it the Lindelöf property. And worst of all, the only finely compact sets are the finite sets.

Gradually it was realized that the fine topology should be truly accepted as a topology in its own right, and that some of the pathologies could be compensated for. Thus Doob [6] discovered in 1966 the "quasi Lindelöf" property of the fine topology. It states that the union of any family of finely open sets differs only by a <u>polar</u> set from the union of some countable subfamily. A polar set is, so to speak, a potential theoretic null set. It is the same as a set of outer capacity 0 (hence, in particular, a Lebesgue null set). From the point of view of the fine topology, a polar set is the same as a finely discrete set.

In 1969 I showed that the fine topology is locally connected [7], and this became the starting point for developing a theory of <u>finely harmonic</u> functions, defined in finely open sets [8]. Such sets need not have any interior points in the usual sense. While compactness arguments play a crucial role in usual potential theory, for instance in establishing the boundary maximum principle for subharmonic functions, they had to be replaced by other methods in the study of finely harmonic functions, the lack of finely compact sets being irreparable as such. Instead were used lattice properties of subharmonic functions, and approximation arguments based ultimately on the discovery by Choquet [26] that any finely closed set differs from suitable closed subsets (in the usual sense) only by sets of arbitrarily small outer capacity. And subsequently, over the past 10 years, a large portion of classical potential theory has been extended to "fine potential theory".

As an illustration of the use of this fine potential theory let me mention the generalization of the classical theorem of F. Iversen to general subharmonic functions. In his thesis from 1914 [16], Iversen proved that, given a non-constant entire function $f$, there exists a continuous path in the complex plane along which $f(z) \to \infty$, that is, $\log|f(z)| \to +\infty$ as $z \to \infty$. He obtained this result in the course of his study of the inverse to a meromorphic function. It has long been known that a simpler proof can be based on the Phragmén-Lindelöf principle. As noted by Hayman, this latter proof carries over so asto produce an asymptotic path (even an infinite polygonal line) for any <u>continuous</u> subharmonic function $u$ defined in $\mathbb{R}^n$, thus replacing $\log|f|$ in Iversen's theorem. The main step was to show that $u$ must be unbounded in each component of any of the sets $[u > \lambda]$, and it was for this a result of Phragmén-Lindelöf type was needed.

The question arose whether such an asymptotic path exists for a general (not necessarily continuous) subharmonic function $u$ in the plane, or even in higher dimensions. The difficulty lies in the fact that the sets $[u > \lambda]$ ($\lambda$ constant) and their connectivity components are no longer open. Hayman and Talpur therefore replaced them by the sets $[u \geq \lambda]$, which are closed, owing to the upper semicontinuity of $u$, and so their components are continua. Using some deep analytical results by Hayman they settled the plane case in 1967, again with an infinite polygonal path, see [15].

In higher dimensions the problem was solved in 1974 by means of fine potential theory [10], but the asymptotic path was at the time highly non-rectifiable, being composed of arcs of Brownian paths. The subharmonic function $u$ is of course <u>finely continuous</u> by the very definition of the Cartan fine topology. The crucial sets $[u > \lambda]$ are therefore automatically finely open, and so are their fine components on account of the local connectedness of the fine topology. The argument involving the Phragmén-Lindelöf principle turned out to carry over, by consistent use of the fine topology and fine potential theory. To finish the proof it remained only to appeal to a result about arcwise connectedness of fine domains. In 1972 Nguyen-Xuan-Loc and T. Watanabe [19] had proved that there is a positive probability for a Brownian particle, starting at a point $x$ of a fine domain $U$, to reach into any prescribed fine neighbourhood of another point $y$ of $U$ before leaving $U$. And this property (which actually characterizes the fine domains among all finely open sets) easily implies the arcwise connectedness (in the usual sense) of any fine domain, and hence completes the proof of the existence of an asymptotic path for any subharmonic function on $\mathbb{R}^n$ in any dimension $n$.

Soon after this, Carleson [3] devised a direct proof, in principle elementary, but technically quite complicated, leading to an infinite polygonal path. His method, however, does not seem to extend to other cases of asymptotic paths which have been settled by use of fine potential theory.

The missing link in the complete success of treating questions about existence of asymptotic paths for subharmonic functions by the natural method of fine potential theory, was found in 1978 by

Lyons [17] (see also [13]), who proved that any two points of a fine domain $U$ can be joined by a usual finite polygonal line contained in $U$. Invoking this result we immediately obtain infinite polygonal asymptotic paths in the various generalizations of Iversen's theorem and similar results, see [10].

Let me pass to another application of fine potential theory, now in the complex plane. It is possible to introduce a notion of _finely holomorphic_ functions, defined in finely open sets, and extending the category of classical holomorphic functions in one complex variable. I should like to present here the newest and I believe most natural definition of finely holomorphic functions, and to describe some of their properties.

In the sequel $U$ will always denote a _fine domain_ in the complex plane $\mathbb{C}$, that is, a set which is open and connected in the fine topology in the plane.

A function $f: U \to \mathbb{C}$ is called _finely differentiable_ (in the complex sense) at a point $z_0 \in U$ if the fine limit

$$f'(z_o) = \underset{z \to z_o}{\text{fine lim}} \frac{f(z) - f(z_o)}{z - z_o}$$

exists. This is understood naturally as follows: To any $\varepsilon > 0$ there shall correspond a fine neighbourhood $V$ of $z_o$ in $U$ such that

$$\left| \frac{f(z) - f(z_o)}{z - z_o} - f'(z_o) \right| < \varepsilon \quad \text{for all} \quad z \in V .$$

This implies of course that $f$ is finely continuous at $z_o$. - Note that we use the _fine_ topology for the independent variable $z \in U$, but the _Euclidean_ topology on the target $\mathbb{C}$, that is, for the dependent variable.

By a known result on fine limits, this definition of fine differentiability amounts to saying that $z_o$ has a fine neighbourhood $V$ such that

$$f'(z_o) = (\text{usual}) \lim_{z \to z_o,\, z \in V} \frac{f(z) - f(z_o)}{z - z_o} .$$

<u>Definition</u>. A function $f: U \to \mathbb{C}$ is <u>finely holomorphic</u> if $f$ is finely differentiable as above at every point of $U$, and if the fine derivative $f': U \to \mathbb{C}$ is finely continuous (again from $U$ with the fine topology to $\mathbb{C}$ with the usual topology).

Actually, it turns out later that it would amount to the same to use the fine topology for both variables, that is, both on $U$ and on the target $\mathbb{C}$. We may express this by saying that $f$ is finely holomorphic if and only if $f$ is fine-to-fine differentiable and its derivative is fine-to-fine continuous. - The above definition thus copies the standard definition by Cauchy of a usual holomorphic function.

To get started we need a few lemmas.

<u>Lemma</u> 1. Let $f: U \to \mathbb{C}$ be finely holomorphic, and suppose that $f$ and $f'$ are continuous even in the <u>usual</u> topology. Then

$$f(b) - f(a) = \int_K f'(z)dz,$$

the integral being taken along any rectifiable arc $K \subset U$ from a point $a \in U$ to $b \in U$.

The proof depends on the fact that every fine neighbourhood of a point in the plane contains arbitrarily small circles (not disks) centered at the point. This property goes back to Lebesgue and Beurling. It does not generalize to higher dimensions, as shown by the complement of a so-called Lebesgue thorn. In view of this property the fine differentiability of $f$ at the generic point $z$ of the arc $K$ tells us something about the behavior of $f$ at certain points of $K$ near $z$, namely the points common to $K$ and the small circles contained in $V$, and this information turns out to be sufficient to carry out the proof of the lemma by a standard procedure.

The hypothesis in the lemma that $f$ and $f'$ be continuous in the usual topology is not a serious restriction in view of the following known lemma [12] relating the fine topology to the usual topology, and applied in the present case to a single function:

Lemma 2. If $f_n : U \to \mathbb{C}$, $n = 1,2,\cdots$, are finely continuous then every point of $U$ has a fine neighbourhood $U_o$ such that the restriction of each $f_n$ to $U_o$ is continuous in the usual (relative) topology on $U_o$.

Also, the consideration of rectifiable arcs $K$ contained in $U$ is acceptable in view of the polygonal connectedness of the fine domain $U$. We shall need the following stronger property valid in $\mathbb{R}^n$ for any $n$ and due to Lyons [17] (see also [13]):

Lemma 3. Every point $z$ of a fine domain $U$ has a fine neighbourhood $V$ in $U$ such that any two points $a,b$ of $V$ can be joined by a finite polygonal line $K$ contained in $U$ (not necessarily in $V$) and of length less than $\alpha|b-a|$, $\alpha > 1$ being given.

Lyons even showed that $K$ can be taken as the union of just two segments of equal length.

Theorem 4. A function $f: U \to \mathbb{C}$ is finely holomorphic if and only if every point of $U$ has a fine neighbourhood $V$ in $U$ on which $f$ coincides with some usual $C^1$-function $\tilde{f}: \mathbb{C} \to \mathbb{C}$ satisfying the Cauchy-Riemann equation $\bar{\partial}\tilde{f} = 0$ in $V$.

Proof: The "if part" is immediate. Conversely, suppose that $f$ is finely holomorphic. Since $f$ and $f'$ are finely continuous, the given point $z$ of $U$ has, by Lemma 2, a fine neighbourhood $U_o$ in $U$ such that the restrictions of $f$ and $f'$ to $U_o$ are continuous in the usual sense. Next, choose a fine neighbourhood $V$ of $z$ as in Lemma 3 (now with $U_o$ in place of $U$). Using Lemma 1 it is easy to verify the conditions for applying Whitney's extension theorem [25], and the result follows.

Obviously, by choosing $V$ bounded, we may arrange that $\tilde{f}$ has compact support: $\tilde{f} \in C^1_c(\mathbb{C})$. By Pompeiu's identity it then follows that

$$\tilde{f}(z) = \frac{1}{\pi}\int \frac{1}{z-\zeta}\,\bar{\partial}\tilde{f}(\zeta)\,d\lambda(\zeta), \quad \text{all} \quad z \in \mathbb{C},$$

where $\lambda$ denotes 2-dimensional Lebesgue measure. Since $\tilde{f} = f$ in $V$, and $\bar{\partial}\tilde{f} = 0$ in $V$, we have thus obtained the "only if part" of

Theorem 5. A function $f: U \to \mathbb{C}$ is finely holomorphic if and only if every point of $U$ has a fine neighbourhood $V$ in $U$ in which $f$ is representable as the Cauchy-Pompeiu transform

$$f(z) = \int \frac{1}{z-\zeta} \varphi(\zeta) d\lambda(\zeta), \qquad z \in V,$$

of a function $\varphi \in C_c(\mathbb{C})$ (or $L_c^p(\mathbb{C})$, $p > 1$) with $\varphi = 0$ on $V$.

The proof of the "if part" is based on the following lemma:

Lemma 6. Let $K \subset \mathbb{C}$ be compact. For any real number $\alpha$ the function $h_\alpha$ defined by

$$h_\alpha(z) = \int_{K \setminus U} |z-\zeta|^{-\alpha} d\lambda(\zeta), \qquad z \in U,$$

is finite and finely continuous in all of $U$. Every point of $U$ therefore has a fine neighbourhood $V$ in $U$ on which each $h_\alpha$ is bounded (viz. continuous, by Lemma 2, taking $V$ compact):

$$\sup_{z \in V} h_\alpha(z) < +\infty .$$

This lemma does not extend to higher dimensions. The proof uses the classical inequality (see [24])

$$\lambda(E) \leq \pi(\text{cap } E)^2$$

between Lebesgue measure and logarithmic capacity of a bounded Borel set $E$ in $\mathbb{C}$. Moreover one applies the following lemma, due to Lyons [17]:

Lemma 7. Let $z_o \in U$ be given. To any $\alpha > 0$ correspond a fine neighbourhood $U_\alpha$ of $z_o$ and a number $N_\alpha$ such that

$$\text{cap}^*(A_n(z) \setminus U) \leq 2^{-n\alpha}$$

for every $z \in U_\alpha$ and every $n \geq N_\alpha$, whereby $A_n(z)$ denotes the annulus

$$A_n(z) = \{\zeta \in \mathbb{C} \mid 2^{-n-1} \leq |\zeta - z| < 2^{-n}\}.$$

The proof of this lemma uses the Wiener criterion (applied to $\complement U$), but a separate argument is needed in order to show that the estimate holds uniformly for all $z$ in some fine neighbourhood $U_\alpha$ of $z_o$. (See also [14].)

In order to study local properties of finely holomorphic functions it suffices, according to Theorem 5, to consider the Cauchy-Pompeiu transform $f$ of a function $\varphi \in C_c(\mathbb{C})$ and to study $f$ in the fine interior of the set where $\varphi$ is $0$. In this way one obtains by application of Lemma 6 the following results:

Theorem 8. Every finely holomorphic function $f: U \to \mathbb{C}$ has fine derivatives of all orders everywhere in $U$, and these derivatives $f', f'', \cdots, f^{(n)}, \cdots$ are finely holomorphic in $U$.

Theorem 9. If $f: U \to \mathbb{C}$ is finely holomorphic then every point of $U$ has a fine neighbourhood $V$ in $U$ such that:

a) There exists a sequence of usual holomorphic functions $f_j$ in open sets $\omega_j \supset V$ such that each derivative $f_j^{(n)}$ approaches $f^{(n)}$ uniformly on $V$ as $j \to \infty$ $(n = 0,1,2,\cdots)$.

b) For any $m = 0,1,2,\cdots,$

$$\left| f(w) - \sum_{k=0}^{m-1} \frac{1}{k!} (w-z)^k f^{(k)}(z) \right| / |w-z|^m$$

is bounded as a function of $(z,w) \in V \times V$, $z \neq w$.

c) There exists a function $\tilde{f} \in C_c^\infty(\mathbb{C})$ such that, for all $m \geq 0$, $n > 0$,

$$\partial^m \tilde{f} = f^{(m)} \text{ in } V, \quad \partial^m \bar{\partial}^n \tilde{f} = 0 \text{ in } V.$$

Property b) thus tells us that the Taylor series of $f$, though generally divergent off the given point $z$, does represent $f$ asymptotically on $V$, even in a uniform sense.

Proof: To prove a) in Theorem 9 in the typical case where $f$ is the Cauchy-Pompeiu transform of a function $\varphi \in C_c(\mathbb{C})$ such that $\varphi = 0$ in $U$, choose $V$ as in Lemma 6. Taking $V$ compact, we may find a decreasing sequence of usual open sets $\omega_j$ shrinking to $V$. Writing

$$\varphi_j = 1_{\complement\omega_j} \varphi,$$

we have $\|\varphi - \varphi_j\|_{L^2} \to 0$ as $j \to \infty$ because $\varphi = 0$ in $V$. The Cauchy-Pompeiu transform $f_j$ of $\varphi_j$ is of course holomorphic in $\omega_j$, and its derivatives can be obtained there by differentiation under the integral sign. Hence a) follows by applying Lemma 6 and the Cauchy-Schwarz inequality, but $f^{(n)}(z)$ is at this stage defined by differentiation of $f(z) = \int (z-\zeta)^{-1}\varphi(\zeta)d\lambda(\zeta)$ under the integral sign.

From a) follows b), e.g. by repeated application of Lemma 3. Actually, this yields b) even with $f$ replaced by $f^{(n)}$ for any $n$. And from this extended version of b) follows c) immediately by Whitney's extension theorem [25]. Finally, c) implies easily Theorem 8 (including the identification of the above $f^{(n)}$ with the fine derivatives of $f$), and also the "if part" of Theorem 5.

At this point, let me mention some other definitions of finely holomorphic functions which have been proposed: A function $f: U \to \mathbb{C}$ (defined in a fine domain $U \subset \mathbb{C}$) was called finely holomorphic if

$1^\circ$. Every point of $U$ has a fine neighbourhood $V$ in $U$ such that the restriction of $f$ to $V$ is in $R(V)$, that is, $f$ can be approximated uniformly on $V$ by restrictions to $V$ of holomorphic functions in open neighbourhoods of V. (Fuglede, 1973 [12]).

$2^\circ$. $f$ is finely complex harmonic in $U$, and $\bar{\partial} f = 0$ holds a.e. in $U$ in the sense of stochastic differentiation. (Debiard and Gaveau, 1974 [4], [5]. Also used by Nguyen-Xuan-Loc [20], [21], [22] and by Lyons [17], [18]).

$3^\circ$. $f$ and $zf(z)$ are finely harmonic in $U$ (Lyons, 1978 [17]).

$4^\circ$. $f$ is finely continuous in $U$, and every point of $U$ has a fine neighbourhood $V$ on which $f$ agrees with some function $\tilde{f}$ in the Sobolev space $W_1^2(\mathbb{C})$ such that $\bar{\partial}\tilde{f} = 0$ a.e. in $V$. (Fuglede, 1979 [14]).

Fortunately, these definitions are all equivalent to the present one. The equivalence between $1^\circ$, $2^\circ$, and $3^\circ$ is due to Lyons (at one point jointly with O'Farrell).

Theorem 10. The set $\mathcal{O}(U)$ of all finely holomorphic functions $U \to \mathbb{C}$ is a subalgebra of the algebra $\mathbb{C}^U$ of all complex-valued functions on $U$. It is closed in the topology of finely local uniform convergence. In a usual domain the usual holomorphic and the finely holomorphic functions are the same.

Proof: $\mathcal{O}(U)$ is an algebra in view of the standard rules for differentiation of sums and products, applicable also in the present case of the fine topology on $U$. It is closed by virtue of Def. $3^o$ since fine harmonicity is preserved under the stated notion of convergence. The finely harmonic and the usual harmonic functions in a usual domain in $\mathbb{C}$ are the same [9], hence the final assertion.

Theorem 11. Every finely holomorphic function $f: U \to \mathbb{C}$ is fine-to-fine continuous and (if non-constant) fine-to-fine open.

Proof: By Def. $2^o$ or $3^o$, $f$ is (complex) finely harmonic, and so are similarly the powers $f^2, f^3, \cdots$ by Theorem 10. This shows that $f$ is a finely harmonic morphism of $U$ into $\mathbb{C}$, whence the two assertions by known properties of such morphisms, see [12].

The first proof of the fine-to-fine openness of a non-constant finely holomorphic function was given by Debiard and Gaveau [5], who used probabilistic methods.

Corollary. a) In the definition of fine holomorphy one could equivalently use the fine topology for both variables.

b) The composite $g \circ f$ of two finely holomorphic functions $f$ and $g$ is finely holomorphic, and $(g \circ f)' = (g' \circ f)f'$.

c) The inverse $g = f^{\circ -1}$ of an injective finely holomorphic function $f$ is finely holomorphic, and $g' = 1/(f' \circ g)$.

Some of the preceding results are due more or less to Debiard and Gaveau and to Lyons. They used stochastic differentiation, Lyons also uniform algebras. Methodically new here is notably the use of the Cauchy-Pompeiu transform, the key being Lemma 6.

Theorem 12. A finely holomorphic function $f: U \to \mathbb{C}$ is uniquely determined by its derivatives $f^{(n)}(z_o)$, $n = 0,1,2,\cdots$, at a given point $z_o$ of $U$.

Proof: We may suppose that $z_o = 0$ and that $f^{(n)}(0) = 0$, $n = 0,1,2,\cdots$. By Theorem 9b) there is a fine neighbourhood $V$ of $0$ such that $f(z)/z^n$ is bounded on $V$ for each $n$:

$$|f(z)| \leq a_n |z|^n, \qquad z \in V.$$

We may suppose in addition that $V$ is a fine domain contained in the unit disk $\Omega$, and further that $|f| \leq 1$ in $V$. Denote by

$$G^V(z,0) = \log(1/|z|) - \hat{R}^{CV}_{\log(1/|\zeta|)}(z)$$

the <u>Green function</u> on $V$ with pole at $0$, (the balayage operation $\hat{R}$ pertaining to $\Omega$), see [11]. Since $G^V(z,0) \leq \log(1/|z|)$, we get

$$\log(1/|f(z)|) \geq n\, G^V(z,0) - \log a_n .$$

Since $G^V(z,0)$ is finely continuous in $z$ and $G^V(0,0) = +\infty$ there is a finely closed fine neighbourhood $W_n$ of $0$ in $V$ such that

$$\tfrac{1}{2} n\, G^V(z,0) \geq \log a_n \quad \text{for } z \in W_n .$$

The function

$$u(z) := \log(1/|f(z)|) - \frac{1}{2}\, n\, G^V(z,0)$$

is therefore $\geq 0$ on $W_n$, in particular on the fine boundary of the finely open set $V \setminus W_n$ relative to $V$. Moreover $u$ is finely hyperharmonic in $V$, and $u \geq -p$ in $V$, where $p(z) := \frac{1}{2}\, n\, G^V(z,0)$ is a fine potential on $V$, see [11]. By a known "relative fine boundary minimum principle" [8, th. 10.8] we conclude that $u \geq 0$ in $V \setminus W_n$ and hence in all of $V$.

Since $G(z,0) > 0$ in $V$ we infer, letting $n \to \infty$, that $\log(1/|f(z)|) = +\infty$, i.e. $f(z) = 0$ for all $z \in V$. This implies that $f$ vanishes in all of $U$ since otherwise $f$ would be non-constant, and hence fine-to-fine open according to Theorem 11.

Theorem 13. A finely holomorphic function $f: U \to \mathbb{C}$ has at most countably many zeros (unless $f = 0$ identically).

Proof: It is easy to reduce the proof to the case where the closure $\overline{U}$ is contained in a disk $\Omega$ of diameter 1 and where $|f| \leq 1$ in $U$. For any finite set of zeros $z_j$ with corresponding multiplicities $m_j$ $(j = 1,\cdots,n)$ the function

$$g(z) := f(z)(z - z_1)^{-m_1} \cdots (z - z_n)^{-m_n}$$

extends to a function defined and finely holomorphic in all of $U$ on account of a result on removable singularities for finely holomorphic functions derived from a corresponding result for finely harmonic functions, see [8, §9.15]. The Green functions $G^U(z,z_j)$ with poles $z_j$ are fine potentials on $U$, and so is the affine combination $p(z) = \sum_{j=1}^{n} m_j G^U(z,z_j)$. From the relations

$$0 \leq \log(1/|f(z)|) = \sum_{j=1}^{n} m_j \log(1/|z - z_j|) + \log(1/|g(z)|)$$

follows therefore easily the Jensen type inequality

$$\sum_{j=1}^{n} m_j G^U(z,z_j) \leq \log(1/|f(z)|).$$

Since $G^U(z,z_j) > 0$, this inequality leads to the desired conclusion when applied to some $z \in U$ such that $f(z) \neq 0$.

There is no difficulty in extending the classification of a (finely) isolated singularity for a finely holomorphic function into regular points, poles, and essential singularities.

Theorem 14. A finely holomorphic function maps every (punctured) fine neighbourhood of an essential singularity onto the whole complex plane less a polar set.

This result of Picard type (which arose in conversation with Lyons and O'Farrell) extends and strengthens a result by Af Hällström and Kametani (see Noshiro 23,p.10]). The proof - like that of Kametani - uses Myrberg's theorem on the existence of a Green function on any plane domain of non-polar complement. It follows easily that $f^{-1}(w)$ is (countably) infinite for any $w \in \mathbb{C}$ off some polar set.

It is easy to see that Theorem 14 is sharp in the sense that every polar set can occur as the complement of the range of a finely holomorphic function with an isolated essential singularity.

Towards the end of my talk let me turn the eyes backward in time and touch upon the earliest and hitherto only attempt to extend 1-dimensional complex analysis beyond the usual domains in the complex plane or on a Riemann surface. I am thinking of Émile Borel's monogenic functions.

In 1892 - at the age of 21 - Borel had conceived the idea that it must be possible to continue certain analytic functions of a complex variable beyond their classical maximal domain of existence - their Weierstrass domain. This idea was almost a heretic one, for Weierstrass' beautiful concept of a global analytic function - das analytische Gebilde - had such a definitive character that it was hard to imagine any extension beyond it.

Poincaré had even constructed certain analytic expressions with the aim of disproving any possibility of such extensions. In his thesis from 1894 - at which Poincaré was his rapporteur - Borel opposed to this negative view, without yet being able to effectively construct a proper extension of the classical concept of an analytic function while retaining its distinctive property of unique continuation.

Borel illustrated his idea by simple examples like this: Consider on the unit circle $|z| = 1$ a dense sequence of points $z_n$, and form the series

$$\sum_n \frac{a_n}{z - z_n} ,$$

where $\Sigma|a_n| < +\infty$. This series clearly converges locally uniformly for $|z| < 1$ as well as for $|z| > 1$, and hence determines a holomorphic function $f_1$ in the open unit disk and a holomorphic function $f_2$ in the open exterior of the disk. It follows from a general result by Goursat that neither $f_1$ nor $f_2$ can be continued analytically across any arc of the unit circle. In fact, an easy direct estimate shows that each of the two functions has radial limit $\infty$ at each of the points $z_n$.

Nevertheless, Borel showed that if the coefficients $a_n$ decrease sufficiently rapidly then there exists a dense subset of the unit circle such that $f_1$ and $f_2$ are both representable, on any given finite segment from 0 through a point of that set, as a uniform limit of a certain sequence of polynomials derived from the given series in a natural way - a kind of over-convergence. In particular, this common extension of $f_1$ and $f_2$ is continuous, presumably even smooth. Borel therefore felt that the two holomorphic functions $f_1$ and $f_2$ - which from the Weierstrass point of view are totally unrelated - should be conceived as restrictions of one and the same function, defined and - in Cauchy's terminology - monogenic in the "domain" consisting of the complement of the unit circle plus a certain dense subset of the circle itself.

In the years which followed, Borel strived to give his ideas a more general form and a solid foundation. A first difficulty was to define a suitable notion of domain of definition for monogenic functions - a Cauchy domain, in Borel's terminology. Starting from a (possibly dense) sequence $(z_n)$ in a usual (say bounded) domain $A$ in the complex plane he constructed an increasing sequence of compact connected subsets $D_h$ of $A$, each obtained from $A$ by removing a sequence of disjoint disks with very rapidly decreasing radii. Under specific conditions as to the construction of these "swiss cheeses" $D_h$, he called their union $D$ a Cauchy domain.

Borel next introduced the notion of a monogenic function $f$ defined in such a Cauchy domain $D = \cup D_h$. He required that the restriction of $f$ to each $D_h$ should be continously differentiable in the complex sense, thus extending Cauchy's definition. He proved that every monogenic function in this sense is infinitely differentiable everywhere in its Cauchy domain $D$, this being understood to apply to the restriction of $f$ to each $D_h$.

His main result was that a monogenic function $f$ is uniquely determined from the sequence of its derivatives of all orders at a single point. Not every point of the Cauchy domain $D$ could serve as such a point of uniqueness, however, but Borel constructed a certain dense subset - a so-called reduced domain - such that every point of that is a point of uniqueness for every monogenic function defined on $D$. It was this result which above all justified his theory.

After 25 years, Borel had developed his ideas to a coherent theory, as exposed in his book "Leçons sur les fonctions monogènes" [1], published in 1917 with the assistance of his pupil Gaston Julia. Yet this part of Borel's extensive contributions to mathematics never met wide response, owing perhaps to the complexity and rather special character of the Cauchy domains. It did, however, give inspiration to the creation of the beautiful theory of quasi-analytic functions (of a real variable) by Denjoy, Carleman, Mandelbrojt, and Bang.

Borel declares in his book that there undoubtedly exist domains more general than his Cauchy domains on which a still further generalization of monogenic functions is possible, and that one will probably never succeed in fixing the exact limits beyond which such an extension is no longer possible.

The finely holomorphic functions, defined in a domain for the fine topology, do constitute such a extension - and a very natural one - of Borel's monogenic functions. It is in fact easily verified, by use of the Wiener criterion, that:

- Every Cauchy domain is also a fine domain, indeed a very special kind of fine domain since it differs only by a polar set from the initial, usual domain $A$.

- Every monogenic function on a Cauchy domain $D$ is finely holomorphic in $D$ less some polar set independent of the function (e.g. finely holomorphic in the union of the fine interiors of the sets $D_h$ used in the construction of $D$). From the point of view of the theory of finely holomorphic functions, the need for removing such a polar set explains why not every point of a Cauchy domain is a point of uniqueness for monogenic functions, although every point of a fine domain is a point of uniqueness for finely holomorphic functions.

## References

1. É. Borel: Leçons sur les fonctions monogènes uniformes d'une variable complexe.
Gauthier Villars, Paris 1917.

2. M. Brelot: Points irréguliers et transformations continues en théorie du potentiel.
J. de Math. 19 (1940), 319-337.

3. L. Carleson: Asymptotic paths for subharmonic functions on $R^n$.
Ann. Acad. Sci. Fennicae (A.I.) 2 (1976), 35-39.

4. A. Debiard & B. Gaveau: Potentiel fin et algèbres de fonctions analytiques.
J. Funct. Anal. 16 (1974), 289-304.

5. A. Debiard & B. Gaveau: Potentiel fin et algèbres de fonctions analytiques II.
J. Funct. Anal. 17 (1974), 296-310.

6. J.L. Doob: Applications to analysis of a topological definition of smallness of a set.
Bull. Amer. Math. Soc. 72 (1966), 579-600.

7. B. Fuglede: Connexion en topologie fine et balayage des mesures.
Ann. Inst. Fourier 21, 3 (1971), 227-244.

8. B. Fuglede: Finely harmonic functions.
Springer Lecture Notes in Mathematics 289. Berlin-Heidelberg-New York 1972.

9. B. Fuglede: Fonctions harmoniques et fonctions finement harmoniques.
Ann. Inst. Fourier 24, 4 (1974), 77-91.

10. B. Fuglede: Asymptotic paths for subharmonic functions.
Math. Ann. 213 (1975), 261-274.

11. B. Fuglede: Sur la fonction de Green pour un domaine fin.
Ann. Inst. Fourier 25, 3-4 (1975), 201-206.

12. B. Fuglede: Finely harmonic mappings and finely holomorphic functions.
Ann. Acad. Sci. Fennicæ (A.I.) 2 (1976), 113-127.

13. B. Fuglede: The fine topology in potential theory.
Séminaire de Théorie du Potentiel, Paris, No. 5, Springer Lecture Notes in Mathematics 814, 97-116. Berlin-Heidelberg-New York 1980.

14. B. Fuglede: Sur les fonctions finement holomorphes.
KUMI Preprint Series 1980.

15. W. Hayman & P.B. Kennedy: Subharmonic functions I.
Academic Press, London-New York 1976.

16. F. Iversen: Recherches sur les fonctions inverses des fonctions méromorphes.
Thèse, Helsingfors 1914.

17. T. Lyons: Finely holomorphic functions.
J. Funct. Anal. 37 (1980), 1-18.

18. T. Lyons: A theorem in fine potential theory and applications to finely holomorphic functions.
J. Funct. Anal. 37 (1980), 19-26.

19. Nguyen-Xuan-Loc & T. Watanabe: A characterization of fine domains for a certain class of Markov processes with applications to Brelot harmonic spaces.
Z. Wahrscheinlichkeitstheorie u. verw. Geb. 21 (1972), 167-178.

20. Nguyen-Xuan-Loc: Sur la théorie des fonctions finement holomorphes.
Bull. Sci. Math. 102 (1978), 271-308.

21. Nguyen-Xuan-Loc: Sur la théorie des fonctions finement holomorphes (II).
Complex Analysis Joensuu 1978, 289-300. Springer Lecture Notes in Mathematics 747. Berlin-Heidelberg-New York 1979.

22. Nguyen-Xuan-Loc: Singularities of locally analytic processes (with applications to the study of finely holomorphic and finely harmonic functions).
Potential Theory Copenhagen 1979, 267-288. Springer Lecture Notes in Mathematics 787. Berlin-Heidelberg-New York 1980.

23. K. Noshiro: Cluster Sets. Ergebnisse d. Math. (Neue Folge) 28. Springer, Berlin-Göttingen-Heidelberg 1960.

24. G. Polya & G. Szegö: Isoperimetric Inequalitites in Mathematical Physics.
Annals of Mathematics Studies No. 27. Princeton 1951.

25. H. Whitney: Analytic extensions of differentiable functions defined in closed sets.
Trans. Amer. Math. Soc. 36 (1934), 63-89.

26. G. Choquet: Sur les points d'effilement d'un ensemble.
Ann. Inst. Fourier 9 (1959), 91-102.

# ON THE FUNDAMENTAL GROUP OF THE COMPLEMENT OF A PLANE CURVE

William Fulton*

§1 A fruitful way to study an n-dimensional complex algebraic variety $X^n$ is to realize the variety as a branched covering of complex projective n-space $\mathbf{P}^n$. For any branched covering, i.e. finite morphism $f: X \to \mathbf{P}^n$, there is a hypersurface $H \subset \mathbf{P}^n$ such that the induced mapping from $X_\circ = X - f^{-1}(H)$ to $\mathbf{P}^n - H$ is a finite-sheeted topological covering.

When $n = 1$, $\mathbf{P}^1 = \mathbf{C} \cup \{\infty\}$ is the Riemann sphere, and $H$ is a finite set. Since the fundamental group of a punctured disk is $\mathbb{Z}$, any finite covering of a punctured disk is isomorphic to a covering given in local coordinates by $z \to z^e$; $e$ is the ramification index. Therefore any finite sheeted covering $X_\circ$ of $\mathbf{P}^1 - H$ may be completed to a compact Riemann surface $X$ and a branched covering $X \to \mathbf{P}^1$. Since compact Riemann surfaces are algebraic curves, the covering is given by some algebraic function.

This construction of Riemann generalizes to higher dimensions. By the Riemann-Enriques-Grauert-Remmert existence

---

* Research partially supported by NSF Grant MCS78-04008

theorem, any finite topological covering $X_{\circ}$ of $\mathbb{P}^n - H$ has a canonical structure of an algebraic variety, and can hence be extended to a branched covering $X \to \mathbb{P}^n$ . In case $n = 1$ , $X$ is determined by $X_{\circ}$ if $X$ is taken to be non-singular. In higher dimensions, $X$ is taken to be a normal variety.

The fundamental group of $\mathbb{P}^n - H$ , denoted $\pi_1(\mathbb{P}^n - H)$ , is the group of homotopy classes of loops in $\mathbb{P}^n - H$ with some fixed base point. Classically this was called the Poincaré group of $H$ . Finite topological coverings correspond to subgroups of finite index in the fundamental group. Thus to each branched covering $X \to \mathbb{P}^n$ whose branch locus is contained in $H$ corresponds a subgroup $N$ of finite index in $\pi_1(\mathbb{P}^n - H)$ , and conversely. The covering is Galois, i.e. the extension of fields of rational functions $R(\mathbb{P}^n) \subset R(X)$ is a Galois extension, if and only if $X_{\circ} \to \mathbb{P}^n - H$ is a regular covering, or equivalently, $N$ is a normal subgroup of $\pi_1(\mathbb{P}^n - H)$ . In this case

$$\mathrm{Gal}(R(X)/R(\mathbb{P}^n)) = \mathrm{Aut}(X/\mathbb{P}^n) = \pi_1(\mathbb{P}^n - H)/N \quad .$$

To study varieties as branched coverings of $\mathbb{P}^n$ , one therefore wants to calculate $\pi_1(\mathbb{P}^n - H)$ , or at least its profinite completion

$$\pi_1^{alg}(\mathbb{P}^n - H) = \varprojlim \pi_1(\mathbb{P}^n - H)/N$$

which classifies all finite coverings.

In case $n = 1$, $\pi_1(\mathbf{P}^1 - H)$ is a free group on $d$ generators $\sigma_1, \ldots, \sigma_d$, with the relation $\sigma_1 \cdot \ldots \cdot \sigma_d = 1$:

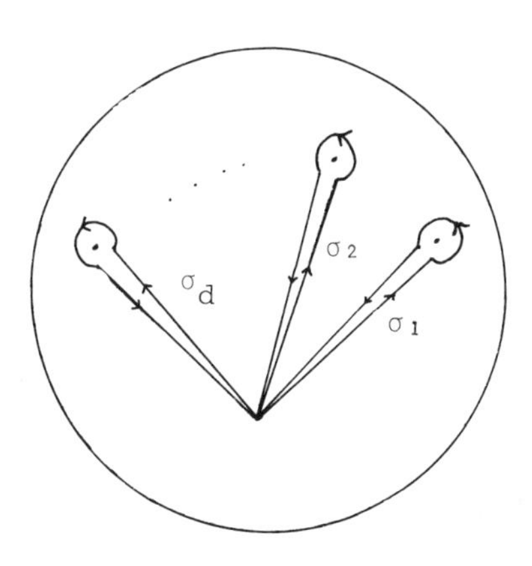


The action of $\pi_1(\mathbf{P}^n - H)$ on the fiber over the base point is the familiar data for gluing together the sheets of a Riemann surface. Here is one classical application of these ideas. To a variation of the branch points corresponds a unique variation of the covering. It is not hard to calculate how the gluing data changes as two adjacent branch points in the above picture move around each other. For simple coverings, i.e. coverings with one ramification point of index two over each branch point of $H$, it becomes a combinatorial problem, solved by Clebsch and Lüroth, to show that any two simple coverings with the same number of sheets and the same branch set $H$ become isomorphic after a finite number of such elementary moves. It follows that the space of m-sheeted simple coverings with $d$ branch points is connected [8]. Since any curve of genus $g$ can be written as a simple covering of $\mathbf{P}^1$ with g+1 sheets, one has a simple proof that the moduli space of curves of genus $g$ is connected [13].

When $n=2$, $H=C$ is a curve. Any non-singular surface $X^2$ admits a branched covering to $\mathbf{P}^2$ whose branch curve has only nodes (with local analytic equation $zw=0$) and cusps (equation $w^2=z^3$). In this case the relation between variation of $C$ and variation of $X$ is more subtle, but has been studied profitably by Wahl [14]. For modern studies relating invariants of $X$ to invariants of $C$ see Iversen [9] and Libgober [11].

The case $n \geq 3$ can be reduced to the case $n=2$. A fundamental theorem of Zariski [17] asserts that for a generic hyperplane $L^{n-1} \subset \mathbf{P}^n$, the homomorphism

$$\pi_1(L - L \cap H) \to \pi_1(\mathbf{P}^n - H)$$

is an isomorphism for $n>2$, and a surjection for $n=2$. Complete proofs of this fact have been given in recent years by Varchenko, Cheniot, and Hamm and Lê.

The basic problem is therefore to calculate $\pi_1(\mathbf{P}^2 - C)$ for $C$ a plane curve of degree $d$. Even when $C$ has only nodes and cusps, the problem remains very difficult. In this lecture we concentrate on the case when $C$ is a node curve, i.e., $C$ has only nodes as singularities.

§2 The foundations for work in this subject were laid by Zariski in 1929 [15]. He considered a pencil of lines $L_t$ through a general point $o$ in $\mathbf{P}^2$, parametrized by $t \in \mathbf{P}^1$.

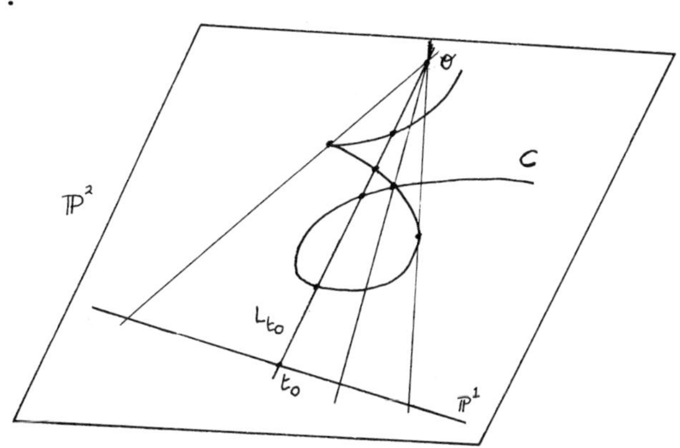


For general $t$, $L_t$ meets $C$ in $d$ points. Fix such a point $t_o$. The $d$ generators for $\pi_1(L_{t_o} - L_{t_o} \cap C)$ are generators for $\pi_1(\mathbf{P}^2 - C)$. The relations come from going around the points $t$ where $L_t \cap C$ has fewer than $d$ distinct points. When the generators are taken in standard form as in §1, so that adjacent points come together in an exceptional fibre, these relations can be worked out explicitly. For example, (i) at a simple tangent, a relation $\sigma_i = \sigma_{i+1}$ arises; (ii) at a node, $\sigma_i\sigma_{i+1} = \sigma_{i+1}\sigma_i$; (iii) at a cusp, $\sigma_i\sigma_{i+1}\sigma_i = \sigma_{i+1}\sigma_i\sigma_{i+1}$. The problem, however, is that standard generators chosen for a fixed $L_{t_o}$ may be carried to quite complicated generators when $t$ is near a point where points come together; one can only deduce that some conjugates of the standard generators satisfy relations (i), (ii), (iii). Nevertheless, Zariski stated the following as a theorem:

(*) <u>If $C$ is a node curve, then</u> $\pi_1(\mathbb{P}^2 - C)$ <u>is abelian</u>.

If $C_1, \ldots, C_m$ are the irreducible components of $C$, with $d_i$ the degree of $C_i$, the homology $H_1(\mathbb{P}^2 - C)$ is free abelian on $m$ generators, modulo the one relation $(d_1, \ldots, d_m)$; this follows for example from the duality isomorphism of $H^3(\mathbb{P}^2, C)$ with $H_1(\mathbb{P}^2 - C)$, and holds regardless of the singularities of $C$. Thus Zariski's theorem is equivalent to

$$(*)' \qquad \pi_1(\mathbb{P}^2 - C) = \mathbb{Z}^m/(d_1, \ldots, d_m) \quad .$$

For $C$ irreducible, it says that $\pi_1(\mathbb{P}^2 - C)$ is cyclic of order $d$. If $F(x,y,z) = 0$ is a homogeneous equation for $C$, let $V$ be the non-singular affine surface in $\mathbb{C}^3$:

$$V = \{(x,y,z) \in \mathbb{C}^3 \mid F(x,y,z) = 1\} \quad .$$

The canonical projection $(x,y,z) \to (x:y:z)$ determines a d-sheeted topological covering $V \to \mathbb{P}^2 - C$. Hence for irreducible curves, (*)' is asserting that $V$ is the universal cover of $\mathbb{P}^2 - C$, i.e.

$$(*)'' \qquad \pi_1(V) = 0 \quad .$$

Zariski reasoned as follows:

1) If $C$ specializes to a curve $C'$ of degree $d$, then there is a surjection of $\pi_1(\mathbb{P}^2 - C')$ onto $\pi_1(\mathbb{P}^2 - C)$.

2) If $C'$ consists of $d$ lines in general position, then $\pi_1(\mathbb{P}^2 - C')$ is abelian.

His proof therefore concluded by appealing to an assertion of Severi [13]:

(**) *If* $C$ *is a node curve of degree* $d$ , *then* $C$ *specializes to* $d$ *lines in general position*.

However, (**) has never been proved; it is the famous "Anhang F gap". It is essentially equivalent with an affirmative answer to the important question:

(**)' *For fixed* $d, \delta$ , *is the variety parametrizing irreducible plane node curves of degree* $d$ *with* $\delta$ *nodes an irreducible variety*?

Work of Zariski, Popp, and Alibert and Maltsiniotis have proved that this is true if $\delta$ is sufficiently large, e.g. $\delta > d^2/2 - 9d/4 + 1$ [2]. The proofs use the irreducibility of the moduli space of curves of genus $g$ . Severi intended his proof of (**)' to give an algebraic-geometric proof of the irreducibility of the moduli spaces of curves, but even this has never been achieved.

Zariski also gave an example which indicated how subtle the problem can become when cusps are allowed. Consider a cubic surface $X^2 \subset \mathbb{P}^3$ with homogeneous equation $w^3 + A(x,y,z)w + B(x,y,z) = 0$ , with $A$ and $B$ general homogeneous polynomials of degree 2 and 3 . Projection to the plane with homegeneous coordinates $(x : y : z)$ gives a branched covering $X \to \mathbb{P}^2$ branched along the curve $C$ defined by the discriminant $4A^3 + 27B^2 = 0$ . This curve $C$ is a curve of degree 6 with 6 cusps (where $A = B = 0$) on a conic $(A = 0)$ . This covering is not Galois, so $\pi_1(\mathbb{P}^2 - C)$ is not abelian; in fact it is $\mathbb{Z}/2 * \mathbb{Z}/3$ . Zariski showed that there are curves of degree 6 with 6 cusps not on a conic, and for these, the Poincaré group is abelian. Moishezon and Dolgachev have made some progress in understanding other non-abelian Poincaré groups.

§3 In 1959 Abhyankar began a study of the algebraic fundamental group, primarily to extend the "known" results to characteristic $p$ [1] . He proved the following basic fact.

Proposition. *If $f : X^2 \to \mathbb{P}^2$ is a Galois covering branched along a node curve $C$, and if $f^{-1}(C_i)$ is irreducible for each irreducible component $C_i$ of $C$, then $\mathrm{Aut}(X/\mathbb{P}^2)$ is abelian.*

In characteristic zero, Abhyankar's proof may be sketched as follows. Let $G = \mathrm{Aut}(X/\mathbb{P}^2)$ be the Galois group. For each component $C_i$ , let $I_i$ be the inertia group of $f^{-1}(C_i)$ :

$$I_i = \{\sigma \in G \mid \text{the restriction of } \sigma \text{ to } f^{-1}(C_i) \text{ is the identity}\}.$$

Since $f^{-1}(C_i)$ is irreducible, $I_i$ is a *cyclic*, *normal* subgroup of $G$ . In general each irreducible component of $f^{-1}(C_i)$ would have its own inertia group, and they would be conjugate. Inertia groups are always cyclic (in characteristic zero); near a point in $f^{-1}(C_i)$ mapping to a non-singular point $P$ in $C$ , the covering must be local analytically isomorphic to the cyclic covering $(z,w) \to (z^e,w)$ ; indeed, if $U$ is a small ball around $P$ , then $\pi_1(U - U\cap C) = \mathbb{Z}$ . In addition, any two inertia groups $I_i$ and $I_j$ must <u>commute</u>; at a point over a point $P$ where $C_i$ and $C_j$ intersect (any two plane curves do intersect!), the covering is dominated by a covering $(z,w) \to (z^{e_1},w^{e_2})$ , since $\pi_1(U - U\cap C) = \mathbb{Z} \oplus \mathbb{Z}$ for $U$ a small neighborhood of $P$ . Finally, the inertia groups <u>generate</u> $G$ , for if $K$ is the subgroup they generate, then $X/K \to \mathbf{P}^2$ is a covering unramified along any curve, and hence unramified, so $X/K = \mathbf{P}^2$ $(\pi_1(\mathbf{P}^2) = 0)$ , and $K = G$ . Thus $G$ is generated by commuting cyclic groups, so is abelian.

One case where the proposition applies is when each $C_i$ is non-singular. For then the local analysis described above shows that each $f^{-1}(C_i)$ must be non-singular. Now a theorem of Bertini implies that for any finite morphism $f : X^n \to \mathbf{P}^n$ , $X$ any irreducible variety, and any hypersurface $D \subset \mathbf{P}^n$ , then $f^{-1}(D)$ is connected. Therefore $f^{-1}(C_i)$ is connected and non-singular, so irreducible. Thus Abhyankar proved that $\pi_1^{alg}(\mathbf{P}^2 - C)$ was abelian in this case. (Over $\mathbf{C}$ , of course, Zariski's original argument is valid, (**) being obvious for non-singular curves.)

Abhyankar and later Prill [12] carried this farther, and proved that the Poincaré group of an irreducible node curve of degree $d$ is abelian if the number $\delta$ of nodes is small, e.g. $\delta < d^2/4$. They also proved generalizations for simply connected surfaces other than $\mathbb{P}^2$. For references to other previous work on this problem see [16], [18], [6].

§4 Zariski's assertion (*) was finally proved last year. Rather than requiring the development of new techniques, the proof depended on posing a series of new questions. It is perhaps most instructive to describe them in historical order.

In 1975-76, K. Johnson, following A. Holme's revival of classical work of Severi, Todd, and Segre, calculated Chern class formulae for the double point locus and ramification locus of a generic projection $X^n \to \mathbb{P}^m$ of a possibly singular variety $X^n \subset \mathbb{P}^N$. The algebraic formulae thus obtained for the double point cycle and for the ramification cycle are quite similar, and he noticed an identity relating them. From this followed a striking geometric conclusion [10]:

<u>If</u> $X^n \subset \mathbb{P}^{2n}$, <u>and a projection</u> $f$ <u>from</u> $X^n$ <u>to</u> $\mathbb{P}^m$ <u>is unramified</u>, $m < 2n$, <u>then</u> $f$ <u>is one-to-one</u>.

It was natural to ask if the same holds for $X^n \subset \mathbb{P}^N$, with $N > 2n$; this had been proved by Mumford and Moishezon, for

$n=2$ , and by J. Harris for $n=2,3$ . More generally, we asked if <u>any</u> unramified morphism from an n-dimensional projective variety to $\mathbb{P}^m$ , with $m<2n$ , must be one-to-one. An affirmative answer implies, for example, that there is no singular surface in $\mathbb{P}^3$ whose only singularities are normal crossings (local analytic equations $z_1z_2 = 0$ or $z_1z_2z_3 = 0$).

In 1976-77, I had the good fortune to spend a year at the Matematisk Institut in Århus. There J. Hansen suggested that one should look at the product mapping

$$X \times X \xrightarrow{f \times f} \mathbb{P}^m \times \mathbb{P}^m$$

and that it is enough to prove that $(f \times f)^{-1}(\Delta)$ is <u>connected</u>, where $\Delta$ is the diagonal in $\mathbb{P}^m \times \mathbb{P}^m$ . For then if $f$ were not one-to-one, there would be a path $(x_1(t), x_2(t))$ of double points, i.e. $f(x_1(t)) = f(x_2(t))$ with $x_1(t) \neq x_2(t)$ , converging to a point $(x,x)$ on the diagonal of $X$ ; but it is clear that $f$ would ramify at such a point $x$ .

In 1977-78 these questions were answered affirmatively.

<u>Theorem</u> [5]. <u>If</u> $Z^n$ <u>is an irreducible compact variety, and</u> $F : Z \to \mathbb{P}^m \times \mathbb{P}^m$ <u>is a morphism with</u> $\dim F(Z) > m$ , <u>then</u> $F^{-1}(\Delta)$ <u>is connected</u>.

Although the authors of [5] were unaware of it, Barth had proved a special case of this in 1969, and his proof is essentially the same as that given in [5]. A geometric construction replaces the diagonal subspace of $\mathbb{P}^m \times \mathbb{P}^m$ by a linear subspace of a projective space. For generic linear spaces, Bertini's theorem applies; for limiting linear spaces, Zariski's holomorphic functions are used. In addition to answering the above questions, this connectedness theorem has several immediate consequences, of which we mention two:

Corollary 1. If $X^n$ is an irreducible subvariety of $\mathbb{P}^m$, $m < 2n$, then $\pi_1^{alg}(X) = 0$.

Indeed, a non-trivial finite covering $X' \to X$ would give an unramified morphism $X' \to \mathbb{P}^m$ which has double points.

Corollary 2. If $f : X \to \mathbb{P}^m$, $g : Y \to \mathbb{P}^m$ are finite morphisms, with $X, Y$ irreducible compact varieties and $\dim(X) + \dim(Y) > m$ then

$$X \times_{\mathbb{P}^m} Y = \{(x,y) \in X \times Y \mid f(x) = g(y)\}$$

is connected.

For this, apply the theorem to $Z = X \times Y$. It is this corollary, a strengthening of Bertini's theorem, that is used in most applications.

In 1978-79, T. Gaffney and R. Lazarsfeld proved a striking result on branched coverings of $\mathbb{P}^n$.

Theorem [7]. If $f : X \to \mathbb{P}^n$ is a branched covering with $k$ sheets, then $\min(k, n+1)$ sheets must come together at some point.

For their proof they applied the connectedness theorem to $Z = X \times S$ , where $S$ is a subvariety of the locus when $r$ sheets come together; approaching the diagonal on $S$ produces points where $r + 1$ sheets come together, as long as $k$ and $\dim(S)$ are sufficiently large. They deduce as a corollary that $\pi_1^{alg}(X) = 0$ if $k \leq n$ and $X$ is normal.

This line of reasoning led in 1979 to a proof of Zariski's assertion.

<u>Theorem</u> [4]. <u>If</u> $C$ <u>is a node curve, then</u> $\pi_1^{alg}(\mathbf{P}^2 - C)$ <u>is abelian</u>.

For the proof, by Abhyankar's proposition, it suffices to see that $f^{-1}(C_i)$ is irreducible for a branched covering $f : X \to \mathbf{P}^2$ . Let $\tilde{C}_i$ be the non-singular model of $C_i$ , and consider the product morphism

$$F : X \times \tilde{C}_i \to \mathbf{P}^2 \times \mathbf{P}^2$$

Then $F^{-1}(\Delta) \cong f^{-1}(C_i) \times_{C_i} \tilde{C}_i$ ; the problem to circumvent is indicated in the following diagram:

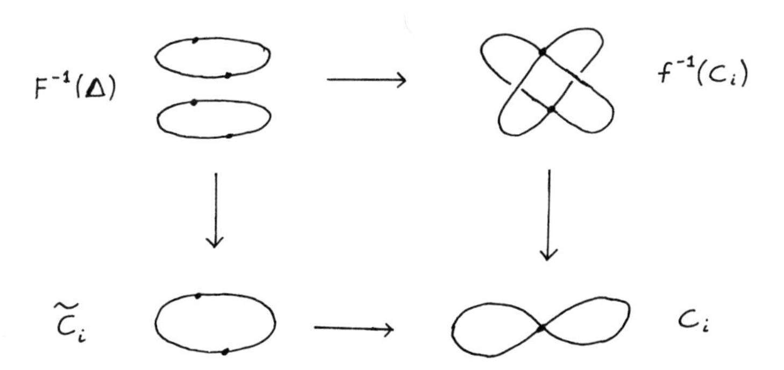

The local analysis done before shows that $F^{-1}(\Delta)$ is non-singular. The connectedness theorem implies that $F^{-1}(\Delta)$ is connected. Therefore $F^{-1}(\Delta)$ is irreducible, so its image $f^{-1}(C_i)$ is also irreducible.

Deligne saw how to improve the connectedness theorem to deduce Zariski's assertion on the topological fundamental group.

Theorem [3]. *Let $Z^n$ be a non-singular algebraic variety (not necessarily compact), $F : Z^n \to \mathbb{P}^m \times \mathbb{P}^m$ a morphism with finite fibres, and $n > m$. Then for arbitrarily small neighborhoods $\Delta_\varepsilon$ of $\Delta$, $F^{-1}(\Delta_\varepsilon)$ is connected, and the canonical homomorphism*

$$\pi_1(F^{-1}(\Delta_\varepsilon)) \to \pi_1(Z)$$

*is surjective.*

He applies this to

$$Z = (\mathbb{P}^2 - C) \times \tilde{C}_i \to \mathbb{P}^2 \times \mathbb{P}^2$$

to prove (*):

Theorem [3]. *If $C$ is a node curve, then $\pi_1(\mathbb{P}^2 - C)$ is abelian.*

The theorem of Deligne can also be used to prove that the other results of this section of the form "$\pi_1^{alg} X = 0$" can be replaced by "$\pi_1(X) = 0$". Libgober [11] has recently used it to study branched coverings of $\mathbb{P}^2$ when the branch curve has singularities worse than nodes and cusps.

§5 The circle of ideas described in §4 has already led to a number of new discoveries. A most striking result is a theorem of F. Zak:

> <u>If</u> $X^n \subset \mathbf{P}^N$ <u>is non-singular, and not a linear subspace, then no linear space</u> $L^m \subset \mathbf{P}^N$ <u>can be tangent to</u> $X$ <u>along a subvariety of dimension greater than</u> $m - n$ .

In particular, the Gauss map is always finite! Other applications of and generalizations to the connectedness theorem are discussed in Hansen's lecture in these proceedings, and in [6].

Deligne in [3] also conjectured a higher homotopy analogue of his theorem, which has since been proved - and strengthened from non-singular to local complete intersection varieties - by M. Goresky and R. MacPherson. From their theorem one can prove:

> <u>If</u> $X^n$ <u>is a compact local complete intersection,</u> $f : X^n \to \mathbf{P}^m$ <u>a finite morphism,</u> $Y^{m-d} \subset \mathbf{P}^m$ <u>a local complete intersection, then</u>
>
> $$\pi_i(X, f^{-1}(Y)) \to \pi_i(\mathbf{P}^m, Y)$$
>
> <u>is an isomorphism for</u> $i \leq n - d$ , <u>and surjective for</u> $i = n - d + 1$ .

For $Y$ a linear space, one finds a strong version of the Lefschetz hyperplane theorem. On the other hand, if one sets $X = Y$ , the Barth-Larsen theorem is generalized to local complete intersections $X^n \subset \mathbb{P}^m$ :

$$\pi_i(\mathbb{P}^m, X) = 0 \quad \text{for} \quad i \leq 2n - m + 1 \quad .$$

We refer to [6] for further discussion.

## References


1. S. Abhyankar, Tame coverings and fundamental groups of algebraic varieties, I; II, Amer. J. Math. 81 (1959), 46-94; 82 (1960), 120-178.

2. D. Alibert and G. Maltsiniotis, Groupe fondamental du complémentaire d'une courbe à points doubles ordinaires, Bull. Soc. Math. France 102 (1974), 335-351.

3. P. Deligne, Le groupe fondamental du complément d'une courbe plane n'yant que des points doubles ordinaires est abélian, Séminaire Bourbaki n° 543, Nov. 1979.

4. W. Fulton, On the fundamental group of the complement of a node curve, Ann. of Math. 111 (1980), 407-409.

5. W. Fulton and J. Hansen, A connectedness theorem for projective varieties, with applications to intersections and singularities of mappings, Ann. of Math. 110 (1979), 159-166.

6. W. Fulton and R. Lazarsfeld, Connectivity and its applications in algebraic geometry, Proceedings of the Midwest Algebraic Geometry Conference, 1980.

7. T. Gaffney and R. Lazarsfeld, On the ramification of branched coverings of $\mathbb{P}^n$ , Inventiones Math. 59 (1980), 53-58.

8. A. Hurwitz, Über Riemannsche Flächen mit gegebenen Verzweigungspunkten, Math. Ann. 39 (1891), 1-61.

9. B. Iversen, Numerical invariants and multiple planes, Amer. J. Math. 92 (1970), 968-996.

10. K. Johnson, Immersion and embedding of projective varieties, Acta Math 140 (1978), 49-74.

11. A. Libgober, Alexander polynomial of plane algebraic curves and cyclic multiple planes, preprint.

12. D. Prill, The fundamental group of the complement of an algebraic curve, Manuscripta math. 14 (1974), 163-172.

13. F. Severi, Verlesungen über algebraische Geometrie, Leipzig, 1921.

14. J. Wahl, Deformations of plane curves with nodes and cusps, Amer. J. Math. 96 (1974), 529-577.

15. O. Zariski, On the problem of existence of algebraic functions of two variables possessing a given branch curve, Amer. J. Math. 51 (1929), 305-328.

16. O. Zariski, Algebraic Surfaces, Ergebnisse der Mathematik, vol. 3, no. 5, (1935); second supplemented edition, with appendices by S. S. Abhyankar, J. Lipman, and D. Mumford, Ergebnisse der Mathematik, vol. 61, Springer-Verlag, Berlin-Heidelberg-New York (1971).

17. O. Zariski, A theorem on the Poincaré group of an algebraic hypersurface, Ann. of Math. 38 (1937), 131-141.

18. O. Zariski, Collected Papers, Part III, Introduction by M. Artin and B. Mazur, The MIT Press, Cambridge, Mass. and London, Eng., 1978.

Brown University
Providence, RI

# Symbolic calculus and differential equations

Lars Hörmander

By a symbolic calculus one means a correspondence between a class of operators and a class of functions such that computations involving the operators (composition, passage to the adjoint, change of variables) correspond to simple algebraic manipulations of the symbols. A classical example is the Heaviside calculus where the symbols are rational functions with no poles in a half plane. The basis for it is of course that the Laplace transformation converts an ordinary differential operator to multiplication by a polynomial.

A differential operator with constant coefficients in $\mathbb{R}^n$ can be written in the form $P(D)$ where $P$ is a polynomial and $D = -i\partial/\partial x = -i(\partial/\partial x_1, \ldots, \partial/\partial x_n)$. The Fourier transformation converts it to a multiplication operator

$$P(D)u(x) = (2\pi)^{-n}\int P(\xi)\,\hat{u}(\xi)\, e^{i\langle x,\xi\rangle}d\xi, u \in C_0^\infty; \hat{u}(\xi) = \int u(x)e^{-i\langle x,\xi\rangle}dx.$$

Of course the composition $P(D)Q(D)$ corresponds to $P(\xi)\,Q(\xi)$. Let us now examine to what extent this remains true if

$$Q(x,D) = \Sigma\, a_\alpha(x)D^\alpha \quad (\alpha=(\alpha_1,\ldots,\alpha_n)),\ D^\alpha = D_1^{\alpha_1}\ldots D_n^{\alpha_n})$$

has variable coefficients, which we shall always put to the left of the derivatives. We call the polynomial $Q(x,\xi)$ in $\xi$ the symbol of $Q(x,D)$ and can still write

$$(1) \qquad Q(x,D)\ u(x) = (2\pi)^{-n}\int Q(x,\xi)\ \hat{u}(\xi)\ e^{i\langle x,\xi\rangle}d\xi,\ u \in C_0^\infty$$

If we let another operator $P(x,D)$ act under the integral sign, we find that the composition $P(x,D)\ Q(x,D)$ has the symbol

(2) $P(x,D_x+\xi)\ Q(x,\xi) = \Sigma_\alpha (iD_\xi)^\alpha P(x,\xi)\ D_x{}^\alpha Q(x,\xi)/\alpha!.$

Here $\alpha! = \alpha_1!\ldots\alpha_n!$. This is just a way of writing Leibniz' formula.

Now (1) also makes perfectly good sense if $Q$ is not a polynomial in $\xi$. If we just know that $Q(x,\xi)$ and its derivatives do not grow faster than a polynomial as $\xi \to \infty$, then $Q(x,D)$ is a continuous map $C_0^\infty \to C^\infty$. Let us now see if we can find $Q$ so that $Q(x,D)$ is a right inverse (fundamental solution) of $P(x,D)$. This means that the expression (2) should be equal to 1. Writing

$$P(x,\xi) = P_m(x,\xi) + P_{m-1}(x,\xi) + \ldots$$

where $P_j$ is homogeneous in $\xi$ of degree $j$, we try to find $Q$ of the form

$$Q(x,\xi) = Q_{-m}(x,\xi) + Q_{-m-1}(x,\xi) + \ldots$$

where $Q_j$ is homogeneous in $\xi$ of degree j. Our first equation is then $P_m(x,\xi)\ Q_{-m}(x,\xi) = 1$, the next one is

$$0 = P_m(x,\xi)Q_{-m-1}(x,\xi) + P_{m-1}(x,\xi)Q_{-m}(x,\xi) -$$

$$-i\ \Sigma_j \partial P_m(x,\xi)/\partial\xi_j \partial Q_{-m}(x,\xi)/\partial x_j$$

and so on. If $P$ is <u>elliptic</u>, that is, $P_m(x,\xi) \neq 0$ for all real $\xi \neq 0$, these equations can be solved successively. Of course we cannot expect the series $\Sigma Q_j$ to converge, but it is easy to show that there is a function $Q \in C^\infty$ such that

$$Q(x,\xi) \sim \sum_{-\infty}^{-m} Q_j(x,\xi)$$

in the sense that for every $N$ we have

$$Q(x,\xi) - \sum_{j>-N} Q_j(x,\xi) = O(|\xi|^{-N}),\ \xi \quad \infty$$

as well as corresponding estimates for the derivatives. Of course Q does not quite make the expression (2) equal to 1, but

$$P(x,D_x+\xi)\ Q(x,\xi) = 1+R(x,\xi)$$

where R decreases faster than any power of $1/|\xi|$ as $\xi \to \infty$. Thus

$$P(x,D)\ Q(x,D) = \mathrm{Id}+R(x,D)$$

where $R(x,D)$ has an infinitely differentiable kernel

$$r(x,y) = (2\pi)^{-n} \int R(x,\xi) e^{i\langle x-y,\xi\rangle} d\xi .$$

The problem of finding a right inverse of $P(x,D)$ is therefore reduced to that of inverting $\mathrm{Id}+R(x,D)$, which at least on a compact set is a very simple Fredholm equation. Error terms such as $R$ can therefore usually be ignored in questions concerning existence and smoothness of solutions, although they may be important in other contexts such as index theory.

The roles played by $P$ and $Q$ above can be made perfectly symmetric. If

$$(3) \qquad P(x,\xi) \sim \sum_0^\infty P_{m-j}(x,\xi), \qquad Q(x,\xi) \sim \sum_0^\infty Q_{m'-j}(x,\xi)$$

where subscripts denote degrees of homogeneity in $\xi$, then $P(x,D)\ Q(x,D) = S(x,D)$ where

$$(2)' \qquad S(x,\xi) \sim \sum_\alpha (iD_\xi)^\alpha P(x,\xi) D_x^\alpha\ Q(x,\xi)/\alpha! .$$

In fact, (2)' remains valid for considerably larger classes of symbols. The simplest and most useful example is perhaps the class $S^m$ consisting of all $a \in C^\infty(\mathbb{R}^{2n})$ with

$$(4) \qquad |D_x^{\alpha} D_\xi^\beta\ a(x,\xi)| \leqq C_{\alpha\beta}(1+|\xi|)^{m-|\beta|} \quad \text{for all } \alpha,\beta .$$

(This condition is fulfilled if $a$ is homogeneous in $\xi$

of degree $m$ for large $|\xi|$). If $P \in S^m$ and $Q \in S^{m'}$ then (2)' is valid in the sense that for all $N$

$$S(x,\xi) - \sum_{|\alpha|<N} (iD_\xi)^\alpha P(x,\xi)\, D_x{}^\alpha Q(x,\xi)/\alpha! \in S^{m+m'-N} .$$

In an analogous sense the usual formulas for computing the adjoint of a differential operator or making a change of variables remain true. Note that this implies that the adjoint of $a(x,D)$ maps $C_0^\infty$ continuously to $C^\infty$, so we can extend $a(x,D)$ to a continuous map from $\mathcal{E}'$ to $\mathcal{D}'$. Together with the simple fact that $a(x,D)$ is $L^2$ continuous if $a \in S^0$, this is all the basic calculus of pseudo-differential operators as it was first established by Kohn and Nirenberg [22] (see also Hörmander[13,14]).

In the case where $Q$ is homogeneous of degree $0$ in $\xi$ the study of $Q(x,D)$ has a long history under the name of singular integral operator theory. The point is that the inverse Fourier transform of $Q(x,\xi)$ with respect to $\xi$ is then a distribution $K(x,z)$ which is homogeneous of degree $-n$ in $z$, so it just barely fails to be locally integrable in $z$. We can write (1) in the form

$$Q(x,D) = \int K(x,x-y)\, u(y)\, dy$$

where the integral has to be interpreted in the sense of distribution theory or more precisely as a principal value. (A delta function may also occur.) This representation has some advantages particularly when $Q$ is not assumed to be smooth as we have tacitly done here. However, the singular integral operator representation tends to hide the simple composition rules which were discovered only in a very indirect way by Tricomi, Giraud, Michlin and others in the 1920's and 1930's. The next step in the development was taken by Calderón and Zygmund in the 1950's, but their symbolic calculus was still crude in the sense that only the leading term in (2)' was considered. An interesting history of this subject with many references can be found in Seeley [31]; a recent exposition is given in Michlin and Prössdorf [28].

If $\varphi, \psi \in C_0^\infty$ and the supports are disjoint, then the symbol of $\varphi a(x,D)\psi$ is in $S^{-\infty}$ so the kernel is a $C^\infty$ function. (Here $\varphi$ and $\psi$ denote multiplication by $\varphi$ and $\psi$.) Thus the kernel of $a(x,D)$ is a $C^\infty$ function outside the diagonal, which implies the <u>pseudo-local property</u>

$$\text{sing supp } a(x,D)u \subset \text{sing supp } u, u \in \mathcal{E}' .$$

Here sing supp $u$ denotes the smallest closed set such that $u$ is in $C^\infty$ in the complement. Since we are ignoring $C^\infty$ kernels the only interesting feature of $a(x,D)$ is therefore the behavior of the kernel on the diagonal, and we can define the notion of pseudo-differential operator on a manifold by demanding that the kernel shall be in $C^\infty$ off the diagonal and that the restrictions to sufficiently small coordinate patches shall be defined by pseudo-differential operators in the sense of (1). The symbol will of course depend on the choice of local coordinates, but for classical symbols, that is, symbols of the form (3), the leading term (the principal symbol) is invariantly defined on the cotangent bundle just as in the case of differential operators.

Let us now see how the classical theory of elliptic and hyperbolic differential equations looks from the point of view of pseudo-differential operator calculus. First consider an elliptic (pseudo-) differential operator $P(x,D)$ on a compact manifold $X$. This means that the principal symbol $P_m$, defined on the cotangent bundle $T^*X$ outside the zero section, is never equal to $0$ there. As seen above, it is then possible to find a pseudo-differential operator $Q$ of order $-m$ such that

$$P(x,D)\ Q(x,D) = \mathrm{Id} + R(x,D)$$

where $R$ is of order $-\infty$. The equation

$$g + R(x,D)g = f$$

has a solution $g \in \mathcal{D}'(X)$ for every $f \in \mathcal{D}'(X)$ orthogonal to the finitely many functions $v \in C^\infty$ such that $v + R^*v = 0$, and $g - f \in C^\infty$. Thus the equation

$$P(x,D)u = f$$

has the solution $u = Q(x,D)g$ if $f$ is in a space of finite codimension, and since $u - Q(x,D)f \in C^\infty$ the regularity properties of $u$ can be determined from those of $f$ and the continuity properties of $Q$. On the other hand, we can also find $Q_1(x,D)$ with

$$Q_1(x,D)\ P(x,D) = \mathrm{Id} + R_1(x,D)$$

where $R_1$ is of order $-\infty$. (Multiplication by $Q(x,D)$ to the right shows that $Q - Q_1$ has a $C^\infty$ kernel so we could in fact have taken $Q_1 = Q$.) Hence $P(x,D)u = 0$ implies $u + R_1(x,D)u = 0$, thus $u \in C^\infty$, so $u$ is in a finite dimensional subspace of $C^\infty$. The equation

$$P(x,D)u = f$$

can therefore be solved if and only if $f$ is in a space

of finite codimension with $C^\infty$ annihilator, and the solutions of the homogeneous equation form a finite dimensional subspace of $C^\infty$.

The preceding arguments work with no real change for elliptic systems. Also boundary problems for elliptic equations can easily be handled. For the sake of simplicity we just consider a boundary problem for the homogeneous Laplace equation in a bounded open set $X \subset \mathbb{R}^n$ with $C^\infty$ boundary,

$$\Delta u = 0 \quad \text{in} \quad X, \quad Bu = f \quad \text{on} \quad \partial X,$$

where $B$ is a differential operator. Using the equation $\Delta u = 0$ we can always write the boundary condition in the form

$$Bu = B_0 u + B_1 du/d\nu = f \quad \text{on} \quad \partial X,$$

where $du/d\nu$ is the outer normal derivative and $B_0$, $B_1$ are differential operators in $\partial X$. Let $u_0$ be the restriction of $u$ to $\partial X$ and let $u_1 = du/d\nu$. If $E$ is the Newton kernel, the fundamental solution of the Laplacean, then

$$(5) \qquad u(x) = \int_{\partial X} (u_0(y)\, dE(x-y)/d\nu_y - u_1(y)\, E(x-y))\, dS(y), \; x \in X,$$

where dS is the surface measure on $\partial X$. In the classical theory Green's formula could not be used to solve the boundary problem because (5) does not usually give a function with boundary value $u_0$ and normal derivative $u_1$. However, using coordinate systems where the boundary is flat (which is legitimate by the invariance of pseudo-differential operators) it is easily seen that the limit of the right hand side of (5) when $x$ approaches the boundary is of the form

$$A_0u_0 + A_1u_1$$

where $A_0$ is a pseudo-differential operator of order 0 and $A_1$ is elliptic of order $-1$. Thus the harmonic function $u$ defined by (5) has boundary value $u_0$ if and only if

$$(I - A_0)u_0 = A_1u_1.$$

If we compare (5) with Green's formula for $u$, it follows that $A_1(u_1-du/d\nu) = 0$. Assuming for the sake of simplicity that $n > 2$ we obtain $u_1 = du/d\nu$. In fact, the equation $A_1u_1 = 0$ has no solution $u_1$ other than 0 since (5) with $u_0 = 0$ would then give a harmonic function $u$ in $\mathbb{R}^n\setminus\partial X$ vanishing on both sides of $\partial X$ and at $\infty$; thus $u = 0$ so $u_1dS = -\Delta u = 0$. Hence $A_1^{-1}$ is a pseudo-differential operator of order 1. Our boundary problem reduces to

$$(B_0 + B_1A_1^{-1}(I - A_0))u_0 = f \quad \text{on} \quad \partial X.$$

If this pseudo-differential equation is elliptic, the boundary problem is called elliptic, and we have already proved results on existence and smoothness of solutions then. However, it is important that the preceding reduction is completely general. To study non-elliptic boundary problems is thus equivalent to studying non-elliptic pseudo-differential operators on a compact manifold. One should therefore always try to study general pseudo-differential operators rather than restricting oneself to differential operators. In this spirit we note that the preceding discussion is valid with no real change if $B_j$ are pseudo-differential operators. One may also replace $\Delta$ by any elliptic differential operator (Calderón [5], Hörmander [15]) or even any elliptic pseudo-differential operator satisfying the so-called transmission condition (see Boutet de Monvel [4]).

Let us next look at the Cauchy problem for a hyperbolic differential equation with $C^\infty$ coefficients, of order $m$ in $n+1$ variables,

$$P(x,t,D_x,D_t)u = \sum_0^m a_{m-j}(x,t,D_x)D_t^j u = 0, \tag{6}$$

$$D_t^j u = \Psi_j \quad \text{when} \quad t = 0, j = 0,\dots,m-1. \tag{7}$$

Here $\Psi_j$ are prescribed functions in $\mathbb{R}^n$. We assume strict hyperbolicity, that is, $a_0 = 1, a_{m-j}$ is of degree $m-j$ and if $a^o_{m-j}$ is the homogeneous part of degree $m-j$ then the equation

$$\sum_0^m a^o_{m-j}(x,t,\xi)\tau^j = 0$$

has $m$ distinct real roots $\tau$ for every $\xi \in \mathbb{R}^n \setminus 0$. Let the roots in decreasing order be $\lambda_k(x,t,\xi)$, $k = 1, \dots, m$. Under natural uniformity conditions with respect to $(x,t)$ one can for every $k$ find a pseudo-differential operator $\Lambda_k(x,t,D_x)$ with principal symbol $\lambda_k$ such that

$$P(x,t,D_x,D_t) = Q_k(x,t,D_x,D_t)(D_t - \Lambda_k(x,t,D_x)) + R_k(x,t,D_x,D_t)$$

where $R_k$ is of the form (6) with $a_{m-j}$ replaced by an operator of order $-\infty$ and $Q_k$ is of the same form with $m$ replaced by $m-1$ and pseudo-differential operators as coefficients. The proof follows the same pattern as the construction of an inverse of an elliptic operator. Ignoring $R_k$ as usual (which can be justified by solving a Volterra integral equation with smooth kernel) we note that (6) is satisfied by $u = u_1 + \ldots + u_m$ if

$$(D_t - \Lambda_k(x,t,D_x))u_k = 0.$$

The boundary conditions (7) mean that

$$\sum_k \Lambda_k(x,t,D_x)^j u_k(0) = \Psi_j, j = 0,\ldots,m-1.$$

This is an elliptic system since the determinant of the principal symbols is a non-vanishing van der Monde determinant. (We encounter here a slightly more general class of elliptic systems where the different equations have different degrees.)

Thus we can reduce the Cauchy problem to the Cauchy problem for an equation

$$(D_t - \Lambda(x,t,D_x))u = 0 \tag{8}$$

which is a first order differential equation in $t$ and a pseudo-differential equation with real first order principal symbol in $x$. The solution of (8) is very elementary. In fact, the equation gives

$$\frac{d}{dt} \int |u(x,t)|^2 \, dx = (iD_t u,u) + (u,iD_t u) = i((\Lambda(x,t,D_x)- -\Lambda^*(x,t,C_x)u,u).$$

Here $\Lambda - \Lambda^*$ is of order $0$, thus $L^2$ continuous, so we obtain

$$\left|\frac{d}{dt} \|u(.,\ t)\|^2\right| \leqq C\|u(.,\ t)\|^2 .$$

Hence one can easily conclude existence and uniqueness of a solution for initial data in $L^2$, satisfying the estimate

$$(9)\quad e^{-C|t|}\,\|u(.,0)\|^2 \leqq \|u(.,t)\|^2 \leqq e^{C|t|}\,\|u(.,0)\|^2 .$$

(Note that for $t>0$ the essential point is to have an upper bound for $i(\Lambda-\Lambda^*)$. For a more general study of the Cauchy problem it is therefore important to establish precise onesided bounds for pseudo-differential operators which are of positive order.)

Let now $a(x,t,D_x)$ be a pseudo-differential operator depending on $t$. If $a(x,t,D_x)$ commutes with $D_t-\Lambda(x,t,D_x)$ then (8) implies the same equation for $a(x,t,D_x)u$, so (9) is also valid for this function. If

$$a \sim a_\mu + a_{\mu-1} + \dots$$

then commutativity (mod $C^\infty$) means that

$$\partial a_\mu(x,t,\xi)/\partial t = \{\lambda, a_\mu\} = \langle \partial\lambda/\partial\xi, \partial a_\mu/\partial x\rangle - \langle \partial\lambda/\partial x, \partial a_\mu/\partial\xi\rangle$$

and that $a_{\mu-1}, a_{\mu-2}, \dots$ satisfy similar inhomogeneous equations where the inhomogeneous part is determined by the preceding terms. Thus $a_\mu$ is constant along the orbits of

$$(10)\qquad dx/dt = -\partial\lambda/\partial\xi,\ d\xi/dt = \partial\lambda/\partial x$$

and the other terms $a_{\mu-1}, \dots$ are successively determined by solving ordinary differential equations (transport equations) along these orbits. If a vanishes outside a narrow cone when $t=0$, this will remain true for all $t$, and the cone moves with $t$ according to the Hamilton equations (10).

At this point it is useful to introduce a refinement of the notion of singularity. We shall say that $u$ is in $C^\infty$ at $(x_0,\xi_0)$, where $\xi_0 \in \mathbb{R}^n \setminus 0$, (or rather on the ray through $(x_0,\xi_0)$ which can be thought of as a point in the cosphere bundle) if $a(x,D)u \in C^\infty$ for some pseudo-differential operator with principal symbol $a^0$ different from $0$ at $(x_0,\xi_0)$. Then we have in fact $b(x,D)u \in C^\infty$ for all pseudo-differential operators $b(x,D)$ such that $b(x,\xi)$ vanishes outside a sufficiently small conic neighborhood of $(x_0,\xi_0)$, for the discussion of elliptic operators above shows that we can find $q$ so that

$$b(x,D) = q(x,D)\ a(x,D) + r(x,D)$$

where $r$ is of order $-\infty$, thus smoothing. The definition is quite analogous to the fact that $u \in C^\infty$ at $x_0$ if and only if $\Psi u \in C^\infty$ for some $\Psi \in C_0^\infty$ with $\Psi(x_0) \neq 0$ (or equivalently for all $\Psi \in C_0^\infty$ with support close to $x_0$). If $u \in C^\infty$ at $(x_0,\xi)$ for every $\xi \neq 0$, then $u \in C^\infty$ at $x_0$ for using the Borel-Lebesgue lemma and a partition of unity in the $\xi$ variable on the unit sphere we can find $b_j(x,\xi)$ with $b_j(x,D)u \in C^\infty$ and $\sum b_j(x,\xi) = \Psi(x), \Psi(x_0) \neq 0$. If we denote by sing spec $u$ the closed cone in $\mathbb{R}^n \times (\mathbb{R}^n \setminus 0)$ where $u$ is not in $C^\infty$, then dropping the second argument gives a surjective map

$$\begin{array}{ccc} \text{sing spec } u \subset \mathbb{R}^n \times (\mathbb{R}^n \setminus 0) & \to & \mathbb{R}^n \times S^{n-1} \\ \downarrow \qquad\qquad \searrow & & \swarrow \\ \text{sing supp } u \subset \qquad \mathbb{R}^n & & \end{array}$$

For other basic properties of this notion we refer to Hörmander [17] where the notation WF(u) was used instead of sing spec u.

Now it follows from the construction of a commuting operator above that the singular spectrum of $u(.,t)$ is equal to that of $u(.,0)$ transported by the Hamilton-Jacobi flow (10). Thus the singularities in our refined sense propagate along the rays of geometrical optics for the Hamiltonian $\Lambda(x,t,\xi)$ (or rather $\tau-\Lambda(x,t,\xi)$ if one thinks of $t$ as one of the variables).

If $A$ is a pseudo-differential operator it is clear that

$$\text{sing spec } Au \subset \text{sing spec } u, \; u \in \mathcal{E}',$$

which improves the pseudo-local property. Thus $A$ induces an operator in the space $\mathcal{D}'_{x_0,\xi_0}$ of germs of singularities at $(x_0,\xi_0)$, defined as $\mathcal{E}'$ (or $\mathcal{D}'$) modulo the subspace for which $(x_0,\xi_0)$ is not in the singular spectrum. This induced operator is 0 if the symbol of $A$ is of order $-\infty$ in a conic neighborhood of $(x_0,\xi_0)$. The study of the germs $\mathcal{D}'_{x_0,\xi_0}$ is called microlocal analysis. In the microlocal study of singularities of solutions of pseudo-differential equations only the local properties of the symbols are thus important.

Now any real valued smooth function $p(x,\xi)$ which is homogeneous in $\xi$ and satisfies the conditions $p=0, \partial p/\partial\xi \neq 0$ at $(x_0,\xi_0)$, can with suitable labelling of the coordinates be factored in the form

$$p(x,\xi) = q(x,\xi)(\xi_n-\lambda(x,\xi')), \xi' = (\xi_1,\ldots,\xi_{n-1}),$$

where $q$ is homogeneous in $\xi$ and not 0 at $(x_0,\xi_0)$. If $P(x,D)$ has principal symbol $p$ it follows as for hyperbolic operators that one can find a pseudo-differential operator $E$ and a pseudo-differential operator $\Lambda$ in $x'$ with symbol $\lambda(x,\xi')$ such that the symbol of

$$E(x,D)\ P(x,D) - (D_n - \Lambda(x,D'))$$

is of order $-\infty$ in a conic neighborhood of $(x_0, \xi_0)$. (Here it does not really matter that $\Lambda$ is a pseudo-differential operator in $\mathbb{R}^{n-1}$ involving $x_n$ only as a parameter; cf. Sjöstrand [32, appendix].) It is now easy to conclude from the discussion of (8) above that if $P(x,D)u \in C^\infty$ then

a) sing spec $u \subset \{(x,\xi);\ p(x,\xi) = 0\}$ = characteristic set of $P$;

b) sing spec $u$ is invariant under the Hamilton flow, that is, the flow along the vector field $H_p = \partial p/\partial\xi \partial/\partial x - \partial p/\partial x\ \partial/\partial\xi$ .

Statement a) is of course valid in general by the definition of sing spec u. In a modified form b) can be extended to quite general pseudo-differential operators with $dp(x,\xi) \neq 0$ when $p(x,\xi) = 0$ (principal type condition). For such results we refer to Hörmander [18,19,20] and Dencker [6]. The proofs require several additional technical tools, which we shall now discuss.

Let us first return to the solutions of the equation (8). Let $F_t$ be the map $u(.,0) \to u(.,t)$, which is continuous in $S$ and in $S'$ as well as in $L^2$. We have seen that if $A$ is a pseudo-differential operator then there is a pseudo-differential operator $A_t$ of the same order such that

$$F_t A u = A_t F_t u$$

(apart from errors of order $-\infty$). Thus

$$F_t A F_t^{-1} = A_t$$

(with an error of order $-\infty$ which can be absorbed in $A_t$). The principal symbol $a_t$ of $A_t$ is the composition of the principal symbol $a$ of $A$ and the inverse of the map $\chi_t$ obtained by integrating (10) from 0 to t. Thus $\chi_t(x,\xi)$ is the value at time t of the solution with initial condition $(x,\xi)$ when $t = 0$. The map $\chi_t^{-1}$ has two essential properties: it is homogeneous, that is, commutes with multiplication by positive scalars in the $\xi$ variable, and it is symplectic, that is, preserves the symplectic form

$$\sum d\xi_j \wedge dx_j .$$

This is a classical and basic fact in mechanics. Now $F_t$ has for small t an analytic representation which is very close to that of a pseudo-differential operator,

$$F_t u(x) = (2\pi)^{-n} \int e^{i\varphi(x,t,\xi)} F(x,t,\xi)\, \hat{u}(\xi)\, d\xi, \quad u \in C_0^\infty. \tag{11}$$

Here $\varphi$ is determined by solving the "eiconal equation"

$$\partial\varphi/\partial t = \lambda(x,t,\partial\varphi/\partial x), \quad \varphi(x,0,\xi) = \langle x,\xi \rangle$$

which requires that t is small enough. Since $F_0$ = Identity we pose the initial condition $F(x,0,\xi) = 1$ and can then write down transport equations for $F \in S^0$ along the curves (10) to make sure that $F_t u$ satisfies (8).

Operators of the form

$$Fu(x) = (2\pi)^{-n} \int e^{i\varphi(x,\xi)} F(x,\xi)\, \hat{u}(\xi)\, d\xi$$

where $F \in S^m$, $\det\ \partial^2\varphi/\partial x \partial\xi \neq 0$ and $\varphi$ is a real valued function homogeneous of degree 1 in $\xi$ are called Fourier integral operators. They are associated with a homogeneous symplectic map

$$\chi\colon (\partial\varphi/\partial\xi, \xi) \to (x, \partial\varphi/\partial x).$$

With a suitable choice of local coordinates every homogeneous symplectic map (i.e., contact transformation) has a representation of this form. If $F$ is elliptic in the obvious sense then given any pseudo-differential operator $A$ we can find another $B$ such that just as in the special case already discussed

$$F^{-1} AF = B. \tag{12}$$

The principal symbol of $B$ is the composition of that of $A$ with $\chi$ (Egorov's theorem [9]). Fourier integral operators can therefore be used to simplify the principal symbol microlocally when this can be done geometrically by composition with a homogeneous symplectic map. For example, if the principal symbol $a$ of $A$ vanishes at $(x_0,\xi_0)$ and $\partial a/\partial\xi \neq 0$ there we can after multiplication by an elliptic operator conjugate $A$ to an operator with symbol of the form

$$\xi_n + if(x,\xi')$$

where $f$ is real. (We can make $f$ vanish if $a$ is real.) This gives a good start for a further study of operators of principal type. Another important case is the induced Cauchy-Riemann system on the boundary of a strictly pseudo-convex domain. Micro-locally it turns out to be independent of the domain - which is what remains of the Riemann mapping theorem for several variables.

Fourier integral operators are useful not only when one wants to simplify pseudo-differential operators by conjugation but also when one wants to give precise descriptions of, say, fundamental solutions. They have a symbolic calculus just as precise as that of pseudo-differential operators but it is somewhat more complicated because of the presence of the symplectic transformation $\chi$. For the technical details of the theory we must refer to Hörmander [17] (see also the summary in Hörmander [16] ), Duistermaat-Hörmander [8] , Duistermaat-Guillemin [7] . An extension to complex phase functions, corresponding to canonical transformations which pass into the complex domain, has been developed by Melin and Sjöstrand [23], [24]. For another class involving a combination of two different transformations $\chi$ see also Melrose and Uhlman [27].

So far we have only considered pseudo-differential and Fourier integral operators with the simplest kind of symbols. However, the study of operators of principal type already forces one to extend the class of admissible symbols a great deal. The point is that an operator with principal symbol $\xi_n+if(x,\xi')$ behaves essentially as a Cauchy-Riemann operator when $df \neq 0$ whereas the Cauchy problem is well behaved for increasing (decreasing) $x_n$ if $f \geqq 0$ $(f \leqq 0)$. In general one has a mixture of these cases which must be disentangled by means of an infinite partition of unity, adapted to the operator at hand. (We assume here that $f(x,\xi')$ does not change sign for fixed $(x',\xi')$.)

By such analysis Beals and Fefferman [2] were able to remove simplifying analyticity assumptions made by Nirenberg and Treves [29] and were led to the study of very general classes of symbols (cf. Beals-Fefferman [3], Beals [1]).

If $m = 0$ then condition (4) means that each derivative of $a$ has a uniform bound provided that the derivatives are calculated with a scale of the order of magnitude $1 + |\xi|$ in the $\xi$ direction at $(x,\xi)$. This observation leads us to the following general definition of symbol classes. Let $g$ be a Riemannian metric in $\mathbb{R}^{2n}$, that is, for every $(x,\xi) \in \mathbb{R}^{2n}$ we assume given a positive definite quadratic form $g_{x,\xi}$. We assume that it varies slowly with $(x,\xi)$ so that for some $c > 0$ and $C > 0$

$$g_{x,\xi}(y,\eta) < c \Rightarrow g_{x,\xi}/C \leqq g_{x+y,\xi+\eta} \leqq C\, g_{x,\xi}.$$

If $m$ is a function such that

$$g_{x,\xi}(y,\eta) < c \Rightarrow m(x,\xi)/C \leqq m(x+y,\xi+\eta) \leqq C\, m(x,\xi)$$

we can then define a space $S(m,g)$ of symbols $a$ such that for every $k$ the norm of the $k^{th}$ differential $a^{(k)}$ at $(x,\xi)$ with respect to $g_{x,\xi}$ is bounded by $C_k m(x,\xi)$. Thus we require uniform bounds for the derivatives divided by $m(x,\xi)$ if the derivatives are calculated in terms of a $g_{x,\xi}$ orthonormal system at $(x,\xi)$.

To establish a pseudo-differential operator calculus with symbols in $S(m,g)$ we first modify (1) to the Weyl prescription

$$a^w(x,D)u(x) = (2\pi)^{-n} \iint e^{i\langle x-y,\xi\rangle} a((x+y)/2,\xi)\, u(y)\, d\xi dy . \tag{13}$$

This is a symmetric compromise between putting coefficients to the left or to the right of differentiations, which would mean replacing $(x+y)/2$ by $x$ or $y$ respectively. The different possibilities reflect of course the fact that we are defining functions of non-commuting operators. An obvious drawback of (13) is that the interpretation is no longer quite elementary, but on the other hand there are important advantages. First we note that it is clear at least formally that (13) makes $a^w$ symmetric if $a$ is real valued. This is the reason why Weyl proposed (13) in his book on quantum mechanics. What is more important for us here is the invariance properties of (13). To every linear symplectic transformation $\chi: \mathbb{R}^{2n} \to \mathbb{R}^{2n}$ there corresponds a unitary operator $U$ such that

$$U^{-1} a(x,D) U = (a\circ\chi)(x,D) \tag{14}$$

if $a(x,\xi)$ is linear in $(x,\xi)$; apart from a constant factor of modulus one this determines $U$ uniquely. The Weyl formula (13) is the only way to make (14) remain valid for general $a$ and $a^w(x)$ equal to multiplication by $a(x)$.

We must now assume that the Riemannian metric satisfies the familiar uncertainty relations of quantum mechanics. Thus let $g^\sigma_{x,\xi}$ be the dual of $g_{x,\xi}$ with respect to the symplectic form which identifies $^{2n}$ with its dual, and let $h(x,\xi)$ be the smallest positive number such that

$$g_{x,\xi} \leqq h(x,\xi)^2 g^\sigma_{x,\xi}.$$

The condition is then that $h \leqq 1$ (or any other fixed constant since equivalent metrics define the same symbols). Note that if

$$g_{x,\xi} = \sum a_j^2 dx_j^2 + \sum b_j^2 d\xi_j^2$$

then

$$g^\sigma_{x,\xi} = \sum b_j^{-2} dx_j^2 + \sum a_j^{-2} d\xi_j^2$$

so the condition means that $a_j b_j \leqq 1$. This is the familiar Heisenberg condition for the local unit ball at $(x,\xi)$. Finally we assume that $g$ has at most polynomial growth over large distances in the sense that for some $N$ and $C$

$$g^\sigma_{x+y,\xi+\eta} \leqq C\, g^\sigma_{x,\xi}(1+g^\sigma_{x,\xi}(y,\eta))^N \quad \text{for all} \quad x,\xi,y,\eta \in \mathbb{R}^n .$$

A similar condition is placed on $m$. (In some contexts this condition can be relaxed (cf. Unterberger [33]) but it is not entirely superfluous.) Under these conditions (13) defines for every $a \in S(m,g)$ an operator $a^w(x,D)$ which is continuous from $\mathcal{S}$ to $\mathcal{S}$ and from $\mathcal{S}'$ to $\mathcal{S}'$; the symmetry condition (14) is valid. If $m = 1$ (resp. $m \to 0$ at $\infty$) then $a^w$ is continuous (resp. compact) in $L^2$. Finally, if $a_j \in S(m_j,g)$ for $j = 1,2$ then $a_1{}^w(x,D)\, a_2{}^w(x,D) = a^w(x,D)$ where $a \in S(m_1 m_2,g)$ and

$$a(x,\xi) - \sum_{j<N} (i(\langle D_\xi,D_y\rangle - \langle D_x,D_\eta\rangle)/2)^j a_1(x,\xi)a_2(y,\eta)/j!_{/x=y,\xi=\eta}$$

is in $S(h^N m_1 m_2,g)$ for every integer $N$. If the metric $g$

contains no cross products between $dx$ and $d\xi$ then we could also use the definition (1) of pseudo-differential operators and replace the differential operator in the sum by $i\langle D_\xi, D_y\rangle$. The result is then completely parallel to (2)' and a substantial generalization. Note that the composition formula in the Weyl calculus starts with

$$a_1(x,\xi)\, a_2(x,\xi) + \{a_1,a_2\}(x,\xi)/2i$$

where $\{a_1,a_2\}$ is the classical Poisson bracket in classical mechanics.

A systematic development of the Weyl calculus was given in Hörmander [21]. It extends the earlier calculus of general pseudo-differential operators by Beals and Fefferman [3], Beals [1]. In particular a thorough discussion of lower bounds for pseudo-differential operators was given. This important topic has now been greatly advanced by Fefferman and Phong [10,11] but unfortunately there is no time to discuss their results here. Instead we shall give a few observations on localization by means of the general symbolic calculus outlined above.

Let $\Psi_j \in C_0^\infty(\mathbb{R}^{2n})$ have support in a small ball

$$\{(x,\xi);\ g_{x_j,\xi_j}(x-x_j,\xi-\xi_j) < c^2\}$$

and be chosen so that we have bounds independent of $j$ for the derivatives with respect to $g_{x_j,\xi_j}$ orthonormal coordinates. If there is a fixed bound for the number of overlapping supports, one can then regard $\{\Psi j\}$ as a symbol in $S(1,g)$ with values in $l^2 = L(\mathbb{C}, l^2)$. The calculus outlined above is still valid for symbols with values in $L(H_1,H_2)$ where

$H_1$ and $H_2$ are Hilbert spaces; it gives operators from $H_1$ valued functions to $H_2$ valued ones. Hence we obtain from the $L^2$ continuity

$$(15) \qquad \Sigma \, \| \Psi_j^w(x,D)u \|^2 \leqq C \| u \|^2, \, u \in C_0^\infty$$

where the norms are $L^2$ norms. If $\Psi_j$ is real valued and $\Sigma \Psi_j^2 = 1$, the composition formula with $\{\Psi_j\}$ also regarded as a symbol with values in $L(l^2, \mathbb{C})$ gives

$$\Sigma \Psi_j^w \Psi_j^w = 1 + R^w, \quad R \in S(h^2, g) \ .$$

If $h$ is small enough this allows us to prove an estimate opposite to (15). Thus we have essentially succeeded in localizing $u$ to the balls with center $(x_j, \xi_j)$ and $g_{x_j, \xi_j}$ radius $c$, so it is quite natural that the uncertainty relations had to be assumed. If $u$ is a solution of a (pseudo-)differential equation $P(x,D)u=f$ one can often obtain a reduction to manageable local problems if the partition of unity is suitable chosen in terms of the properties of the symbol of $P$. However, we can go no further here in discussing these fairly technical matters.

In addition to the classes of pseudo-differential operators just discussed, the study of mixed boundary problems has led to the introduction of pseudo-differential operators with other, more complicated, conditions on the symbols (see e.g. Melrose [25,26]). The theory of Fourier integral operators can also be extended to far more general symbols (cf. Hörmander [21]) and also far more general symplectic maps $\chi$ (i.e. more general phase functions $\varphi$). Many such results are scattered in the literature but as far as we know there is no systematic presentation available.

Finally it should be mentioned that the microlocal analysis discussed here has an analogue in the analytic category, developed mainly by the Japanese school of M. Sato and his students Kashiwara and Kawai. Indeed, the term "microlocal" was coined by them. We must content ourselves here with referring to Sato-Kawai-Kashiwara [30] and to Guillemin-Kashiwara-Kawai [12].

In conclusion we wish to point out that the literature on pseudo-differential operators is already so vast that it has been impossible to attempt giving a complete bibliography here. We have only given some references related to the historical development as outlined here. They should be supplemented by the references in the quoted papers but will still remain a rather incomplete bibliography.

# REFERENCES


1. R.Beals, Spatially inhomogeneous pseudo-differential operators II. Comm. Pure Appl. Math. 27(1974), 161-205.

2. R.Beals and C.Fefferman, On local solvability of linear partial differential equations. Ann. of Math. 97(1973), 482-498.

3. R.Beals and C.Fefferman, Spatially inhomogeneous pseudo-differential operators I. Comm. Pure Appl. Math. 27(1974), 1-24

4. L.Boutet de Monvel, Boundary problems for pseudo-differential operators, Acta Math. 126(1971), 11-51.

5. A.P.Calderón, Boundary value problems for elliptic equations. Outline Soviet-American Symposium on Partial Differential Equations, Novosibirsk 1963, 303-304.

6. N.Dencker, On the propagation of singularities for pseudo-differential operators of principal type. In preparation.

7. J.J.Duistermaat and V.Guillemin, The spectrum of positive elliptic operators and periodic geodesics. Invent. Math. 29(1975), 184-269.

8. J.J.Duistermaat and L.Hörmander, Fourier integral operators II. Acta Math. 128(1972), 183-269.

9. Yu.V.Egorov, On canonical transformations of pseudo-differential operators. Uspehi Mat. Nauk 25(1969), 235-236.

10. C.Fefferman and D.H. Phong, On positivity of pseudo-differential operators. Proc. Nat. Acad. Sci. USA 75(1978), 4673-4674.

11. C.Fefferman and D.H.Phong, The incertainty principle and sharp Gårding inequalities. Mimeographed manuscript.

12. V.Guillemin, M.Kashiwara and T.Kawai, Seminar on micro-local analysis. Annals of Mathematical Studies 93, Princeton University Press 1979.

13. L.Hörmander, Pseudo-differential operators. Comm. Pure Appl. Math. 18(1965), 501-517.

14. L.Hörmander, Pseudo-differential operators and hypoelliptic equations. Amer. Math. Soc. Symp. Pure Math. 10 (1966), Singular integral operators, 138 - 183.

15. L.Hörmander, Pseudo-differential operators and non-elliptic boundary problems. Ann. of Math. 83 (1966), 129 - 209.

16. L.Hörmander, The calculus of Fourier integral operators. Conference on Prospects in Mathematics, Princeton University Press, 1971, 33 - 57.

17. L.Hörmander, Fourier integral operators. Acta Math. 127 (1971), 79 - 183.

18. L.Hörmander, Linear differential operators. Actes Congr. Int. Math. Nice 1970, 1, 121 - 133.

19. L.Hörmander, On the existence and the regularity of solutions of linear pseudo-differential equations. Ens. Math. 17 (1971), 99 - 163.

20. L.Hörmander, Propagation of singularities and semi-global existence theorems for (pseudo-) differential operators of principal type. Ann. of Math. 108 (1978), 569 - 609.

21. L.Hörmander, The Weyl calculus of pseudo-differential operators. Comm. Pure Appl. Math. 32 (1979), 359 - 443.

22. J.J.Kohn and L.Nirenberg, On the algebra of pseudo-differential operators. Comm. Pure Appl. Math. 18 (1965), 269 - 305.

23. A.Melin and J.Sjöstrand, Fourier integral operators with complex phase functions and parametrix for an interior boundary value problem. Comm. in Partial Diff. Eq. 1 (1976), 313 - 400.

24. A.Melin and J.Sjöstrand, Fourier integral operators with complex phase functions. Springer Lecture Notes 459 (1974), 120 - 223.

25. R.Melrose, Airy operators. Comm. in Partial Diff. Eq. 3 (1978), 1-76

26. R.Melrose, Differential boundary value problems of principal type. Seminar on singularities of solutions of linear partial differential Equations. Princeton University Press 1979, 81 - 112.

27. R.Melrose and G.A.Uhlmann, Lagrangian intersection and the Cauchy problem. Comm. Pure Appl. Math. 32 (1979), 483 - 519.

28. S.G.Michlin and S.Prössdorf, Singuläre Integraloperatoren. Akademie Verlag, Berlin 1980.

29. L.Nirenberg and F.Treves, On local solvability of linear partial differential equations; Part I: Necessary conditions, part II: Sufficient conditions, Correction. Comm. Pure Appl. Math. 23 (1970), 1-38 and 459-509; 24 (1971), 279-288.

30. M.Sato, T.Kawai and M.Kashiwara, Microfunctions and pseudo-differential equations. Lecture Notes in Mathematics 287 (1973), 263 - 529.

31. R.T.Seeley, Elliptic singular integral equations. Amer. Math. Soc. Symp. Pure Math. 10 (1966), Singular integral operators, 308 - 315.

32. J.Sjöstrand, Operators of principal type with interior boundary conditions. Acta Math. 130 (1973), 1 - 51.

33. A.Unterberger, Oscillateur harmonique et opérateurs pseudo-différentiels. Ann. Inst. Fourier 19 (1979), 201 - 221.

# SPHERICAL SPACE FORMS: A SURVEY

Ib Madsen

## 1. Introduction.

Suppose $\phi: \pi \to O(n)$ is an orthogonal representation of the finite group $\pi$. It defines an action of $\pi$ on the unit sphere $S^{n-1}$ in euclidian n-space. If the action is free in the sense that $\phi(g)$ acts without fixed points on $S^{n-1}$ for every $g \neq 1$ then the orbit space is a Riemannian manifold, $L(\pi,\phi) = S^{n-1}/\pi$. It is locally indistinguishable from the sphere but globally it is different. It has constant positive Gaussian curvature; we call it an orthogonal space form. In fact, every complete Riemannian manifold with constant curvature is isometric to some $L(\pi,\phi)$ by celebrated results of H. Hopf and W. Killing.

We can forget all metric properties in the orbit space and just remember that it is a manifold covered by a sphere. Then we get the concept of a topological (or differentiable) spherical space form. It is the orbit space of a free action of $\pi$ on $S^{n-1}$ defined by a representation $\phi: \pi \to \mathrm{Homeo}(S^{n-1})$ (or $\phi: \pi \to \mathrm{Diffeo}(S^{n-1})$) of $\pi$ in the group of homeomorphisms (or diffeomorphisms) of the unit sphere.

Classification of orthogonal space forms is equivalent to questions in linear representation theory. In a very roundabout manner classification of topological and differentiable spherical space forms depends on results in the theory of integral representation theory.

The methods we use in the paper are typical for dealing with classification problems in topology.

One starts with a concrete geometric/topological question; reduce it to an abstract algebraic setting, which can be overlooked. The steps back from the abstract algebra to the concrete geometry is then governed by layers of obstructions. The obstructions lie in groups whose explicit structure depend on arithmetic, e.g. the algebraic K-groups and their unitary analogues. This is one of the virtues of the subject: solutions to natural geometric questions require an interplay with other mathematical disciplines.

The process above, usually called surgery theory, contains contributions from many mathematicians, and I have attempted to give the relevant references throughout the paper. The main conclusions, Theorem A and B in paragraph 4 represent joint work with C.B. Thomas and C.T.C. Wall and work of R.J. Milgram and the author.

The lecture contained results, conjectures and speculations about periodic homeomorphisms on spheres. With the limited space available I have not been able to include this part of the lecture in the writeup.

## 2. From topology to algebra.

In this paragraph we associate to a free topological action $\phi: \pi \to \mathrm{Homeo}(S^{n-1})$ a finite object called a *triangulation of the action*. A triangulation of $\phi$ is a graded, partially ordered finite set on which $\pi$ acts by permutations. The induced integral permutation representation implies a "periodic resolution" of $\mathbb{Z}$ by free modules over the integral group ring $\mathbb{Z}\pi$. The existence of such a resolution has implications for the structure of $\pi$: we get necessary conditions for the existence of a free action.

Before giving the definitions it might be of help to consider some (otherwise irrelevant) examples. The surfaces of the five regular polyhedra are triangulations of the 2-dimensional sphere. Their symmetry groups (the tetrahedral group, the octahedral group and the icosahedral group) are examples of triangulated (but not free) actions. Now the definitions.

Let $X$ be a compact topological space. A homeomorphic image in $X$ of the standard i-dimensional simplex $\Delta^i \subseteq \mathbb{R}^{i+1}$ is called an i-simplex of $X$ and is denoted $\sigma^i$. It has $i+1$ faces $\partial_0\sigma^i,\ldots,\partial_i\sigma^i$ which are simplices of dimension $i-1$. A triangulation of $X$ is a finite collection $S$ of simplices in $X$ subject to the conditions: if $\sigma$ belongs to $S$ then so does any (iterated) face, and distinct elements of $S$ have disjoint interiors (interior $(\sigma) = \sigma - \cup\, \partial_i\sigma$).

Given a triangulated space $X$, an action $\phi: \pi \to \mathrm{Homeo}(X)$ is called triangulated, if each $\phi(g)$ acts on $X$ via a permutation of the elements in the triangulation $S$.

(In the examples above, the cube and the dodecahedron are,

strictly speaking, not triangulated. However, they can easily be subdivided further into triangulated spaces).

Given a free action $\phi: \pi \to \text{Homeo}(S^{n-1})$. The orbit space $S^{n-1}/\pi$ is locally homeomorphic with $S^{n-1}$ and hence a topological manifold. The celebrated results of Kirby and Siebenmann on the structure of topological manifolds imply that $S^{n-1}/\pi$ can be triangulated. If we choose the simplices in $S^{n-1}/\pi$ small enough their inverse images in $S^{n-1}$ are disjoint. They become permuted by the action of $\pi$, so we obtain a triangulated action. If the action to start with is differentiable, $\phi: \pi \to \text{Diffeo}(S^{n-1})$ then the orbit space is a differentiable manifold, and the triangulation exists by old results of J.H.C. Whitehead.

From a triangulation $S$ of the free action $\phi: \pi \to \text{Homeo}(S^{n-1})$ we pass to a more managable object by forming the induced permutation representations. They fit together into a chain complex of free, finitely generated $\mathbb{Z}\pi$-modules. More precisely, let $C_i = C_i(S)$ denote the free abelian group generated by the i-simplices of $S$. There is a homomorphism

2.1 $$\partial_i: C_i \to C_{i-1}, \quad \partial(\sigma^i) = \sum_{j=0}^{i} (-1)^j \partial_j(\sigma^i)$$

and $\partial \circ \partial = 0$. Each $C_i$ is a free module over $\mathbb{Z}\pi$; its basis is in one to one correspondance with the i-simplices in $S/\pi$.

The homology groups $H_i = \text{Ker}(\partial: C_i \to C_{i-1})/\text{Im}(\partial: C_{i+1} \to C_i)$ are independent of the chosen triangulation; they only depend on the topological type of the underlying space. In our case where the underlying space is $S^{n-1}$ the only non-zero groups are $H_0 = \mathbb{Z}$ and $H_{n-1} = \mathbb{Z}$. Thus we obtain an extension (exact sequence)

$$0 \to \mathbb{Z} \xrightarrow{\mu} C_{n-1} \to \dots \to C_0 \xrightarrow{\varepsilon} \mathbb{Z} \to 0 \tag{2.2}$$

of free (in fact based) f.g. modules $C_i$ over $\mathbb{Z}$.

The action of $\pi$ on the right hand $\mathbb{Z}$ is trivial. This need not be the case for the left hand $\mathbb{Z}$ $(=H_{n-1})$. We call the action *oriented* if the action on $H_{n-1}$ is trivial. *In the rest of the paper we only consider oriented actions, taking place on odd dimensional spheres.* This restriction might look severe but in fact it is not. The only cases excluded are $\pi = \mathbb{Z}/2$, acting on even dimensional spheres (the "fake $RP^{2n}$"). This concludes the transition from topology to algebra.

We can now draw conclusions from the algebraic data in 2.2. This will give necessary conditions for the existence of spherical space forms.

Given the extension 2.2, we can splice it together an infinite number of times to get a periodic resolution of $\mathbb{Z}$ by free $\mathbb{Z}\pi$-modules. Therefore the groups $\mathrm{Ext}^k_{\mathbb{Z}\pi}(\mathbb{Z},\mathbb{Z})$ are periodic in $k$ of period $n$. We shall follow tradition in topology and denote $\mathrm{Ext}^k_{\mathbb{Z}\pi}(\mathbb{Z},\mathbb{Z})$ by $H^k(\pi;\mathbb{Z})$, and call it the k'th cohomology group of $\pi$.

The point is, that group cohomology is calculable. For example, it is periodic for a cyclic group but not for a product of cyclic groups. We have recovered P.A. Smith's result

2.3 $\mathbb{Z}/p \times \mathbb{Z}/p$ cannot act freely on a sphere.

In conclusion, if $\pi$ acts freely on a sphere then all its subgroups of order $p^2$ are cyclic: $\pi$ satisfies the $p^2$-conditions. The only p-groups which satisfy this criteria are the cyclic groups $\mathbb{Z}/p^n$ and the quaternion groups $H2^k = \{x,y \mid x^{2^{k-2}} = y^2,\ yxy^{-1} = x^{-1}\}$. Cartan and Eilenberg proved in [CE] that $\pi$ satisfies the $p^2$-conditions if and only if it has periodic cohomology groups. It is fair to say that this reformulation of P.A. Smith' result marked the beginning of the modern development of spherical space forms.

A concise tabulation of the groups with periodic cohomology can be found in J.A. Wolf's book on spaces of constant curvature. We give an indication of the tabulation. Let $O(\pi) \triangleleft \pi$ be the maximal normal subgroup of odd order. This is metacyclic and the quotient $\pi/O(\pi)$ has one of six types:

2.4

| | | | |
|---|---|---|---|
| I. | $\mathbb{Z}/2^k$ (cyclic), | II. | $H2^k$ (quaternion), |
| III. | $T^*$ (binary tetrahedral), | IV. | $O^*$ (binary octahedral), |
| V. | $Sl_2(\mathbb{F}_p)$, | VI. | $Tl_2(\mathbb{F}_p)$. |

Not all the groups listed in 2.4 have free, orthogonal actions on spheres; for example the groups $Sl_2(\mathbb{F}_p)$ of $2 \times 2$ matrices over the field with $p$ elements only allow free orthogonal actions for $p \leq 5$.

The spherical space form problem is to decide which of the groups above admit free, topological (or differentiable) actions on spheres, and to classify such actions.

## 3. Inside algebra: Two invariants.

This paragraph deals with the problem initially considered by R. Swan, of reversing the implication which from the based exact sequence 2.2 deduced periodic cohomology. There are two basic and interrelated invariants: The *Swan obstruction* and the *Reidemeister torsion*.

The invariants live in the algebraic K-groups $K_0(R)$ and $K_1(R)$ associated with a ring $R$ with unit. I recall their definition, and refer the reader to [Mil 2] for further details.

Two R-modules $P$ and $Q$ are called stably equivalent if $P \oplus R^n$ is isomorphic to $Q \oplus R^m$ for suitable $n$ and $m$. We consider the stable equivalence classes of *finitely generated* projective modules. They form an abelian semi-group under direct sum. The associated group is the *reduced projective class group* and is denoted $\tilde{K}_0(R)$. The group is easy to define but not easy to calculate. For example, if $R$ is a Dedekind domain then the order of $\tilde{K}_0(R)$ is known as the class number of the associated field of fractions. This is a classical and difficult invariant of the field.

The other classical algebraic K-group $K_1(R)$ is the universal value group for determinants. It was introduced by J.H.C. Whitehead. Let $Gl_n(R)$ be the group of invertible n×n-matrices over $R$. Then $Gl_n(R) \subset Gl_{n+1}(R)$; we let $Gl(R)$ be the limit. It's maximal abelian quotient is $K_1(R)$. The class in $K_1(R)$ represented by an isomorphism $f\colon C_1 \to C_0$ between based R-modules is denoted $\det(f)$. When $R$ is a field, and in many other commutative cases $K_1(R)$ is equal to the invertible elements $R^{\times}$ in the ring, and $\det(f)$ becomes the usual determinant.

It is a basic result due to R.Swan that if a finite group $\pi$ has periodic cohomology groups then there exists an exact sequence

3.1 $$0 \to \mathbb{Z} \to P_{n-1} \to \dots \to P_0 \overset{\varepsilon}{\to} \mathbb{Z} \to 0 \quad (P_*)$$

where each $P_i$ is a finitely generated and projective $\mathbb{Z}\pi$-module. Here $n$ can be taken to be any period for the cohomology groups of $\pi$, that is, any $n$ such that $H^k(\pi;\mathbb{Z}) \cong H^{k+n}(\pi;\mathbb{Z})$ for all $k > 0$. The reader should compare with 2.2; the only difference is that the modules above are only projective while in 2.2 we have free (in fact based) modules.

The Swan invariant of 3.1 is defined as

3.2 $$\sigma(P_*) = \sum_{i=0}^{n-1} (-1)^i [P_i] \in \tilde{K}_0(\mathbb{Z}\pi) .$$

It is a tautology that $\sigma(P_*) = 0$ if and only if 3.1 can be replaced by an exact sequence $C_*$ with each $C_i$ a finitely generated free $\mathbb{Z}\pi$-module. The replacement is achieved by succesively adding to 3.1 elementary complexes of the form $0 \to P \to P \to 0$ with $P$ projective.

The Swan invariant depends not only on the group in question and the length of the sequence but also on the equivalence class of the sequence as an element of $\mathrm{Ext}^n_{\mathbb{Z}\pi}(\mathbb{Z},\mathbb{Z})$. However, in a certain quotient group of the reduced projective class group it becomes independent of the particular projective sequence considered. We write $\sigma_n(\pi)$ for this reduced invariant. (It lives in $\tilde{K}_0(\mathbb{Z}\pi)/S(\pi)$, where $S(\pi)$ is the subgroup generated by the projective ideals $\langle r,N\rangle \subseteq \mathbb{Z}\pi$ where $(r,|\pi|) = 1$ and $N$ is the norm element).

The reduced projective class group is finite by results from [S2]. On the other hand, it follows from 3.2 that $\sigma_{kd}(\pi) = k\sigma_d(\pi)$. Hence the Svan obstruction vanishes for suitable sequences. But the length of the sequences are left uncontrolled. In particular it need not be equal to the minimal period $d = d(\pi)$ for $H^k(\pi;\mathbb{Z})$. In [W3], Swan's result was improved to give $\sigma_{2d}(\pi) = 0$, but the question remained unanswered if $\sigma_d(\pi) = 0$ for certain groups of type II in the classification list 2.4. This was settled in the negative in [Mg1]. Subsequently the Swan obstruction has been calculated in many cases and with the recent results in [Mg3] it is fair to say that it is well under control. For certain type II groups its actual value depends on the arithmetic of the number fields in the center of the rational group ring. More details are given in the next paragraph.

Our second invariant, the torsion, is an invariant of *based* exact sequences. The sequence in 2.2 is not of this form, since the outer terms $\mathbb{Z}$ are not even free. So first we must associate to 2.2 a based exact sequence over a related ring. This requires we introduce denominators. Given 2.2 we define

$$D_i = (1-\varepsilon)(C_i \otimes \mathbb{Q})$$

where $\varepsilon$ is the usual central idempotent in $\mathbb{Q}\pi$, present in any group ring, $\varepsilon = 1/|\pi| \sum \{g \mid g \in \pi\}$.

We get an exact sequence

$$D_{n-1} \xrightarrow{\partial} D_{n-2} \xrightarrow{\partial} \cdots \xrightarrow{\partial} D_0$$

of based modules over the ring $(1-\varepsilon)\mathbb{Q}\pi$ ($=I\pi$, the augmentation ideal in $\mathbb{Q}\pi$). There exist maps in the other direction $\delta: D_i \to D_{i+1}$ forming a contracting homotopy (cf. [C]) and

$$\partial + \delta: \bigoplus D_{2i} \to \bigoplus D_{2i-1}$$

is an isomorphism. Its determinant in $K_1(I\pi) \subseteq K_1(\mathbb{Q}\pi)$ is denoted $\Delta(C_*;\mathbb{Q})$. This is the Reidemeister torsion, cf. [Mil3].

The Swan invariant and the Reidemeister torsion are related in a somewhat complicated algebraic fashion, which I now attempt to outline; but it is rather rough going. First, we need a procedure for constructing projective modules over group rings, alias for calculating $\tilde{K}_0(R)$. The point of departure is that $\tilde{K}_0(R)$ is easy for semi-simple rings. This follows from Wedderburns structure theorems. Integral group rings $\mathbb{Z}\pi$ are not semi-simple, but are surrounded by rings which are, or which can be reduced to semi-simple rings.

More precisely, let $\hat{\mathbb{Z}} = \Pi\hat{\mathbb{Z}}_p$ where $\hat{\mathbb{Z}}_p$ denotes the p-adic integers and let $\hat{\mathbb{Q}} = \hat{\mathbb{Z}} \otimes \mathbb{Q}$. Then

$$\mathbb{Z}\pi = \hat{\mathbb{Z}}\pi \cap \mathbb{Q}\pi$$

where the intersection takes place in $\hat{\mathbb{Q}}\pi$. The fundamental point can now be stated as follows: to every isomorphism $f: (\hat{\mathbb{Q}}\pi)^m \to (\hat{\mathbb{Q}}\pi)^m$ the $\mathbb{Z}\pi$-module

$$P(f) = \{(\hat{a},a^0) \in (\hat{\mathbb{Z}}\pi)^m \times (\mathbb{Q}\pi)^m \mid f(\hat{a}) = a^0\}$$

is projective. This leads to the exact sequence

$$3.3 \qquad K_1(\mathbb{Z}\pi) \to K_1(\hat{\mathbb{Z}}\pi) \oplus K_1(\mathbb{Q}\pi) \overset{J}{\to} K_1(\hat{\mathbb{Q}}\pi) \to \tilde{K}_0(\mathbb{Z}\pi) \to 0,$$

where $\partial([f]) = [P(f)]$. Such types of sequences were originally constructed in [Mil 2].

The gluing together process giving $P(f)$ above can be generalized to extensions: instead of constructing one projective module one constructs an exact sequence of the form 3.1. More precisely, one starts out with exact sequences (and they can be *explicitly* constructed)

$$0 \to \hat{\mathbb{Z}} \to \hat{C}_{n-1} \to \dots \to \hat{C}_0 \to \hat{\mathbb{Z}} \to 0$$

$$0 \to \mathbb{Q} \to C^0_{n-1} \to \dots \to C^0_0 \to \mathbb{Q} \to 0$$

of based $\hat{\mathbb{Z}}\pi$- and $\mathbb{Q}\pi$-modules, respectively. They are glued together over $\hat{\mathbb{Q}}\pi$ to give an exact sequence $P_*$ as in 3.1. The promised relationship between our two invariants can be stated in

3.4 $$\sigma(P_*) = \partial(\Delta(C^0_*;\hat{\mathbb{Q}}) - \Delta(\hat{C}_*;\hat{\mathbb{Q}})).$$

This process was used in [W3], and a variant of it in [Mg1].

The two middle terms in 3.3 can be calculated in terms of the centers in the respective rings via the reduced norm homomorphism; but the map $J$ between them is not so easy. This is one place where the arithmetic comes in. In any case, the results of Milgram demonstrate forcefully that actual calculations can be carried out: $\sigma(P_*)$ can be determined from 3.4 and 3.3 in many cases.

If $\sigma(P_*) = 0$ then the sequence can be replaced with a sequence $C_*$ consisting of free modules, rather than projective ones. We have lost the bases, so $C_*$ does not have a well-defined torsion invariant in $K_1(\mathbb{Q}\pi)$. But there is left an invariant $\Delta(P_*) \in K_1(\mathbb{Q}\pi)/K_1(\mathbb{Z}\pi)$, the reduced torsion, which indicates the Reidemeister torsion after some choice of basis.

It is a secondary invariant, defined only if the Swan obstruction vanishes. Its value explicates the vanishing of $\sigma$:

3.5 $$\Delta(P_*) = a^0 - \Delta(C_*^0,\mathbb{Q}),$$

where $J(\hat{a},a^0) = \Delta(\hat{C}_*;\hat{\mathbb{Q}}) - \Delta(C_*^0,\hat{\mathbb{Q}})$.

## 4. From algebra to topology.

Starting with an extension

$$0 \to \mathbb{Z} \to C_{n-1} \to \dots \to C_0 \to \mathbb{Z} \to 0 \quad (C_*)$$

consisting of f.g. based $\mathbb{Z}\pi$-modules, the object is to construct a free action $\phi\colon \pi \to \mathrm{Diffeo}(S^{n-1})$. There are three steps in the construction.

The *first step* is to construct a triangulated space $\Sigma$ with a free, triangulated action $\phi\colon \pi \to \mathrm{Homeo}(\Sigma)$ such that the associated chain complex $C_*(\Sigma)$ is isomorphic to $C_*$ above. This uses classical theorems of Hurewicz and Whitehead. In fact when $n \geq 6$ one can assume that $\Sigma$ is homotopy equivalent (that is, contineously deformable) to a sphere. Details can be found in [S1].

The *second step* is to construct over the orbit space $X = \Sigma/\pi$ a vector bundle $\nu$ which can play the role of the normal bundle. More precisely, let $N(X)$ denote equivalence classes (under cobordism) of pairs $(c,\nu)$ where $\nu$ is a $(k+1)$-dimensional vector bundle and $c\colon S^{n+k} \to T(\nu)$ is a map of "degree one". Here $T(\nu)$ denotes the one point compactification of $\nu$ and $k$ is large. If $X$ is a manifold, then $\nu$ is the normal bundle to an embedding $X \subseteq \mathbb{R}^{n+k}$, and $c$ is the map which collapses points of $\mathbb{R}^{n+k}$ outside a small tube around $X$ to one point.

Every equivalence class $[c,\nu]$ contains a representative $c\colon S^{n+k} \to T(\nu)$ which is transverse regular ("in general position") with respect to the zero section $\Sigma/\pi \subseteq T(\nu)$. Then

$M = c^{-1}(\Sigma/\pi)$ is a smooth manifold. In general $M$ is "larger" than $\Sigma/\pi$ (it has more homology).

The *third step* is to examine if $[c,\nu]$ contains a representative for which $M = c^{-1}(\Sigma/\pi)$ is homotopy equivalent to $\Sigma/\pi$. In general this is not the case but there is a map (the surgery obstruction)

4.1 $$\lambda\colon N(\Sigma/\pi) \to L_{n-1}(\mathbb{Z}\pi)$$

and the question has a positive solution iff $\lambda([c,\nu]) = 0$. Here $L_{n-1}(\mathbb{Z}\pi)$ is the socalled surgery obstruction group; it is related to the algebraic K-groups considered in the previous paragraph, and "calculated" for finite $\pi$ in [W2].

If we can find $[c,\nu] \in N(\Sigma/\pi)$ with $\lambda([c,\nu]) = 0$ and hence a manifold $M^{n-1}$ homotopy equivalent to $\Sigma/\pi$ then we are done. Indeed, the universal cover $\tilde{M}^{n-1}$ of $M^{n-1}$ is a manifold on which $\pi$ acts freely as a group of covering transformations, and it is homotopy equivalent to $\Sigma$, hence to $S^{n-1}$. The h-cobordism theorem proved by S. Smale implies that $\tilde{M}^{n-1}$ is homeomorphic to $S^{n-1}$, if $n \geq 6$. Off hand, $\tilde{M}^{n-1}$ need not be diffeomorphic to $S^{n-1}$, it could be an exotic smoothing. However, the arguments can be sharpened to actually give $\tilde{M}^{n-1} = S^{n-1}$ and hence a free action $\phi\colon \pi \to \mathrm{Diffeo}(S^{n-1})$.

The above machinery, called surgery theory, can be found in the books [B] and [W1]. The main contributors to the theory are Kervaire, Milnor, Browder, Novikov, Sullivan and Wall.

The key to the solution of the spherical space form problem is to apply a Brauer-type induction theory to both $N(\Sigma/\pi)$ and $L_{n-1}(\mathbb{Z}\pi)$. Keeping $\Sigma$ fixed, both groups are contravariant functors on the lattice of subgroups of $\pi$, and $\lambda$ in 4.1 is a natural transformation. We single out the set of 2-hyperelementary subgroups $H(\pi)$ of $\pi$ to induct from. In our case, an element $\rho \in H(\pi)$ is an extension of an odd order cyclic group by a cyclic 2-group or by the quaternion group,

$$4.2 \quad \begin{array}{ll} \text{(i)} & 0 \to \mathbb{Z}/m \to \rho \to \mathbb{Z}/2^r \to 1 \\ \text{(ii)} & 0 \to \mathbb{Z}/m \to \rho \to H2^k \to 1 \end{array}$$

($m$ is odd).

The 'induction' result we need is the following commutative diagram

$$4.3 \qquad \begin{array}{ccc} N(\Sigma/\pi) & \xrightarrow{\lambda} & L_{n-1}(\mathbb{Z}\pi) \\ N \downarrow \cong & & L \downarrow \cong \\ \varprojlim N(\Sigma/\rho) & \xrightarrow{\lambda} & \varprojlim L_{n-1}(\mathbb{Z}\rho) \end{array}$$

The $\varprojlim$ is taken over the category of subgroups $\rho \in H(\pi)$ with maps induced by conjugation. It is a subgroup of the direct product (over $\rho \in H(\pi)$), consisting of families of elements which satisfy certain compatibility conditions. The elements of $\varprojlim$ are called stable elements in [CE]. The left hand isomorphism in 4.3 follows because by a theorem of Boardmann and Vogt, $N(X)$ is a generalized cohcmology theory in X. The right hand isomorphism is due to A.Dress.

Let p and q be primes. A group is said to satisfy the pq-condition if each subgroup of order pq is cyclic. J.A.Wolf proves in [Wo] that a solvable group admits a free orthogonal action on a sphere iff it satisfies all pq-conditions. The groups in 4.2(ii) do satisfy Wolf's criteria: free actions exist on (8k-1)-dimensional spheres, but the groups in 4.2(i) satisfies the criteria only if the unique element of order 2 is central.

Fortunately, back in 1957 Milnor proved that the 2p-conditions are necessary even to have a free topological action. The argument is a generalization of the well-known Borsuk-Ulam theorem and is based on Lefschetz duality in the homology groups of manifolds, see [Mil 1] .

In conclusion, for the relevant groups $\pi$ each $\rho \in H(\pi)$ admits a free orthogonal action on $S^{n-1}$ as long as $n$ is divisible by twice the cohomological period of $\pi$, $2d(\pi) | n$. Here is the main result:

Theorem A. (i) A finite group $\pi$ admits a free, topological action on a sphere if and only if it satisfies all 2p- and $p^2$-conditions. In this case there actually exists a free $\phi: \pi \to \mathrm{Diffeo}(S^{2d-1})$, $d = d(\pi)$.

(ii) Up to conjugacy, $\phi$ is determined uniquely by the value on all $\rho \in H(\pi)$, and there exists $\phi$ such that the restrictions $\phi | \rho$ are orthogonal.

The theorem is an application of the theory outlined above. It is the outcome of a collaboration with C.B.Thomas and C.T.C.Wall. Partial results about the existence of spherical space forms for certain odd order groups were obtained earlier by Petrie and R.Lee. The uniqueness result in (ii) uses a recent result in algebraic K-theory proved by R.Oliver: the group $SK_1(\mathbb{Z}\pi) = 0$, or equivalently Torsion $K_1(\mathbb{Z}\pi) = \langle \pm 1 \rangle \oplus \pi/[\pi,\pi]$ for metacyclic groups $\pi$, [O]

Spherical space forms are generalizations of (complete)

Riemannian manifolds of positive *constant* curvature (orthogonal space forms). Theorem A above tells us that all groups listed in 2.4 except some of type I (which do not satisfy the 2p-conditions) have spherical space forms $S^{n-1}/\pi$. The class of groups $\pi$ which have orthogonal space forms is more restricted: $\pi$ must satisfy all pq-conditions, and if $\pi$ has type V or VI then $p \leq 5$. So we know from the outset that many of the space forms from Theorem A cannot have a constantly curved Riemannian metric. It is natural to ask:

Question: Do the spherical space forms from Theorem A admit Riemannian metrics with positive curvature? (The group of smallest order for which the question is relevant is the non-abelian group of order 21. It is generated by elements A and B of order 7 and 3, respectively. We can make Theorem A (ii) explicit as follows: there exists a unique free differentiable action $\phi$ on $S^5$ conjugate to $\psi_A$ and $\psi_B$ on the subgroups <A> and <B>, where

$$\psi_A(A)(z_1,z_2,z_3) = (\zeta_7 z_1, \zeta_7^2 z_2, \zeta_7^4 z_3)$$

$$\psi_B(B)(z_1,z_2,z_3) = (\zeta_3 z_1, \zeta_3 z_2, \zeta_3 z_3)$$

Here $\zeta_3$, $\zeta_7$ are roots of 1, $\zeta_p = e^{2\pi i/p}$).

Theorem A leaves open the question of existence of spherical space forms in the period dimension. The problem occurs when $\pi$ contain groups of the form

$$0 \to \mathbb{Z}/pq \to \tau_k(pq) \to H2^k \to 1, \quad k \geq 3$$

where the centralizer of $\mathbb{Z}/pq$ in $\tau_k(pq)$ is cyclic of order $2^{k-2}pq$. Such groups cannot act freely on spheres in dimensions

$8l+3$ by orthogonal maps, and the question becomes if there exist topological (or differentiable) free actions in these dimensions.

In 1973, R.Lee proved that $\tau_k(pq)$ cannot act freely on any $S^{8l+3}$ if $k > 3$. Reinterpreted in our terminology, he assumed the Swan obstruction vanished and proved that the surgery obstruction did not. More precisely, the composition

$$N(\Sigma/\tau_k(pq)) \xrightarrow{\lambda} L_{8l+3}(\mathbb{Z}[\tau_k(pq)]) \longrightarrow L_{8l+3}(\hat{\mathbb{Z}}_2[\tau_k(pq)])$$

is injective. We are left with $\tau(pq) = \tau_3(pq)$ of order $8pq$. There are two problems, namely to calculate the Swan obstruction and to calculate the surgery obstruction.

Explicit calculation of $\sigma_{8l+4}(\tau(pq))$ was initiated by R.J.Milgram in [Mg 1]. Subsequently, in [Mg 3] the calculations were perfected and a solid understanding of the number theory involved were obtained. For example, if $p \equiv 3(4)$ then $\sigma_{8l+4}(\tau(pq)) = 0$ iff $q \equiv 1(8)$ or $q \equiv 5(8)$ and $p$ has odd order in $(\mathbb{Z}/q)^{\times}/\langle\pm 1\rangle$.

The surgery obstruction was treated in [M2] and a little later (using somewhat different methods) in [Mg2]. Roughly speaking the surgery obstruction is calculated from the Reidemeister torsion invariant of 3.5. It can be non-zero only because the torsion of manifolds are more restricted than the torsion of triangulated spaces. More precisely, consider orbit spaces $\Sigma/\pi$ of triangulated actions. Let $\Delta(\Sigma/\pi) \in K_1(\mathbb{Q}\pi)$ be the Reidemeister torsion of the associated chain complex $C_*(\Sigma)$. (This is a topological invariant by a result of T.A.Chapman)

There is an action of $K_1(\mathbb{Z}\pi)$ on orbit spaces $\Sigma/\pi$ and $\Delta((\Sigma/\pi)^{\alpha}) = i(\alpha) + \Delta(\Sigma/\pi)$ where $i: K_1(\mathbb{Z}\pi) \to K_1(\mathbb{Q}\pi)$. This action does not restrict to an action an manifolds: in particular, it is not possible to change the Reidemeister torsion of a given spherical space form by an arbitrary element of $K_1(\mathbb{Z}\pi)$.

The following sample results are combinations of results from [M2], [M3] and [Mg3].

<u>Theorem B</u>. (i) Let $p \equiv 3(4)$. If $q \not\equiv 1(8)$ then $\tau(pq)$ cannot act freely on any sphere in dimension $\equiv 3(8)$.

(ii) If $p \equiv 3(4)$ and $q \equiv 1(8)$ then $\tau(pq)$ admits a free differentiable action on some sphere in dimension $\equiv 3(8)$. The same is true if $p \equiv 1(8)$, $q \equiv 1(8)$ and the quadratic symbol $(\frac{p}{q}) = -1$

The proof is similar in spirit to the proof of A in that it follows the prescription: "reduce to subgroups"; but in details it is different. The group $\tau = \tau(pq)$ has three interesting subgroups $\tau_i$, $i = 1,2,3$ of order $8p$, $8q$ and $4pq$, respectively. Each subgroup has a free orthogonal action on $S^{8l+3}$. Assuming the Swan obstruction vanishes (i.e. $q \equiv 1(8)$ or $q \equiv 5(8)$ with $p$ of odd order in $(\mathbb{Z}/q)^{\times}/\langle\pm 1\rangle$) we have a triangulated space $\Sigma/\tau$ with $\Sigma/\tau_i$ deformable to an orthogonal space form $S^{8l+3}/\tau_i$ for $i = 1,2$ and 3. A diagram similar to 4.3 implies an element $[c,\nu] \in N(\Sigma/\tau)$ whose surgery obstruction $\lambda[c,\nu]$ lies in the kernel of

4.4 $$L: L_{8l+3}(\mathbb{Z}\tau) \to \bigoplus_{1}^{3} L_{8l+3}(\mathbb{Z}\tau_i)$$

This kernel can be computed explicitly in favorable cases, and

$\lambda([c,\nu])$ can be identified.

A few words about the arithmetic involved in carrying out this program is in order. Consider the Dedekind domains $A_{pq} = \mathbb{Z}(\zeta_p+\zeta_p^{-1},\zeta_q+\zeta_q^{-1}) \subseteq \mathbb{R}$ and $B_{pq} = \mathbb{Z}(\zeta_{pq}+\zeta_{pq}^{-1}) \subseteq \mathbb{R}$. One (particular difficult) summand of Ker(L) is given by the subgroup of elements of order 2 in $\mathrm{Kernel}(\tilde{K}_0(A_{pq}) \to \tilde{K}_0(B_{pq}))$. In [M3], I show that this summand is mapped monomorphically into $\Pi\{K_1(\mathbb{F}_l[\pi])/\langle 2\rangle \mid l \text{ prime } (l,|\pi|)=1\}$ by a 'modular' Reidemeister torsion invariant. Moreover, for the cases of B(ii) it is proved that $\lambda([c,\nu])$ maps trivially, and hence has no component in the above part of Ker(L).

The rest of the kernel in 4.4 depends on the maps $\rho_A\colon A_{pq}^\times \to F_{pq}^\times$ and $\rho_B\colon B_{pq}^\times \to F_{pq}^\times$, where $F_{pq}$ is the common residue ring of $A_{pq}$ and $B_{pq}$ obtained by dividing out the principal ideal generated by pq. For example, if $\rho_A\colon A_{pq}^\times \to F_{pq}^\times$ maps onto all elements of order 2 then this part of Ker(L) vanishes, so $\lambda([c,\nu]) = 0$. (e.g. $p=3$, $q=89$). More generally, $\lambda([c,\nu]) = 0$ iff a certain element $\rho_B(\Delta)$, whose value depends on the reduced torsions invariant in 3.5 belongs to the image of $\rho_A$. This is the case for the groups in B(ii).

The results above are (unpleasantly) irregular, and one could be tempted to change the question. One can ask instead for actions on $\mathbb{R}^{8l+4}$, free outside the origin. Of course, a free action on $S^{8l+3}$ implies by radial extensions an action on $\mathbb{R}^{8l+4}$ of this type.

Conjecture. If $\sigma_d(\pi) = 0$ then there exists an action of $\pi$ on $\mathbb{R}^d$ which is free outside the origin.

## REFERENCES.


[B] W. Browder, Surgery on simply connected manifolds, Ergebnisse der Math.62, Springer Verlag (1972)

[BV] J. Boardman, R. Vogt, Homotopy everything H-spaces, Bull.Amer.Math.Soc.,74(1968), 1117-1122.

[CE] H. Cartan, S. Eilenberg, Homological algebra, Princeton University Press, (1956).

[C] M. Cohen, A course in simple-homotopy theory, Springer Graduate Texts in Math.10(1973).

[D] A. Dress, Induction and structure theorems for orthogonal representations of finite groups, Ann.of Math.102(1975)291-326.

[H] H. Hopf, Zum Clifford-Kleinschen Raumproblem, Mathematische Annalen, vol.95(125)113-339.

[K] W. Killing, Über die Clifford-Kleinschen Raumformen, Mathematische Annalen, vol.39(1891).

[KS] R. Kirby, L. Siebenmann, Foundational essays on topological manifolds smoothings and triangulations, Ann.of Math. studies 88, Princeton Univ. Press, 1977.

[L] R. Lee, Semicharacteristic classes, Topology 12 (1973)183-199.

[MTW] I. Madsen, C.B. Thomas, C.T.C. Wall, The topological spherical space form problem, Topology 15(1978)375-382.

[M1] I. Madsen Spherical Space Forms, Proc.of the IMC, Helsinki 1978.

[M2] I. Madsen, Spherical space forms in the period dimen sion: I, Proc. for conference on algebrai topology, Moscow 1979; Preprint, Aarhus 7

[M3] I. Madsen, Spherical space forms in the period dimen sion: II, (to appear).

[Mg1] R.J. Milgram, The Swan finiteness obstruction for periodic groups, Preprint Stanford 1979.

[Mg2] R.J. Milgram, Exotic examples of free group actions on spheres, Preprint Stanford 1979.

[Mg3] R.J. Milgram, Odd index subgroups of units in cyclotomi fields and applications, Preprint Stanfor 1980.

[Mil 1] J. Milnor, Groups which operate on $S^n$ without fixed points, Amer.J.Math.79(1975)623-630.

[Mil 2] J. Milnor, Introduction to algebraic K-theory, Ann. of Math. Studies, 72, Princeton Univ. Press 1971.

[Mil 3] J. Milnor, Whitehead torsion, Bull AMS 72(1966)358-426.

[O] R. Oliver, $SK_1$ for finite group rings: I, Invent. Math. (to appear).

[S] P.A. Smith, Permutable periodic transformations, Proc.Nat.Acad.Sci.U.S.A.30, 105-108.

[S1] R. Swan, Periodic resolutions for finite groups, Ann.of Math.72(1960)267-291.

[S2] R. Swan, Induced representations and projective modules, Ann.of Math.71(1960)552-578.

[P] T. Petrie, Free metacyclic group actions on homotopy spheres, Ann.of Math.94(1971)108-124.

[W1] C.T.C. Wall, Surgery on Compact manifolds, Academic Press, 1970.

[W2] C.T.C. Wall, On the classifications of hermitian forms VI. Group rings, Ann.of Math.103 (1976)1-80.

[W3] C.T.C. Wall, Free actions of finite groups on spheres, Proc.Sym.Pure Math. vol.43(1978)115-124.

[Wh 1] J.H.C. Whitehead, On $C^1$ complexes, Ann.of Math.4(1940) 804-824.

[Wh 2] J.H.C. Whitehead, Simple homotopy types, Amer.J.Math.72 (1952)1-57.

[Wo] J.A. Wolf, Spaces of constant curvature, Publish or Perish, 1977.

/KB

# FORMAL POWER SERIES IN NONCOMMUTING VARIABLES

Arto Salomaa

1. Introduction. The theory of formal power series in noncommuting variables was initiated around 1960 - apart from some scattered work done earlier in connection with free groups. As with most mathematical formalisms, also the formalism of power series is capable of unifying and generalizing known results. However, it is also capable of establishing specific results which are difficult if not impossible to establish by other means. In particular, this has been the case with respect to automata and language theory. That power series constitute a powerful tool in automata and language theory depends on the fact that they in a sense lead to the arithmetization of the theory. Indeed, several important results have been established using power series arguments and, for a number of such results, the only proofs known depend on power series. Some such results are mentioned below.

This paper is mainly expository in nature and discusses some recent work and presents also some open problems. Most of the material selected belongs to the area of rational power series. Algebraic series are briefly discussed in Section 7, where also a new "super normal form" result is established for algebraic series. The attention is focused on results applicable to language theory. Thus, topological aspects are not discussed at all.

Apart from automata and language theory, power series in noncommuting variables have been applied, for instance, to the enumeration of graphs and to other enumeration problems, as well as to nonlinear systems theory. The reader is referred to [2] for further details. Formal power series constitute an important tool also in the theory of developmental systems, see [14] or [18].

Although this paper is mostly self-contained, the reader is referred to [18] for further details and for all proofs of the basic results. As regards unexplained notions in the theories of automata and formal languages, cf. [16].

We conclude this section with some historical remarks. A systematic study of formal power series in noncommuting variables was initiated by M. P. Schützenberger, [19] - [21], to whom also belong some of the major results concerning rational series. However, the history of the fore-runners of the theory has a definite Scandinavian flavor. Some of the basic notions were implicitly present in the work of J. Nielsen concerning free groups. See, for instance, [9]. Th. Skolem presented in the 8th Scandinavian Congress of Mathematicians, [22], the basic idea of the result that became known later as the Skolem-Mahler-Lech Theorem. This result is of fundamental importance for the decidability theory of rational power series. See [18] for details and further references. Scandinavian flavor is present also in the very early history, although in a somewhat different sense: the papers of Fatau and Hadamard,

[5] and [6], appeared in Acta Mathematica! The significance of the papers to the theory of formal power series in noncommuting variables will become apparent below.

2. Basics of rational series. Consider a monoid $M$ and a semiring $A$. Mappings $r$ of $M$ into $A$ are called formal power series. The values of $r$ are denoted by $(r,w)$, where $w \in M$, and $r$ itself is written as a formal sum

$$r = \sum_{w \in M} (r,w)w.$$

The values $(r,w)$ are also referred to as the coefficients of the series. We shall be interested in this paper in the case where $M$ is the free monoid $X^*$ generated by an alphabet $X$. Then we also say that $r$ is a series with (noncommuting) variables in $X$. The identity of $X^*$ (i.e., the empty word) is denoted by $\lambda$.

The collection of all formal power series in this set-up is denoted by $A\langle\langle M\rangle\rangle$ or $A\langle\langle X^*\rangle\rangle$. Given $r$, the subset of $M$ defined by

$$\text{supp}(r) = \{w \mid (r,w) \neq 0\}$$

is termed the support of $r$. The subset of $A\langle\langle M\rangle\rangle$ consisting of all series with a finite support is denoted by $A\langle M\rangle$. Elements of $A\langle M\rangle$ are referred to as polynomials.

In most applications to automata and formal languages, the semiring $A$ will be either $N$ (the semiring of nonnegative integers) or $Z$ (the semiring, in fact a ring, of all integers). For simplicity, we shall assume below that $M = X^*$, although some of the results are valid also in the more general case. However, observe that for instance neither the definition of product given below nor Schützenberger's Representation Theorem is valid for arbitrary monoids.

The sum and product of two series $r$ and $s$ are defined by

$$r + s = \sum_{w \in X^*} ((r,w) + (s,w))\, w$$

and

$$rs = \sum_{w \in X^*} \Big( \Big( \sum_{w_1 w_2 = w} (r,w_1)(s,w_2) \Big) \Big) w \, .$$

Sum, product and the quasi-inverse defined below are referred to as <u>rational operations</u>.

Quasi-inverse $r^+$ is defined for <u>quasi-regular</u> series, i.e.,series $r$ satisfying $(r,\lambda) = 0$, by

$$r^+ = \lim_{k \to \infty} \sum_{i=1}^{k} r^i \, .$$

(We say that a sequence of series $r_1, r_2, \ldots$ converges to a limit $r$, in symbols, $\lim_{j \to \infty} r_j = r$ if for all $n$ there exists an $m$ such that the conditions $lg(w) \leq n$ and $j > m$ imply the condition $(r_j,w) = (r,w)$. )

The family of <u>A-rational</u> series, in symbols $A^{rat} \ll X^* \gg$, is the collection of series obtained from polynomials by (finitely many applications of) rational operations.

For a semiring $S$, we denote by $S^{m \times m}$ the collection (in fact, a semiring) of $m \times m$ matrices with entries in $S$. A series $r$ of $A \ll X^* \gg$ is termed <u>A-recognizable</u> (in symbols $r \in A^{rec} \ll X^* \gg$ ) if

$$r = (r,\lambda)\lambda + \sum_{w \neq \lambda} (\pi h(w) \eta)\, w \, ,$$

where $h : X^* \to A^{m\times m}$, $m \geq 1$, is a homomorphism and $\pi$ (resp. $\eta$) is an m-dimensioned row (resp. column) vector. (Instead of $\pi$ and $\eta$, more general projections can be considered.)

By Schützenberger's Representation Theorem, the families $A^{rat} \langle\langle X^* \rangle\rangle$ and $A^{rec} \langle\langle X^* \rangle\rangle$ coincide. For a proof, we refer to [18].

The family of supports of N-rational series equals the family of regular languages. The family of supports of Z-rational series (referred to as the family of Z-rational languages) is strictly larger. Indeed, there are non-context-free Z-rational languages, for instance, the language

$$\{a^m b^n \mid m,n \geq 1 \text{ and } n \neq m^2\}.$$

We would like to emphasize that very little is known about Z-rational languages although, at least from the mathematical point of view, this family is very natural in formal language theory.

The theory of rational series can be roughly divided into the following three parts:

(i) Study of supports of A-rational, in particular, Z-rational series. For instance, how does the growth rate of the coefficients affect the support?

(ii) Study of properties, in particular arithmetical and growth properties, of the coefficients.

(iii) Decision problems.

Some examples of (i) were already given above. Some typical results concerning (ii) and (iii) are given in Theorems 1 and 2 below. The reader is referred to [18] for proofs and further references. (Indeed, the proof of Theorem 1 is rather straightforward.) Before stating the theorems, we introduce some further terminology and notations.

We denote by R (resp. Q) the semiring of real (resp. rational) numbers. The length of a word $w$ is denoted by $lg(w)$. A series $r$ in $R \ll X^* \gg$ is <u>bounded</u> (resp. <u>polynomially bounded</u>) if there is a natural number $k$ such that, for all words $w$ in $X^*$,

$$|(r,w)| \leq k \quad \text{(resp. } |(r,w)| \leq (lg(w))^k + k).$$

(Observe that the term $k$ is added to take care of words of length 0 and 1.)

<u>Theorem 1.</u> If $r \in R^{rat} \ll X^* \gg$ then there is a positive constant $c$ such that, for all words $w$ in $X^*$,

$$|(r,w)| \leq c^{1+lg(w)}.$$

Consequently, there is a positive constant $c_1$ such that the family

$$\{(r,w) / c_1^{1+lg(w)}\}$$

is absolutely summable.

<u>Theorem 2.</u> Let $X$ be an alphabet containing at least two letters. Then it is undecidable whether or not a given series $r$ in $Z^{rat} \ll X^* \gg$

(i) has at least one qero coefficient,

(ii) has infinitely many qero coefficients,

(iii) has at least one positive coefficient,

(iv) has infinitely many positive coefficients,

(v) has its coefficients ultimately nonnegative,

(vi) has a regular support,

(vii) has two equal coefficients.

On the other hand, it is decidable whether or not $r$ is (i) identically qero, (ii) a polynomial, (iii) ultimately constant, (iv) bounded. Consequently, it is decidable whether or not two given series in $Z^{rat} \ll X^* \gg$ are identical.

The reader is referred to [4], [7], [13], [17] and [18] for further decidability results related to Theorem 2.

<u>3. Rational sequences.</u> The case where the basic alphabet $X$ consists of a single letter, $X = \{x\}$, is of special interest. In this case a power series defines a sequence $a_i$ of elements of the basic semiring $A$, namely, the sequence of coefficients of $x^i$, for $i = 0,1,2,\ldots$ . Thus, we may speak of Z-rational and N-rational sequences. By Schützenberger's Representation Theorem, a sequence of integers $a_i$ is Z-rational (resp. N-rational) if and only if there are a square matrix $M$, a row vector $\pi$ and a column vector $\eta$ with integer (resp. nonnegative integer) entries such that

$$a_i = \pi M^i \eta \quad \text{for all } i.$$

It is easy to see that this is the case if and only if, for all $i \geq 1$, $a_i$ equals the upper right-hand corner of $M^i$, where $M$ is a square matrix with integer (resp. nonnegative integer) entries.

The important problem of characterizing N-rational sequences within the family of Z-rational sequences was solved by Berstel and Soittola, [1] and [23]. The proof of the following characterization result can be found also in [18].

Theorem 3. Assume that $a_i$ is a Z-rational sequence of nonnegative numbers. Then $a_i$ is N-rational if and only if there are natural numbers $m$ and $p$ such that if $0 \leq j \leq p-1$ then, for all $n$,

$$a_{m+j+np} = P_j(n)\alpha_j^n + \sum_i P_{ji}(n)\alpha_{ji}^n$$

where $\alpha_j \geq 0$, $\alpha_j > \max_i |\alpha_{ji}|$ and the $P_j$'s and $P_{ji}$'s are nonzero polynomials.

The characterization of a "dominating pole" contained in Theorem 3 shows, among other things, that not all Z-rational sequences of nonnegative numbers are N-rational. A counter example is provided by the sequence defined as follows:

$$a_{2i} = 30^i \ , \quad a_{2i+1} = 25^i \cos^2 2\pi i\alpha \ ,$$

where $\cos 2\pi\alpha = \frac{3}{5}$ and $\sin 2\pi\alpha = \frac{4}{5}$. The same characterization leads also to the following important decidability result.

Theorem 4. It is decidable whether or not a given Z-rational sequence is N-rational.

Apart from automata theory, N-rational sequences have been applied to the theory of growth functions of developmental systems, cf. [11] and [14]. Our next theorem belongs to this area. We give first the necessary definitions.

A DTOL system is a tuple

$$G = (X, h_1, \ldots, h_m, w),$$

where $X$ is an alphabet, $m \geq 1$ and $h_i : X^* \to X^*$ are homomorphisms for $i = 1, \ldots, m$, and $w \in X^*$. If $m = 1$, $G$ is referred to as a DOL system. A DOL system is a PDOL system if its homomorphism is nonerasing.

A DOL system $G = (X, h, w)$ defines the sequence of nonnegative integers

$$a_i = lg(h^i(w)), \quad i \geq 0.$$

Sequences of this form are referred to as DOL length sequences. PDOL length sequences constitute their subclass, defined by PDOL systems.

It is easy to see that DOL length sequences constitute a proper subclass of N-rational sequences. The exact characterization of this subclass depends on the fact that a DOL length sequence cannot be decomposed (in the sense of Theorem 3) into parts with different growth orders. The reader is referred to [14] or [18] for details. This characterization shows also that it is decidable whether or not a given N-rational sequence is a DOL length sequence. The same is true of the following characterization with respect to PDOL length sequences. The theorem is due to Soittola, cf. [18].

Theorem 5. A sequence $a_i$ is a PDOL length sequence not identically zero if and only if the sequence

$$b_i = a_{i+1} - a_i$$

is N-rational and $a_0$ is a positive integer.

Some of the major open problems concerning rational sequences deal with decidability. In particular, the decidability status corresponding to problems (i), (iii) and (iv) in Theorem 2 is open for Z-rational sequences. On the other hand, problem (ii) in Theorem 2 is decidable for Z-rational sequences: it is decidable whether or not a given Z-rational sequence contains infinitely many zeros. This is essentially due to the Skolem-Mahler-Lech Theorem. (See [18] for details and references, see also [22].)

Let us elaborate to some extent problem (i) in Theorem 2 for Z-rational sequences: Is it decidable whether or not the number 0 appears in a given Z-rational sequence? By Schützenberger's Representation Theorem, this problem can also be formulated in the following very simple way. Consider square matrices $M$ with integral entries. Is it decidable whether or not, given such an $M$, the number 0 appears in the upper right-hand corner of some power $M^i$, for $i = 1, 2, \ldots$ ?

Still an equivalent version of this problem is the following. Is it decidable whether or not a given DOL length sequence contains two consecutive equal numbers? (For this and other related equivalent versions of the problem, the reader is referred to [18].)

Further importance to the problem is brought about by the fact that many other, even purely language-theoretic, problems have been reduced to it. Typical examples are given in [3] and [15]. It is shown that the decidability of the problem we are considering implies the decidability of many language-theoretic problems (whose decidability status is unknown at present). Moreover, it would give new proofs for the decidability of some celebrated problems such as the DOL sequence equivalence problem.

Intuitively, the problem we are considering seems to be "very decidable": only one variable is involved and the process is very deterministic. It seems that the known undecidability tools, such as the Post Correspondence Problem, are just impossible to encode into this problem.

4. DTOL transformations. A recent approach, due to Reutenauer, combines in an interesting way some ideas of L systems with the theory of rational power series. This line of research seems to be very promising, and the possibilities are far from being exhausted. The results in this section are from [13]. (In fact, [13] is a comprehensive thesis, containing also references to papers published by the same author.)

Consider a DTOL system $G = (X, h_1, \ldots, h_m, w)$ as defined in Section 3. (The reader is referred to [14] for motivation and background material concerning DTOL systems.) Consider also the alphabet $Y = \{1, \ldots, m\}$. Then $G$ transforms a series $r$ in $A \ll X^* \gg$ into the series $G(r) = s$ in $A \ll Y^* \gg$, defined as follows. Let $i_1 \ldots i_t$ be an arbitrary nonempty word over $Y$ (each $i_j$ is a letter). Then

$$(s, i_1 \ldots i_t) = (r, (w)h_{i_1} \ldots h_{i_t}).$$

Moreover, $(s,\lambda) = (r,w)$.

Theorem 6. If $r$ is a polynomially bounded N-rational (resp. Z-rational) series and $G$ is a DTOL system with the same alphabet, then $g(r)$ is also N-rational (resp. Z-rational) but not necessarily polynomially bounded.

The main significance of Theorem 6 lies in the fact that decidability results concerning N-rational and Z-rational series become applicable also for the series $G(r)$. For instance, the equation $G(r) = G'(r')$ is decidable for arbitrary given polynomially bounded Z-rational series $r$ and $r'$, and DTOL systems $G$ and $G'$. In particular, if $G = (X, h, w)$ is a DOL system (i.e., $m = 1$) and $u \in X^*$, then the series

$$\sum_n \binom{h^n(w)}{u} x^n ,$$

where $\binom{w}{u}$ is the Eilenberg binomial coefficient, cf. [4], is N-rational. Consequently, the equality of two such series, possibly coming from different DOL systems, is decidable.

We mention, finally, two language-theoretic corollaries of Theorem 6.

Let $G$ be a DTOL system as above and $L \subseteq X^*$ a regular language. Then the language

$$G(L) = \{y \in Y^* \mid (w)y \in L\}$$

is also regular.

Let $G = (X, h, w)$ be a DOL system and $L \subseteq X^*$ a regular language. Then there are integers $q$ and $p$ such that, for all integers $i \geq q$,

$$h^i(w) \in L \text{ if and only if } h^{i+p}(w) \in L.$$

<u>5. Fatou properties.</u> Consider a subsemiring $A$ of a semiring $B$. We say that $B$ is a <u>Fatou extension</u> of $A$ if

$$A \ll X^* \gg \cap B^{rat} \ll X^* \gg = A^{rat} \ll X^* \gg$$

holds for all alphabets $X$.

The notion of a Fatou extension can be defined in a much more general set-up but here we are interested only in the specific structures $A \ll X^* \gg$ and $B \ll X^* \gg$ and the property of being a rational series, associated with these structures.

<u>Theorem 7.</u> A field is always a Fatou extension of its subfield. The semiring $Q_+$ is a Fatou extension of $N$.

The reader is referred to [18] for a proof of Theorem 7, as well as for references and related material. We want to emphasize that $R_+$ is not a Fatou extension of $Q_+$: there are $R_+$-rational series with nonnegative integer coefficients that are not $Q_+$-rational.

Consider the alphabet $X = \{x_1, x_2\}$. For a word $w$ in $X^*$, denote by $lg_i(w)$ the number of occurrences of $x_i$ in $w$, for $i = 1,2$. Then the Z-rational series

$$\Sigma(lg_1(w) - lg_2(w))^2 w$$

has its coefficients in N but is not N-rational. Thus, even if attention is restricted to series with a quadratic growth of the coefficients, Z is not a Fatou extension of N. (Observe that Theorem 3 tells the same fact with respect to series of one variable.) However, the following result, due to [13], shows that a "Fatou property" can be obtained in case the coefficients grow at most linearly with respect to word length.

Theorem 8. A linearly bounded Z-rational series with nonnegative coefficients is N-rational.

Fatou properties are sometimes linked with questions concerning Hadamard product and Hadamard quotient. The Hadamard product $r$ of two series $r_1$ and $r_2$ is defined by the condition

$$(r,w) = (r_1,w)(r_2,w),$$

valid for all $w$. (See [6] where such a product was considered probably for the first time.) It is easy to see that the Hadamard product of two A-rational series is A-rational.

Similarly, we define the notion of Hadamard quotient. It has been shown in [7] that there are N-rational series $r_1$ and $r_2$ such that their Hadamard quotient is not N-rational although its coefficients are in N.

6. Pólya series. Consider a series $r$ in $Q \ll X^* \gg$. The set of prime factors of the coefficients of $r$, in symbols $P(r)$, is defined to be the set of all prime numbers $p$ such that, for some word $w$,

$$(r,w) = p^k \frac{a}{b},$$

where $k, a, b$ are integers distinct from 0, and $p$ divides neither $a$ nor $b$.

For instance, if $r = \Sigma n^2 x^n$ then $P(r)$ equals the set of all prime numbers. For

$$r = \Sigma 2^n x^n,$$

we have $P(r) = \{2\}$.

A series $r$ in $Q \ll X^* \gg$ is termed a Pólya series if $P(r)$ is finite. (For motivation behind this terminology, see [12].)

Rational Pólya series play a special role among all rational series. Of their language-theoretic applications, we mention the following result due to [13].

Theorem 9. The support of a rational Pólya series is regular. If $r$ is a rational Pólya series satisfying $P(r) = \{p\}$, then the language

$$\{w \in X^* \mid (r,w) > 0\}$$

is regular.

We conclude our discussion of rational series with the following two interesting conjectures (i) and (ii) (of which (i) is stronger) from [13]:

(i) Assume that $L_1$ and $L_2$ are disjoint regular languages appearing as supports of some rational series. Then there is a regular language containing $L_1$ and disjoint with $L_2$ (i.e., $L_1$ and $L_2$ are separated by a regular language).

(ii) Assume that $L_1$ is the complement of $L_2$ and that both $L_1$ and $L_2$ appear as supports of some rational series. Then both $L_1$ and $L_2$ are regular.

7. On algebraic series. Consider an alphabet $Z = \{Z_1, \ldots, Z_n\}$ disjoint with $X$. A proper algebraic system (with respect to the pair $(A,X)$ and with variables in $Z$) is a set of equations of the form

$$(1) \qquad Z_i = P_i \ , \quad i = 1, \ldots, n,$$

where each $p_i$ is in $A<(X \cup Z)^*>$ and, for each $i$ and $j$,

$$(p_i, \lambda) = (p_i, Z_j) = 0.$$

An n-tuple $(\sigma_1, \ldots, \sigma_n)$ of quasiregular series in $A<<X^*>>$ is a solution of (1) if (1) is satisfied when $Z_i$ is replaced by $\sigma_i$, for $i = 1, \ldots, n$.

The reader might wonder why special systems (1) are considered instead of general algebraic equations. The reason is two-fold: (i) systems (1) correspond in a natural way to context-free grammars, and (ii) very little is known about more general algebraic series. As regards "transcendental series", there are practically no results.

It is shown in [18] that every proper algebraic system (1) possesses a unique solution. A quasiregular series in $A \ll X^* \gg$ is termed A-algebraic if it appears as a component of the solution of a proper algebraic system. The family of A-algebraic series is denoted by $A^{alg} \ll X^* \gg$.

Shamir's Theorem (cf. [18]) gives a representation result for algebraic series, analogous to Schützenberger's Representation Theorem.

There is a natural correspondence between context-free grammars (having no chain rules and no $\lambda$-rules) and proper algebraic systems of equations: the variables $Z_i$ correspond to nonterminals. In this correspondence, only the supports of the polynomials $p_i$ are significant. (The coefficients correspond to weights in a weighted context-free grammar.) If we begin with a context-free grammar $G$, write the productions as a proper algebraic system of equations (with coefficients equal to 1), and consider the series $r$ in the solution corresponding to the initial letter of $G$, then $\mathrm{supp}(r) = L(G)$ and, moreover, the coefficient of an arbitrary word $w$ in $r$ equals the ambiguity of $r$ according to $G$. (It is assumed that the basic semiring is $N$.)

Assume that $u \geq 3$ and $t_1, \ldots, t_u$ are nonnegative integers. We say that a proper algebraic system (1) is in the <u>$(t_1, \ldots, t_u)$ normal form</u> if

$$\mathrm{supp}(p_i) \subseteq X^+ \cup X^{t_1} Z X^{t_2} Z \ldots X^{t_{u-1}} Z X^{t_u},$$

for every $i = 1, \ldots, n$.

Thus, $(t_1, \ldots, t_u)$ normal form means that the right sides of the equations contain only (i) words over the "terminal" alphabet, and (ii) words involving exactly $u-1$ variables, separated by terminal words of fixed lengths determined by the tuple $(t_1, \ldots, t_u)$.

Theorem 10. For every $u \geq 3$ and nonnegative integers $t_1, \ldots, t_u$, every A-algebraic series can be generated by a proper algebraic system in the $(t_1, \ldots, t_u)$ normal form.

Proof. Let $r$ be generated by the system (1), i.e., $r$ is a component in the solution, say, the component corresponding to $Z_1$. To get the required normal form, we make a sequence of transformations to (1) preserving $r$ (i.e., $r$ will always equal the first component in the solution). Such a transformation may alter the set of variables $Z$: in general, the new set of variables $Z'$ is bigger than the original one. For simplicity, we will denote the set of variables always by $Z$.

It is shown in [18] that $r$ is generated by a proper algebraic system in which the supports of the right sides are included in the set

(2) $$X \cup X^2 \cup XZX \cup XZXZX.$$

Our first transformation consists in bringing the given system into this form.

Thus, we may assume that the supports of the right sides are included in the set (2). Our next step is to show that, given an integer $i \geq 0$, we may assume that the supports of the right sides are included in the set

(3) $$X^+ \cup X^i Z X^i \cup X^i Z X^i Z X^i.$$

(Thus, there may be longer terminal words in the two last terms of the union but then also $X \cup X^2$ might not be sufficient.) The step from (2) to (3) is accomplished exactly as the corresponding step for grammars in [8]. (In [8], this is called "reduction of subgoal 2 to subgoal 1".) The only additional observations needed are the following. (These observations are needed also in the remaining two reduction steps.) (i) Whenever a substitution preserving a language is performed, we have to make sure that the same substitution preserves also the coefficient of each word in the language. But this is easily taken care of by preserving the original coefficient at the starting stage and making the coefficients all equal to 1 at other stages. (As customary when dealing with algebraic series, we assume that $A$ is a commutative semiring with identity.) (ii) Taking the union of languages corresponds to summing up the series.

In the last two reduction steps we first go from (3) to the set

$$X^+ \cup X^j Z X^k Z X^l, \tag{4}$$

for an arbitrary triple $(j, k, l)$ of nonnegative integers and, finally, from (4) to the set

$$X^+ \cup X^{t_1} Z \dots Z X^{t_u},$$

determined by the given tuple $(t_1, \dots, t_u)$. These two reduction steps are established exactly as "reduction of subgoal 3 to subgoal 2" and "reduction of subgoal 4 to subgoal 3" in [8]. This completes the proof of Theorem 10.

The set $X^+$ in the statement of Theorem 10 can be replaced by a finite set

$$X \cup X^2 \cup \dots \cup X^i,$$

where $i$ depends on the tuple $(t_1, \dots, t_u)$. An explicit upper bound for $i$, in terms of the tuple, can be given.

The whole research area concerning trade-offs between the number of variables in the systems and the numbers $u, t_1, \ldots, t_u$ is open.

As an example of a deep result, established using algebraic power series, we mention the fact due originally to Semenov that it is decidable whether or not a given unambiguous context-free language and a given regular language coincide. Cf. [18]. The reader is referred also to [10] for a recent application of this result to a quite different problem.

## References


[1] J. BERSTEL, Sur les pôles et le quotient de Hadamard de séries N-rationnelles. C. R. Acad. Sci. Paris 272, série A (1971) 1079-1081.

[2] J. BERSTEL (ed.), Séries formelles en variables non commutatives et applications. Laboratoire d'informatique théorique et programmation; 2, place Jussieu, Paris (1978).

[3] K. CULIK II, Homomorphisms: decidability, equality and test sets. Proc. Conf. on Formal Languages in Santa Barbara, December 1979, Academic Press, in preparation.

[4] S. EILENBERG, Automata, Languages and Machines, Vol. A-B, Academic Press, New York (1974 and 1976).

[5] P. FATOU, Séries trigonométriques et séries de Taylor. Acta Math. 30 (1906) 335-400.

[6] J. HADAMARD, Théorème sur les séries entières. Acta Math. 22 (1899) 55-63.

[7] J. KARHUMÄKI, Remarks on commutative N-rational series. Theoret. Comput. Sci. 5 (1977) 211-217.

[8] H. MAURER, A. SALOMAA and D. WOOD, On generators and generative capacity of EOL forms. Acta Inform. 13 (1980) 87-107.

[9] J. NIELSEN, Om regning med ikke-kommutative faktorer og dens anvendelse i gruppeteorien. Math. Tidsskrift B (1921) 78-94.

[10] Th. OTTMANN, A. SALOMAA and D. WOOD, Grammar and s-grammar forms. Theoret. Comput. Sci., to appear.

[11] A. PAZ and A. SALOMAA, Integral sequential word functions and growth equivalence of Lindenmayer systems. Inform. and Control 23 (1973) 313-343.

[12] G. PÓLYA, Arithmetische Eigenschaften der Reihenentwicklungen rationaler Funktionen. J. Reine Angew. Math. 151 (1921) 1-31.

[13] C. REUTENAUER, Séries rationnelles et algèbres syntactiques. Thèse, Univ. Paris VI (1980).

[14] G. ROZENBERG and A. SALOMAA, The Mathematical Theory of L Systems. Academic Press, New York (1980).

[15] K. RUOHONEN, Zeros of Z-rational functions and DOL equivalence. Theoret. Comput. Sci. 3 (1976) 283-292.

[16] A. SALOMAA, Formal Languages. Academic Press, New York (1973).

[17] A. SALOMAA, Undecidable problems concerning growth in informationless Lindenmayer systems. Elektron. Informationsverarb. Kybernet. 12 (1976) 331-335.

[18] A. SALOMAA and M. SOITTOLA, Automata-Theoretic Aspects of Formal Power Series. Springer-Verlag, New York, Heidelberg, Berlin (1978).

[19] M.P. SCHÜTZENBERGER, Un problème de la théorie des automates. Séminaire Dubreil-Pisot, 13e année (1959-1960), no. 3, Inst. H. Poincaré, Paris (1960).

[20] M.P. SCHÜTZENBERGER, On a theorem of R. Jungen. Proc. Amer. Math. Soc. 13 (1962) 885-890.

[21] M.P. SCHÜTZENBERGER, On a definition of a family of automata. Inform. and Control 4 (1961) 245-270.

[22] Th. SKOLEM, Ein Verfahren zur Behandlung gewisser exponentialer Gleichungen und diophantischer Gleichungen. C. R. VIII Congr. Mathém. Scandin., Stockholm 14-18 Août 1934; Lund (1935) 163-188.

[23] M. SOITTOLA, Positive rational sequences. Theoret. Comput. Sci. 2 (1976) 317-322.

Mathematics Department
University of Turku, Finland

# QUANTIZATION, STABILITY, AND SCATTERING

Irving Segal*

1. Introduction. This lecture describes work in field quantization going back two or three decades.

There has been considerable development of the concept of quantization itself and for brevity, this is developed here along general algebraic lines. However, the distinction between formal and probabilistic quantization is emphasized and illustrated.

The application of these ideas to non-linear and time-dependent wave equations has only recently become possible, as a result of extensions of the stability theory of the Krein school in the thesis of S.M.Paneitz, and its application along presently relevant lines in joint work. Field quantization involves positivity and causality considerations that have been absent in pure symplectic manifold and group orbit quantization theory, but which are emphasized here, since they underly the connection with stability theory and are important for physical applications. Relevant aspects of scattering and other features of nonlinear wave equations are summarized briefly; it is hoped that at the same time, this gives some impression of a subject of intrinsic interest.

In concrete terms, one of the results described is the introduction of a Lorentz-invariant hermitian structure in the stable solution variety of a nonlinear wave equation; until recently, only the symplectic part of this structure was known. In a purely mathematical way, this may be regarded as in the direction of an extension to partial differential varieties - i.e., such as are defined by partial differential equations - of structures well known for their usefulness in finite-dimensional varieties. Quantum field theory may in large part be considered as function theory in an infinite-dimensional space, and differential geo-

*Research supported in part by The National Science Foundation

metric structures such as those employed in the Hodge theory appear likely to be important, although the basic problems seem integration-and-operator-theoretic, rather than topological or group-theoretic.

I apologize in advance to the many significant contributors to the present developments whose work could not be explicitly described or cited due to space-time limitations.

2. Quantization. Nonlinear quantization theory is highly fragmentary, but linear quantization theory is now well developed and is fundamental for the nonlinear case; consequently, that is what this section is mostly about.

First, the general idea in an historical setting. When Dirac saw how well Heisenberg's scheme of replacing the classical p's and q's by operators satisfying the simplest non-trivial commutation relation, $[p,q] = -i$, worked, he was (apparently) moved to ask: why not do the same thing for classical fields. Formally, the only difference is that there is then an infinite set of p's and q's , - the components of the Cauchy data relative to an orthonormal basis in $L_2$ over space. This worked very well from a physical standpoint if one did not push the theory beyond first-order perturbation theory (if one did, it became riddled with infinities, and remains so today).

Dirac's approach was non-relativistic. A relativistic form was developed by Heisenberg and Pauli. The general idea may be sketched thus. Suppose one is given a classical non-linear relativistic equation, such as

$$\Box \phi + F'(\phi) = 0 , \tag{1}$$

where $F$ is some given function, e.g. a polynomial, and $\Box$

the d'Alembertian ; classical here means that $\phi$ is numerically, say real, valued. The "quantization" $\phi!$ of the classical field $\phi$ is supposed to differ from it fundamentally only in being not real-valued, but Hermitian-operator valued - with however the simplest non-trivial commutation relations. These were analogous to the original Heisenberg relations; at a fixed time, the values of the field throughout space commuted with each other, the same was true of the values of the first partial derivative of the field with respect to time, and the commutator of a field value with a first partial derivative was just a number (formally similar to what Lie's theory of contact transformations would suggest). The same partial differential equation (1) was supposed to be satisfied by the quantized field $\phi!$, in a "suitable sense".

Let us ignore the infinities in this approach and look at the formal theory more closely. The classical equation could be construed as defining an infinite-dimensional Hamiltonian system whose Hamiltonian is the classical energy

$$E = \int \{(\text{grad}\,\phi)^2 + m^2\phi^2 + F(\phi) + (\partial\phi/\partial t)^2\}\, d\vec{x}$$

(this is time-independent, although the integral is only over space). In the quantized case, the Hamiltonian is an operator $E!$, namely that obtained by the replacement of $\phi$ by $\phi!$ in the given expression. In case $F$ is bounded below, as expected from positive-energy (or physical stability) considerations, it would have a "ground" or "vacuum" state, i.e. the lowest eigenstate of the quantized Hamiltonian $E!$ This state $v$ determined the vacuum expectation values of any function of the quantized field; in particular, the so-called n-point function

$$V(x_1,x_2,\ldots,x_n) = \langle \phi(x_1)\phi(x_2)\ldots\phi(x_n)v,v \rangle .$$

(It was recognized that $\phi$ ! and its n-point function were not literal functions, but generalized ones, even in the simplest case).

In a formal algebraic way, the theory appears clearly self-consistent, and with the adjunction of a suitable folk theorem, categorical. The simplicity of the postulated commutation relations guarantees the Jacobi relations, and thereby the formal existence of objects satisfying the commutation relations. These objects only had to be given at one time, say 0; at any other time, they were determined by the Hamiltonian E!, in the usual way, by transformation with exp(it E!). There was a folk theorem to the effect that the commutation relations assumed at time 0 uniquely determined the fields at that time, within an inessential equivalence, so that the initial values of the field (its Cauchy data at time 0) were fully specified. (This "theorem" was actually proved in a rigorous mathematical form by Stone and von Neumann around the same time, only in the case of _finite_ sets of quantum variables, as in Heisenberg's original formulation, and is false in the infinite-dimensional case, except in its weaker C*-algebra formulation). The theory was Lorentz-invariant and enjoyed finite propagation velocity, in "suitable" senses.

This is a theory of compelling conceptual simplicity, which may be why some people still cling to it despite the seeming intractability of the infinities it leads to. In any event the linear aspects of quantum field theory are now fairly well developed; they have turned out to be a bit sophisticated but ultimately algebraically simple; let us consider some of the basic developments in it.

One of the problems with the linear theory in the Dirac-Heisenberg-Pauli form was that one could neither prove nor disprove very much, due to the underlying vagueness. For example, no spaces were given in which the quantized field operators $\phi_i^!(x)$ were supposed to act, so that a question such as the unitary equivalence of given sets of "operators" was inherently moot. It has been necessary to develop new mathematical constructs in order to have a lingua franca for the consideration of rigorous quantum field issues.

One basic one is that of a Weyl system over a given linear symplectic space [1]. The combination of a real linear topological space $L$, and a given non-degenerate continous antisymmetric form $A$ on $L$, may be called a "linear symplectic space". A "Weyl system" over $(L,A)$ is a pair $(K,W)$ in which $K$ is a complex Hilbert space, and $W$ a continous map from $L$ to the unitary group $U(K)$ on $K$ (always taken in the strong operator topology in such connections), that satisfies the Weyl relations:

$$W(z)W(z') = e^{(i/2)A(z,z')} W(z+z') , \tag{2}$$

for arbitrary $z$ and $z'$ in $L$. Essentially, a Weyl system is any non-pathological representation of the Heisenberg commutation relations, in finite group, rather than infinitesimal-group, form. A priori it is not obvious that Weyl systems exist for given $(L,A)$ but this is demonstrable (e.g. by functional integration methods) in case there exists a positive definite symmetric form $S$ on $L$ relative to which $A$ is continous.

Reinterpreted in terms of quantized fields, what this means (more exactly, implies) is that there do exist as mathematical entities fields such as the putative $\phi_i^!$ satisfying the Heisenberg commutation relations at every fixed time, and in addition a given linear wave equation, such as that of Klein-Gordon type:

(3) $\Box\, \phi! + V(x)\phi! = 0$ ,

where $V(.)$ is a given smooth real function on spacetime. To this end one may take $L$ as the classical solution variety of equation (3), with e.g. $C_0^\infty$ Cauchy data, and A as the form

(4) $A(\phi,\psi) = \int (\phi\,\dot{\psi} - \dot{\phi}\,\psi)\,dx$ .

This existence result appears supportive of the quantum field concept, but provides actually an embarassment of riches; There are at least continuum many unitarily inequivalent such solutions $\phi!$ even such as act irreducibly on the representation space $K$ (i.e. the $W(z)$, $z \in L$, do so)[2].

To see how these may arise in a simple manner, let $Sp(L,A)$ denote the group of all continous invertible linear transformations on $L$ that preserve the form A. If T is arbitrary in $Sp(L,A)$, then $W_T$, defined by the equation $W_T(z) = W(Tz)$, gives another Weyl system, with the same representation space $K$; but there is in general no unitary transformation V on $K$ that transforms one Weyl system into the other. More specifically, in case $(L,A)$ "comes from" a complex Hilbert space $H$ by taking $L$ as $H$ with its complex structure deleted, and A to be given by the equation:

$$A(z,z') = \mathrm{Im}\, \langle z,z' \rangle$$

- in this case we speak of a Weyl system over $H$ or of the symplectic group $Sp(H)$ on $H$ - then a unitary operator V such that $V^*W(z)V = W_T(z)$ for all z exists if and only if $TT^{tr} - I$ (or equivalently, the commutator of T with the complex unit i) is Hilbert-Schmidt [3]. (This is of course always the case when $L$ is finite-dimensional).

Which then is the <u>right</u> quantization? One sees that even in the linear case, there appear to be ambiguous issues virtually from the start. At this level, the issue can be "resolved" by judicious use of C*-algebra, which serves to extend the Stone - von Neumann theorem, at the cost of replacing unitary equivalence by C*-isomorphism, which is precisely the extent to which linear quantization is unique, in the absence of additional structure, such as the dynamical features treated below. One defines the "Weyl algebra" of a given Weyl system as the uniform closure of the union of all the W*- (or von Neumann) algebras generated by the $W(z)$ as $z$ ranges over arbitrary finite-dimensional subspaces $M$ of $L$, the restrictions of $A$ to which are non-degenerate. As just seen, there is no spatial isomorphism (i.e. unitary equivalence) of the Weyl algebras that exchanges the corresponding $W(z)$'s, but there is a C*-isomorphism between the Weyl algebras that does, in fact a unique one [1]. One recovers from this the essential unitary equivalence (apart from multiplicity) when $L$ is finite-dimensional, i.e. the Stone - von Neumann conclusion.

In principle, this tells one exactly how the states of the quantized field (3) evolve, independently of the particular quantization employed. For the independence of the given form $A$ from the time means that the temporal evolution defined by the classical equation from one time $t$ to another $t'$, say $V(t',t)$, is in $Sp(L,A)$. Now for any element $T \in Sp(L,A)$, there is a unique corresponding automorphism, say $\Theta(T)$, of the Weyl algebra, as an abstract *-algebra, in view of the C* generalization of the Stone - von Neumann theorem. In particular, $\Theta(V(t',t))$ is an automorphism, say $s(t',t)$ of the Weyl algebra, say $\mathcal{W}$, and the contragredient action of $s(t',t)$ defines in particular the temporal evolution of the states of the Weyl algebra. We are using here the C*- algebra quantum phenomenology, a more operational variant of the conventional Hilbert space phenomenology, and essentially a generalization of it: a state $E$ of $\mathcal{W}$ is a positive linear normalized functional on $\mathcal{W}$. The typical conventional states are of the form $E(A) = \langle A\psi,\psi\rangle$, $\psi$ being the normalized Schrödinger wave function, but $\psi$ is "observed" only through the expectational functional $E$ that derives from it in this fashion, and does not exist in any unique way for $\mathcal{W}$.

But this says nothing about the existence or non-existence of stationary states or of a ground state; above all, it gives no objective means of comparison of the states at one time with the states at another time, that can be correlated even in conceptual principle with the transitions observed experimentally. The states so observed are normally labelled by the eigenvalues of operators in a Hilbert space (so called "quantum numbers"), and there is no preferred Hilbert space in which to represent $\mathcal{W}$, or preferred state from which to derive one by the canonical C*-algebra construction, or otherwise. This is just as it should be physically, in the absence of more structure or more specific dynamical considerations. It is conceivable that a physical system may have no objective, time-independent parametrization for its states, in which case the evolution law of the C*-algebra states tells one all there is to know. (For example, cosmological general relativity is fundamentally devoid of temporal invariance). But the "formal" quantization that one has at this stage is incapable of quantitative theoretical prediction that is susceptible to experimental verification.

For this one needs to specify (at least; but then the rest largely follows) the vacuum (or ground) state. One could define it in a particular representation of $\mathcal{W}$ on a Hilbert space as represented by the vector that is annihilated by all the "annihilation" operators, except that we have no preferred complex structure, such as is required to derive "creation" and annihilation operators from the hermitian field operators that we now have. Another way to look at it is that we have the field commutators, but no n-point, or even 2-point function. This "probabilistic" aspect of the quantization is presently unspecified; if one considers the formal quantization as the first half of the quantization process, this probabilistic second half is much the more difficult. It will be seen that it does not always exist, and, that even when it does, it is highly non-local, unlike the commutation relations and the differential equation from which the formal quantization basically derives.

One natural idea for the fuller specification of the quantization is to use group invariance. In mathematical terms, if the topological group $G$ acts on the linear space $L$ in a continuous symplectic fashion, - i.e. one is given a continuous representation of $G$ into $Sp(L,A)$ , - one defines a "covariant" Weyl system (with respect to $G$ and its action, of course) as a triple $(K,W,\Gamma)$ , where $(K,W)$ is a Weyl system as earlier, and $\Gamma$ is a continuous unitary representation of $G$ on $K$ , satisfying the "intertwining" relation: $\Gamma(g)W(z)\,\Gamma(g)^{-1} = W(gz)$ , where $gz$ denotes the action of $g \in G$ on $z \in L$ . This relation simply extends the way time evolution is supposed to work to the entire "covariance" group $G$; e.g., Lorentz invariance can (in part) be expressed in this way. However, there still remain at least continuum many inequivalent irreducible covariant Weyl systems with respect to the Lorentz group, even in the simplest non-trivial case of the Klein - Gordon equation, $\Box\phi + c\phi = 0$ .

One could try to pin the quantization down further by requiring the existence of a fundamental invariant state - the putative vacuum - which is expected to be Lorentz - invariant in a relativistic theory. By an "equilibrium state vector" for a covariant Weyl system one means a vector $v \in K$ such that $\Gamma(g)v = v$ for all $g \in G$ But the imposition of this requirement is still insufficient. Indeed, it remains so even if the covariance group $G$ is enormously enlarged, for example taken as the full unitary group $U(H)$ on a complex Hilbert space, from which $(L,A)$ "comes" as earlier; the irreducible covariant Weyl systems with equilibrium state vector then have a simple explicit representation, and there are continuum many distinct ones, within unitary equivalence [4].

An idea that does work, and is both mathematically simple conceptually and physically natural, is the exploitation of the positive-energy constraint on the temporal development of the quantum field, in the case of temporal invariance, and one then needs no group beyond the temporal. If $S(.)$ is a continuous one-parameter subgroup of the symplectic group $Sp(L,A)$ on the linear symplectic space $(L,A)$ a quantization $(K,W,\Gamma,v)$ that is covariant with respect to $S(.)$ and has equilibrium state vector $v$, is said to be "positive-energy" in case the infinitesimal generator of the unitary group $\Gamma(t)$ is non-negative. There is a theorem [4,5] to the effect that if a positive-energy quantization exists, and if the subgroup $S(.)$ is non-trivial, then the quantization is unique, and has a simple universal form, exhibiting covariance with respect to a full unitary group on a Hilbert space.

More specifically, it is possible to imbed $L$ as a dense real-linear subspace of a complex Hilbert space $H$ in such a way that A becomes the imaginary part of the inner product and S(.) becomes pre-unitary, its unitary closure having a positive generator. When this is done the quantization becomes the "free field" over $H$ ; more precisely, the free "boson" field, our exposition here being constrained by the need for brevity to neglect "fermions" totally - except to say that all of the results of this section have analogues in that case, with the replacement of the symplectic by the orthogonal group. The free field over $H$ may be characterized intrinsically as the unique $U(H)$ -covariant Weyl system with equilibrium state vector, say $(K,W,\Gamma,v)$ such that either one of the following two equivalent conditions are satisfied: (i) for <u>some</u> strictly positive-energy one-parameter unitary group $U(.)$ on $H$ , the one parameter group $\Gamma(U(t))$ on $K$ has non-negative generator; (ii) the same is true of <u>every</u> one-parameter unitary group on $H$ with non-negative generator. The equivalence of these two conditions is a consequence of the theorem quoted.

It is not obvious that a free field exists, and it was not until some 25 years after Fock's heuristic formulation of a non-relativistic free field in a sense employed by physicists that a pure Hilbert space construction (exhibiting important invariance properties) and a precise mathematical formulation incorporating the features in the present characterization were given [6,7].

As an example of the application of the theorem described, consider the Klein-Gordon equation, $\Box\phi + c\phi = 0$, using Cauchy data in $C_0^\infty$. Using the symmetric form in the solution manifold derived from the positive definite quadratic form $\int\{(\text{grad }\phi)^2 + \dot{\phi}^2 + \phi^2\}d\vec{x}$, relative to which the temporal propagation defined by the equation, as well as the fundamental assymmetric form earlier given, are continuous, a formal quantization can be achieved for any real value of $c$. However, it is only when $c \geq 0$ that the temporal propagation is unitarizable, and a probabilistic quantization, with a vacuum, etc., achieved; one then has the conventional Klein-Gordon free field (within unitary equivalence; there are many other representations). When $c < 0$, there is a well-defined evolution of states of the Weyl algebra, but there is no distinguished reference system for the parametrization of states, in particular no smooth invariant state whatsoever.

The description of the Hilbertization and positive-energy unitarization for the case $c \geq 0$, which are unique, are more readily described for the general abstract equation $u'' + B^2u = 0$, where $u = u(t)$ is a function from $\mathbb{R}^1$ to a real Hilbert space $M$, and $B$ is a positive self-adjoint operator in $M$. If $\mathcal{D}$ is any domain in $M$ that is invariant under $B^2$ and $\sin tB/B$, the solution manifold $L$ of the equation with Cauchy data in $\mathcal{D}$ is invariant under the temporal propagation, and admits the invariant symplectic structure

$$A(u,v) = \langle u(t), v'(t)\rangle - \langle u'(t), v(t)\rangle$$

where $t$ may be chosen arbitrarily. Taking as $H$ the complex Hilbert space $M \oplus iM$, and mapping $L$ into the functions from $\mathbb{R}^1$ to $H$ by the equation

$$(u(t), u'(t)) \to w(t) = Cu(t) - iC^{-1}u'(t) ,$$

where $C = B^{1/2}$, the given equation takes form $w' = iBw$, and the requisite structures are obtained. With $B = (c - \Delta)^{1/2}$ in the space $M = L_2(\mathbb{R}^3)$, the canonical (unique) Lorentz-invariant norm of the solution manifold of the Klein-Gordon equation is obtained.

Thus in a temporally homogeneous context, the positive-energy constraint may suffice to extend a formal quantization to a full probabilistic one.

In a temporally inhomogeneous situation, e.g. for the Klein-Gordon equation on a curved manifold, whose local symplectic structure was established some time ago [27], there has until recently been no available probabilistic quantization - no vacuum, or specification of complex structure needed for the introduction of creation and annihilation operators; not even a 2-point function. But the idea that the vacuum may be characterized as the state invariant under the scattering transformation has proved successful in this connection.

The mathematical machinery involved begins with a single given transformation, rather than a one-parameter group of such, in a linear symplectic space; and the physical interpretation of this transformation, say $S$, is as a scattering transformation, and so quite different from that of a temporal evolution group; in particular, the positive-energy constraint is not involved. Let us say that an element $S \in Sp(L,A)$ is "unitary-quantizable" if there exists a covariant Weyl system, with respect to $S$ (more exactly, the cyclic group it generates) with equilibrium state vector; "C*-quantizable" if there exists a state of the Weyl algebra that is invariant under the automorphism induced by $S$, and regular, in the sense that its restrictions to the Weyl systems over arbitrary finite-dimensional subspaces of $L$ on which A remains non-degenerate should be representable via the trace of the product with a trace-class operator; and that S is unitarizable if $(L,A)$ comes from a complex Hilbert space $H$ in such a way that S is represented by an operator conjugate in $Sp(H)$ to a unitary operator. Theorem [8]:

Assuming that $(L,A)$ comes from a complex Hilbert space, the following conditions on $S \in Sp(L,A)$ are all equivalent:

i) S is unitary-quantizable;

ii) S is C*-quantizable;

iii) S is unitarizable.

As an example, suppose $H$ is one-dimensional, and that S takes the form $e \to \lambda e$, $ie \to \lambda^{-1} ie$, $\lambda$ being real and $>1$. S is then notunitarizable. The Weyl algebra is naturally identifiable with the algebra $B$ of all bounded linear operators on $L_2(\mathbb{R}^1)$; the only invariant states under the induced action of S (which necessarily exist by abstract fixed-point theory) vanish identically on all operators of the form $F(p)$ or $F(q)$, where F is an arbitrary bounded continuous function vanishing at $\infty$ on $\mathbb{R}^1$, and p and q are the usual Heisenberg-Schrödinger operators; or on any compact operator. Note also that the natural action of S on $L_2(\mathbb{R}^1)$, given by the harmonic representation of the symplectic group, has no nonzero fixed vector. If on the other hand, S multiplies both e and ie by $e^{i\theta}$ with $\theta$ real, then S is evidently unitary; the harmonic representation gives a unitary quantization whose vacuum is the lowest hermite function, the induced action of S being e.g. the Fourier transform if $\theta = \pi/2$; and a fortiori, S is C*-quantizable.

The quoted theorem states when there may be a probabilistic quantization, but does not describe the explicit possibilities for such. With the physical interpretation of $S$ as a scattering transformation, it is natural to constrain its quantization to be free-field, and it is then generically unique. Of course the conjugacy to a unitary is not unique; it is only the "hilbertization" that is unique (i.e. invariant positive complex structure, and the real positive definite symmetric bilinear form that it defines in conjunction with the given anti-symmetric form). Moreover, simplicity of the spectrum (i.e. multiplicity-free eigenvalues) is not sufficient for unicity; rather, this requires an asymmetry condition (in finite-dimensional form, in [9,10]) akin to the positive-energy condition. The precise situation is given by the Theorem [8]: <u>If $S$ is a unitarizable element of $Sp(H)$, $H$ being a complex Hilbert space, it has a unique free-field quantization (definable as one such that $\langle W(z)v,v\rangle = \exp(-\|z\|^2/4)$ for arbitrary $z \in H$) if and only if the spectra of $S$ and $S^{-1}$ are disjoint (in the Hilbert space sense; as point sets they may overlap, but spectral absolute continuity classes are mutually singular).</u>

The requisite disjointness condition is satisfied in particular if the spectrum is confined to the upper unit circle, exclusive of the end-points, as will be shown to be the case for a natural general class of scattering transformations. However, the considerations thus far have been somewhat oversimplified from the standpoint of concrete applications, which involve conjugacies via unbounded operators, the introduction of complex structures that are unbounded with respect to the original topology in $L$, and the like. Suffice it here to say that notions of "essential" unitarizability, quantizability, etc. can be introduced that apply to suitable dense domains in $L$ , and that the theorems quoted then remain fully valid, and applicable in concrete scattering-theory contexts [11].

3. Stability of wave equations. The simplest prototype of the results needed to verify that scattering operators S such as occur in connection with wave equations enjoy the unique unitarizability conditions is the classic result of Liapounoff on the stability, in the sense of boundedness for all time, of solutions of Hill's equation. This equation takes the form

(5) $$u'' + p(t)u = 0 ,$$

where $p(t)$ is a real continuous function on $\mathbb{R}^1$ of period 1. It is readily shown that the equation is stable if and only if the "Floquet matrix" $S : (u(0),u'(0)) \to (u(1),u'(1))$ is conjugate to a unitary matrix. Liapounoff proved stability under the condition that p be non-negative and $\int_0^1 p(t)\,dt < 4$ . Many other criteria were later established for the stability of Hill's and related quations, and following the war the theory was greatly extended by M.G.Krein and co-workers [12]. For example, Krein showed that Liapounoff's criterion remained valid in case $u(.)$ had values in a real Hilbert space, $p(.)$ having values that were non-negative hermitian operators, and the "4" being replaced by 4 times the identity operators.

It seems interesting that the key to a deeper understanding and extension of the Liapounoff-Krein criterias was suggested by the field of theoretical astronomy, in which Hill worked. Causality in the form of invariant convex cone fields (defining the future) in space-time structures has led to a reconsideration of the concepts of time and energy that appears cosmologically relevant [12]. However that may be, why not do as Dirac, and extend the idea from the finite-dimensional geometrical context to an infinite-dimensional function-space one ? Illogical, perhaps, but it has been productive. The basic work is the thesis of S.M.Paneitz [10,14].

A starting point is the observation that Hill's equation defines an arc $S(t)$ in the symplectic group $Sp(\mathbb{R}^2)$, relative to the form $A(u,v) = uv' - u'v$, and that the positivity condition of Liapounoff (and similar ones) imply that the tangent $a(t)$ to this arc points into the unique (apart from sign) invariant closed convex cone in the Lie algebra $sp(\mathbb{R}^2)$; moreover, every element of the interior of this cone generates a unitarizable one-parameter subgroup. $S(t)$ differs from $\int_0^t a(s)\,ds$ only by terms of 2nd order in the $a(s)$, so that if $\int_0^t a(s)ds$ is sufficiently distant from the boundary of the cone, it may be expected that $S(t)$ is in the exponential of the cone, and hence unitarizable. Cf. also [12] and other studies by M.G. Krein, V.A. Yakubovich, et al.

The invariant convex cone $C$ in $sp(\mathbb{R}^2)$ takes the form: $C = [a \in sp(\mathbb{R}^2) : A(az,z) \geq 0,\ z \in \mathbb{R}^2]$. This definition generalizes to symplectic groups of arbitrary dimension; for $Sp(H)$, $H$ being a complex Hilbert space, $C$ takes the form $[a \in sp(H) : \mathrm{Im} < az,z > \geq 0, z \in H]$. We mean here by $sp(H)$ only the generators of uniformly continuous one-parameter subgroups, and so bounded real-linear operators on $H$; those in the interior of $C$ are precisely those that are conjugate via an element of $Sp(H)$ to an operator of the form $iH$, where $H$ is a positive hermitian operator. A variety of other groups are similarly causally orientable, and treated by Paneitz, but only the symplectic group will be considered here. It is interesting that while the semigroup $K = \cup_n (\exp C)^n$ generated by $C$ is all of $Sp(H)$, the corresponding semigroup $\tilde{K}$ in the universal cover $\tilde{Sp}(H)$ is a proper semigroup, satisfying $\tilde{K} \cap \tilde{K}^{-1} = \{e\}$ ([10] et seq.). However, here only the linear form of the symplectic group will be considered.

The group-invariant arc length in $Sp(H)$ is closely related to the Hilbert-Schmidt norm; the requirement that the endpoint of the arc in $Sp(H)$ be at a sufficiently small distance from 0 that the exponential mapping is properly applicable then leads to the following extension [10,11] of a beautiful theorem of Krein: If the arc $t \to a(t) \in Sp(H)$ is such that $a(t) \in C$ and is Hilbert-Schmidt for all $t$, $\int a(t)\,dt$ exists in the space of Hilbert-Schmidt operators, and has Hilbert-Schmidt norm $< 2$, then the endpoint of the curve in $Sp(H)$ that starts at $e$ and has $a(t)$ as tangent vector for each $t$, is essentially uniquely unitarizable.

The Hilbert-Schmidt norm, it should be emphasized, is in the space of real-linear operators on $H$, in this connection regarded as a real Hilbert space, with its given complex structure ignored.

Unfortunately, in scattering theory applications, the Hilbert-Schmidt norm of the relevant operators is finite only in 2 space-time dimensions, as is so often the case in quantum-field-theoretic considerations. This makes it necessary to use a non-invariant norm in $sp(H)$, and in addition the norm must be taken under the integral. A further complication of the applications to scattering theory is that in practice, $a(t)$ is not in the interior of the cone $C$ in $sp(H)$, but on its boundary, and not (infinitesimally) unitarizable; it is only after the smoothing effect of time-integration that the endpoint detaches itself from the non-unitarizable boundary of $C$, and even then it becomes only essentially unitarizable. The situation is enormously simpler if $a(t_o)$ is in the interior of $C$ for some value $t_o$. The resulting analogue of the Floquet matrix is then conjugate to a unitary operator directly within $Sp(H)$, and leaves a unique complex Hilbert space structure on $H$, which need agree with the original one only in the imaginary part of the inner product, invariant.

A basic result [11] can be summarized as follows, in the form applicable in scattering contexts. One is there concerned with a differential equation of the form $v' = a(t)v$, where $v$ has its values in a separable complex Hilbert space, and $t \to a(t)$ is a continuous map from $\mathbb{R}^1$ to $sp(H)$ such that $a(t) \in C$ for all $t$; $a(t)$ is in general not a complex-linear operator on $H$. Then if $\int_{\infty}^{\infty} ||a(t)||dt < 2$ and if no nonzero vector in $H$ is annihilated by all $a(t)$, then the limit $S = \lim_{t \to \infty} S(t)$ of the unique continuous solution $S(t)$ of the equation $S(t) = I + \int_{-\infty}^{t} a(s)S(s)ds$ (which solution exists and has values in $Sp(H)$ ) is uniquely essentially unitarizable, and has spectrum contained in the upper unit circle, exclusive of the endpoints.

The constant 2 is best possible. The verification of the hypotheses in concrete cases naturally requires special considerations, to which we now turn.

4. <u>Scattering theory and invariant hermitian structures in partial differential varieties</u>. Now let us return to the time-dependent perturbation of the Klein-Gordon equation

$$\Box\,\phi + m^2\phi + V(x,t)\,\phi = 0\,. \tag{6}$$

The natural complex Hilbert space in which to work here is the space $H$ of solutions of the "free" equation, $\Box\,\phi + m^2\phi = 0$, with its unique Lorentz invariant norm, complex, and symplectic structures. As earlier indicated, these may be described succinctly by introducing

$$\psi = C\,\phi - i\,C^{-1}\dot{\phi}\ ; \qquad \dot{\psi} = iB\psi\ ,$$

where $B = (m^2 I - \Delta)^{1/2}$, $C = B^{1/2}$; for simplicity we assume $m > 0$. Equation (6) defines a linear symplectic motion in $H$ taking the form

$$\dot{\psi} = iB\psi + iC^{-1}M(t)C^{-1}\,\mathrm{Re}(\psi)\ , \tag{7}$$

where $M(t)$ denotes the operation of multiplication by $V(x,t)$.

The scattering operator $S$ may be described as the total motion from time $-\infty$ to time $+\infty$ defined by equation (7), <u>modulo</u> the "free" motion defined by the equation $\dot{\psi} = iB\psi$, which is supposed to be in itself unobservable; $S$ measures the total effect of the interaction, as in the case of free objects entering a "black box", interacting, and then reappearing in altered form. More specifically [15] $S$ is obtained by the method of variation of constants (or transformation to the "interaction representation", in the language of much of the physics literature); setting $u(t) = e^{-itB}\psi(t)$, $u(.)$ satisfies the equation.

$$u' = a(t)u \quad , \quad a(t):\ w \to ie^{-itB}C^{-1}M(t)C^{-1}\mathrm{Re}(e^{itB}w) \tag{8}$$

Applying the criterion of the preceding section, it follows that: <u>if $\int_{-\infty}^{\infty} \| C^{-1} M(t) C^{-1} \| \, dt < 2$, and if there exists $x_o \in \mathbb{R}^n$ and $t_o \in \mathbb{R}^1$ such that $V(x_o, t) \neq 0$ for almost all $t > t_o$, $V(x,t)$ being non-negative and continuous, then $S$ exists and is uniquely essentially unitarizable.</u>

A natural and indeed prototypical source for equations of the form (6) is provided by the tangent spaces to the solution varieties of non-linear wave equations. These tangent spaces are defined by the first-order variational equations of the given non-linear equation. This observation together with the known theory of such equations permits their solution varieties to be given canonical hermitian structures, in a punctured (i.e. with zero deleted) neighbourhood of the zero solution.

The equations of the form

$$\Box \phi + m^2 \phi + g \phi^p = 0 \quad (m > 0 \, , \; g > 0 \, , \quad p \text{ odd}) \tag{9}$$

have had considerable study in the past 20 years, including global existance and scattering (for sufficiently small Cauchy data), the latter to the effect that the nonlinear term $g\phi^p \to 0$ fairly rapidly as $|t| \to \infty$, so that solutions of the non-linear equation are asymptotic to solutions of the free equation as $t \to \pm\infty$, under suitable conditions. In particular, in 3 or more space dimensions, a non-linear S-operator, transforming the asymptotic limit at time $-\infty$ into that at time $+\infty$, is well-defined and regular in a neighbourhood $N$ of $0$ in an appropriate topology [16].

The neighbourhood $N$ becomes a non-linear symplectic variety [17,18] when the undamental 2-form $\Omega$ is defined for any two tangent vectors $\beta$ and $\lambda$ as follows

$$\Omega_\phi(\beta, \lambda) = \int\{\beta(\vec{x},t)\,\dot{\lambda}(\vec{x},t) - \dot{\beta}(\vec{x},t)\,\lambda(\vec{x},t)\}\,d\vec{x} \;,$$

$\beta$ and $\lambda$ being represented by solutions of the variational equation.

(10) $\qquad \Box\beta + m^2\beta + pg\phi^{p-1}\beta = 0 \,.$

To simplify the analytic considerations one may use here $C_0^\infty$ data for $\beta$ and $\lambda$. The resulting form $\Omega$ is non-degenerate, closed, and Lorentz-invariant, generalizing to the nonlinear case the form earlier employed. The same symplectic structure substantially may be defined in terms of the elementary retarded and advanced solutions of the variational equation; the difference of these two solutions is both the commutator in the formal quantization of the variational equation, and the kernel in an integral representation for in terms of general $C_0^\infty$ functions on space-time.

The full, probabilistic quantization of the variational equation can now be effected in $N$, using the earlier theory. However, the fundamental complex structure $J_\phi$ in the tangent space $T_\phi$ at $\phi$ defined by the variational equation is, unlike the symplectic structure itself, highly non-local, as is the corresponding Riemannian structure given by the equation $G_\phi(\beta,\lambda) = \Omega_\phi(J_\phi\beta,\lambda)$. More specifically, $J_\phi$ is a function of the scattering operator $S_\phi$ for equation (10), which is in turn the (Frechet-Gateau) differential of the nonlinear scattering transformation $S$. With the use of the complex structure $J_\phi$, creation and annihilation operators may be defined for the quantization of the variational equation, and the vacuum state vector thereby characterized. This gives the 2-point function in particular, and this is related to the Riemannian structure, being essentially its kernel in an integral representation of $G_\phi$ in terms of general $C_0^\infty$ functions on space-time, analogous to that for the symplectic structure.

The verification of the conditions required for essential unitarization of the $S_\phi$ for $\phi$ in a sufficiently small $N$ depends on fairly delicate decay [16,19] and unique continuation properties [20,21,22] of non-linear wave equations, to show that no nonzero vector is annihilated by all the $a(t)$, or equivalently, left fixed by $S_\phi$. One needs to show, e.g., that if $\phi$ is a solution of the nonlinear wave equation, if $\psi$ is a solution of the Klein-Gordon equation, and if $\phi\psi = 0$ identically, then either $\phi$ or $\psi$ vanishes identically. A more quantitative substitute for the relatively qualitative considerations involved here should be useful. There has indeed been recent progress [23] on the scattering and decay issues, but this has not been adapted to present needs as yet.

What does this tell us about the solution variety $N$ itself? - the term "variety" is used, since $N$ is by no means a Banach manifold in a useful canonical fashion (in part due to the unboundedness of temporal evolution generation), and the situation is somewhat reminiscent of algebraic varieties (cf. [24]). Due to the unicity of the essential unitarization of $S_\phi$, the complex structure $J_\phi$ is left invariant by transformations commuting with $S$. Since all Lorentz transformations do so [25], the complex structure in $N$, and hence the entire hermitian structure, are Lorentz-invariant. It seems likely that the Lorentz group is nearly, perhaps precisely, the full invariance group of the hermitian variety $N$. Thus from a geometrical standpoint, non-linear wave equations serve to define Lorentz-invariant varieties that are non-trivial in the (physically relevant) sense of not being invariant under much larger (e.g. infinite-dimensional) groups; no other means of achieving this is presently known. Note also that a Riemannian structure moreover has the potential to determine a canonical generalized probability measure in the solution variety, needed for physical applications, by a kind of conjunction of the natural Gaussian measures in each tangent space that essentially define the vacuums for the tangential equations (cf. [26, 27]).

In all the cases discussed above, the formal expression of the symplectic structure has been time-independent, but the general method described is not at all dependent on this. An interesting case in which the form of the symplectic structure varies is that of the Klein-Gordon equation in a curved space time [28] that is asymptotically flat; this may be treated in a similar fashion [11] with appropriate restrictions on the perturbation of the Minkowski metric.

The particular program in constructive quantum field theory that Nelson [29] and I [30] initiated, and was later developed, in part in modified form, by Glimm-Jaffe, Gross, and Simon, among others, in addition to ourselves, has not been directly treated here for a variety of reasons, its limitation to lower-than-physical space dimensions, its present quiescence, the relative promise of the alternative program based on the classical solution variety, and particularly, the need for brevity. The latter constraint has also made it impossible to do more than refer to the work of Leray [31] applying largely symplectic considerations to a restructuring of finite-dimensional quantum mechanics.

## REFERENCES:

[1] Segal, I., Foundations of the theory of dynamical systems of infinitely many degrees of Freedom, I. Mat.-Fys. Medd. Danske Vid. Selsk. 31, no.12,39 pp (1959).

[2] Segal, I., Distributions in Hilbert space and canonical systems of operators. Trans. Amer. Math. Soc. 88 (1958), 12-41.

[3] Shale, D., Linear symmetries of free boson fields. Trans. Amer. Math. Soc. 103 (1962), 149-167.

[4] Segal, I., Mathematical characterization of the physical vacuum. Illinois Journ. Math. 6 (1962), 500-523.

[5] Weinless, M., Existence and uniqueness of the vacuum for linear Quantized fields. Journ. Funct. Anal. 4 (1969), 350-379.

[6] Cook, J.M., The mathematics of second quantization. Trans. Amer. Math. Soc. 74 (1953), 222-245.

[7] Segal, I., Tensor algebras over Hilbert spaces I. Trans. Amer. Math. Soc. 81 (1956), 106-134.

[8] Segal, I., Quantization of symplectic transformations. To appear in vol. of Advances in Math. ded. L.Schwartz.

[9] Gelfand, I, and Lidskii, U.B., On the structure of regions of stability of linear canonical systems of differential equation with periodic coefficients. Uspehi Mat. Nauk (N.S.) 10 (1955), 3-40.

[10] Paneitz, S.M., Causal structures in Lie groups and applications to stability of differential equations. Doctoral dissertation, June 1980. MIT, Dept. of Math.

[11] Paneitz, S.M. and Segal, I., Quantization of wave equation and hermitian structures in partial differential varieties. Proc. Nat. Acad. Sci. USA, in press.

[12] Daleckii, J.L. and Krein, M.G., Stability of solutions of differential equations in Banach spaces. Amer. Math. Soc., Providence, R.I. 1974, Translations of Math. Monographs., vol. 43.

[13] Segal, I., Mathematical cosmology and extragalactic astronomy. Academic Press, N.Y. 1976.

[14] Paneitz, S.M., Unitarization of symplectics and stability for differential equations in Hilbert space. Journ. Funct. Anal., in press.

[15] Segal, I., Nonlinear semigroups. Ann. Math. 78 (1963), 339-364.

[16] Segal, I., Dispersion for nonlinear relativistic equations: I., Proc. Conf. Math. Th. Elem. Part., MIT Press, 1966,79-108. II., Ann. Scient. de l'Ecole Norm. Sup. (4) 1 (1968) 459-497. See also W.A.Strauss, Nonlinear scattering theory, in: Scattering theory in Math. Physics, D. Reidel Publ. 1974, 53-78.

[17] Segal, I., Quantization of nonlinear systems. Journ. Math. Phys. 1 (1960), 468-488.

[18] Segal, I., Symplectic structures and the quantization problem for wave equations. Symposia Mathematica, vol. XIV, pp. 93-117. Academic Press, London, 1974.

[19] Nelson, S., $L^2$ asymptotes for the Klein-Gordon equation. Proc. Amer. Math. Soc. 27 (1971), 110-116.

[20] Segal, I., Direct formulation of causality requirements on the S-operator. Phys. Rev. (2) 109 (1958), 2191-2198.

[21] Goodman, R.W., One-sided invariant subspaces and domains of uniqueness for hyperbolic equations. Proc. Amer. Math. Soc. 15 (1964), 653-660.

[22] Morawetz, C.S., A uniqueness theorem for relativistic wave equations. Comm. Pure App. Math. 16 (1963), 353-362.

[23] Strauss, W.A., Nonlinear scattering theory at low energy. Journ. Funct. Anal., in press.

[24] Segal, I., Banach algebras and nonlinear semigroups. To appear in Vol. of Journ. Integral Eq., ded. E. Hille.

[25] Morawetz, C.S. and Strauss, W.A., On a nonlinear scattering operator. Comm. Pure Appl. Math. 26 (1973), 47-54.

[26] Jørgensen, E., The central limit problem for geodesic random walk. Z. Wahrscheinlichkeitstheorie u. Verw. Gebiete 32(1975),1-

[27] Malliavin, P., Diffusions et géometrie differentielle globale. Lecture notes, August 1975, Institut Henri Poincaré, Paris

[28] Lichnerowicz, A., Propagateurs et commutateurs en relativité générale Inst. Hautes Études Sci., Publ.Math. 10 (1961), 1-56.

[29] Nelson, E., A quartic interaction in two dimensions. In: Math. Th. El. Parts. (proc. Conf., Dedham, Mass. 1965) pp. 69-73, MIT Press, 1966.

[30] Segal, I., Notes towards the construction of nonlinear relativistic quantum fields.,
I. Proc. Nat. Acad. Sci.,USA 57 (1967), 1178-1183.
II. Bull. Amer. Math. Soc. 75 (1969), 1383-1389.
III.ibid pp 1390-1395.

[31] Leray, J., Lagrangian analysis and quantum mechanics. (English transl. of French text. publ. C.N.R.S.) MIT Press, Cambridge MAss., in press.

Massachusetts Institute of Technology
Cambridge, MA 02139, USA

# Recent developments in C*-algebras

Erling Størmer

Introduction. There are several different areas of C*-algebra theory that could be treated in an article with the above title. I have decided to concentrate my attention on a class of C*-algebras which has been at the center of developments over the last few years, namely the nuclear ones. Then I shall discuss their representations, whose weak closures are the injective von Neumann algebras. This will enable me to cover the main examples of simple C*-algebras together with a discussion of the probably most important factor - the hyperfinite $\mathrm{II}_1$-factor. However, we will miss any discussion of the order structure of C*-algebras, the general theory of von Neumann algebras, and of the promising interaction with differential geometry that is presently taking place.

The subject matter is the work of many people, so my references will necessarily be scattered and incomplete. The reader who wants a more complete survey of the recent developments and of the references in the area, should study the forthcoming proceedings of the AMS-Summer Institute on Operator Algebras and Applications which took place at Queen's University the summer of 1980.

## Part I. C*-algebras

1. Basic concepts. A C*-algebra is a complex Banach algebra with an involution and a norm which is a C*-norm, i,e, $\|a^*a\| = \|a\|^2$ for all elements a. The celebrated Gelfand-Naimark theorem states that a C*-algebra has an isometric *-preserving representation as a concrete C*-algebra, i.e. as a C*-algebra of bounded operators on a complex Hilbert space with the usual operator norm and involution. We may therefore whenever desired consider our C*-algebras as operator algebras on a complex Hilbert space H. We denote the C*-algebra of all bounded operators on H by B(H). There are several useful topologies on B(H). We shall only be concerned with the weak topology defined as the weakest such that all the linear functionals $a \to (a\xi,\eta)$, $\xi,\eta \in H$, are continuous. A von Neumann algebra on H is a weakly closed C*-subalgebra M of B(H) containing the identity operator 1. We say M is a factor if its center is the scalar multiples of 1. The factors are the basic building blocks for von Neumann algebras, while the simple C*-algebras, i.e. C*-algebras without any norm closed proper non-zero two-sided ideals, are the basic ones for C*-algebras.

A representation of a C*-algebra, or just a *-algebra, A is a *-preserving homomorphism $\pi$ of A into some B(H). We say $\pi$ is a factor representation if the weak closure $\pi(A)^-$ is a factor.

After these basic definitions we shall mainly be concerned with the main examples and constructions of C*-algebras. Our examples will usually be separable.

2. Abelian C*-algebras. If X is a locally compact Hausdorff space we denote by $C_0(X)$ the continuous complex functions on X, vanishing at infinity if X is noncompact, and with norm

$\|f\| = \sup_{x \in X} |f(x)|$. The main result on abelian C*-algebras is that they all have isometric representations as some $C_0(X)$ with $X$ compact if the algebra has identity. In the latter case we often write $C(X)$ instead of $C_0(X)$. Thus the classification theory for abelian C*-algebras is the same as that of locally compact Hausdorff spaces.

3. Finite dimensional C*-algebras. Every finite dimensional C*-algebra $A$ can be represented as a direct sum $\sum_{i=1}^{k} \oplus M_{n_i}$, where $M_n$ denotes the complex $n \times n$ matrices. In particular, $A$ is simple if and only if $A$ is isomorphic to some $M_n$.

4. Type I C*-algebras. Let $\mathcal{K}$ denote the compact operators on the separable Hilbert space. Then $\mathcal{K}$ is a simple C*-algebra without identity. If $A$ is a C*-algebra and $X$ a locally compact Hausdorff space, we extend the example in (2) to $C_0(X,A)$ - the C*-algebra of continuous $A$-valued functions on $X$ vanishing at infinity with pointwise multiplication and involution and norm $\|f\| = \sup_{x \in X} \|f(x)\|$. The type I C*-algebras are those we obtain if we build up from $C_0(X,M_n)$ and $C_0(X,\mathcal{K})$ by ideals, quotients and extensions. A formal but equivalent definition is that $A$ is of type I if for each factor representation $\pi$ we have $\pi(A)^-$ is isomorphic to some $B(H)$. These algebras were mainly studied in the 1950's and early 1960's and are by now very well understood, see [10]. They often appear in group representations, e.g. if $G$ is a semisimple or nilpotent Lie group, the C*-algebra generated by a continuous unitary representation of $G$ is of type I.

5. Tensor products. The algebras $C_0(X,M_n)$ form a special case of the tensor product construction, indeed $C_0(X,M_n) \simeq C_0(X) \otimes M_n$, where the tensor product is defined as follows.

Let $A$ and $B$ be $C^*$-algebras, say acting on Hilbert spaces $H$ and $K$ respectively. $H$ and $K$ have a natural vector space tensor product which we give the inner product $(\xi\otimes\eta, \xi'\otimes\eta') = (\xi,\xi')(\eta,\eta')$, $\xi,\xi' \in H$, $\eta,\eta' \in K$. The completion gives rise to the Hilbert space $H \otimes K$. If $a \in A$, $b \in B$ we can define the operator $a \otimes b$ in $B(H\otimes K)$ by

$$(a\otimes b)(\xi\otimes\eta) = a\xi \otimes b\eta .$$

In this way we can represent the algebraic tensor product $A\odot B$ as operators on $H \otimes K$. In general there are many different $C^*$-cross-norms, viz $\|a\otimes b\| = \|a\|\|b\|$, on $A\odot B$, but the norm inherited from $B(H\otimes K)$ is the smallest such norm. We call the closure of $A\odot B$ in this norm the Hilbert space tensor product of $A$ and $B$ and denote it by $A \otimes B$. An important class of $C^*$-algebras is the one for which this norm is the unique $C^*$-cross-norm on $A\odot B$ for all $C^*$-algebras $B$. We call $A$ <u>nuclear</u> if $A$ has this property. In this case the tensor product $A \otimes B$ is uniquely defined for all $C^*$-algebras $B$. All the algebras described in (2),(3),(4) are nuclear, and so are tensor products of nuclear $C^*$-algebras, see e.g. [14].

<u>6. Inductive limits</u>. If $(A_n)$ is a sequence of $C^*$-algebras such that there is a faithful, hence isometric, representation $\alpha_n$ of $A_n$ into $A_{n+1}$ for all $n$, we can consider the inductive system

$$A_1 \overset{\alpha_1}{\hookrightarrow} A_2 \overset{\alpha_2}{\hookrightarrow} A_3 \hookrightarrow \cdots$$

Since each $\alpha_n$ is an isometry the algebraic inductive limit of the system has a unique $C^*$-norm, hence its closure is a $C^*$-algebra denoted by $\varinjlim A_n$.

It is easy to see that if each $A_n$ is nuclear so is $\varinjlim A_n$. If each $A_n$ is finite dimensional $\varinjlim A_n$ is called an <u>AF-algebra</u> ( = approximately finite dimensional). Special cases are UHF-algebras, in which case each $A_n$ is a full matrix algebra, a special case of which is the CAR or the Fermion algebra arising from the canonical anticommutation relations. Then each $A_n = M_{2^n}$, and $\alpha_n$ is defined by

$$\alpha_n(x) = \begin{pmatrix} x & 0 \\ 0 & x \end{pmatrix} \in M_{2^{n+1}}, \qquad x \in M_{2^n}.$$

AF-algebras, and especially the CAR-algebra, have been extensively studied over the last twenty years, however, there are still obvious questions which remain open. For example, is a simple infinite dimensional C*-subalgebra of the CAR-algebra an AF-algebra ? More specially, is the fixed point algebra of a *-automorphism of order two necessarily an AF-algebra ?

Using inductive limits it is easy to construct infinite tensor products. Let $(B_k)$ be a sequence of C*-algebras with identity $1_k$. Let

$$A_n = B_1 \otimes B_2 \otimes \cdots \otimes B_n$$

and define $\alpha_n : A_n \to A_{n+1}$ by $\alpha_n(x) = x \otimes 1_{n+1}$ (again we have to be careful with the norms unless each $B_k$ is nuclear). We define the infinite tensor product of the B's to be $\bigotimes_{i=1}^{\infty} B_k = \varinjlim A_n$. Infinite tensor products and their tensor product representations were analysed in great detail in the late 1960's when each $B_k$ is a full matrix algebra, so $\bigotimes_{i=1}^{\infty} B_k$ is UHF, and are by now quite well understood. For example, when R. Powers [18] in 1967 first exhibited an infinite number of nonisomorphic factors on separable Hilbert space, he used different tensor product representations of the CAR-algebra $\bigotimes_{i=1}^{\infty} B_k$ with each $B_k = M_2$.

<u>7. Group algebras</u>. The simplest way to construct separable non-nuclear $C^*$-algebras is to use discrete countable groups. Let $G$ be such a group, let $H = l^2(G)$, and let $\lambda$ denote the left regular representation of $G$ on $H$, namely if $g \in G$, $f \in H$, then $\lambda_g f$ is the function

$$(\lambda_g f)(h) = f(g^{-1}h) , \qquad h \in G .$$

Let $C_r^*(G)$ denote the $C^*$-algebra generated by the unitaries $\lambda_g$, $g \in G$. This $C^*$-algebra is called the <u>reduced group algebra</u> as compared with the group algebra $C^*(G)$ defined below. One can show that $C_r^*(G)$ is nuclear if and only if $G$ is amenable [16]. In particular $C_r^*(F_n)$ is nonnuclear if $F_n$ denotes the free group on $n \geq 2$ generators. This $C^*$-algebra can also be shown to be simple [19], and it and its weak closure, which is a factor, were up to recently the canonical counter examples in operator algebras. Quite recently another group has taken over, namely $SL(3,\mathbb{Z})$ and closely related groups. These are discrete groups which satisfy property $T$ of Kazhdan, i.e. the trivial representation is an isolated point in the space of unitary equivalence classes of irreducible representations of the group.

<u>8. Crossed products</u>. We now come to the most important construction of $C^*$-algebras. It is closely related to the study of automorphism groups of $C^*$-algebras, an area which has close connections with physics and ergodic theory. We are given a $C^*$-algebra $A$, a locally compact group $G$, and a representation of $G$ into the group $\text{Aut}\, A$ of $*$-automorphisms of $A$ with the continuity property $\lim_{g \to e} \|\alpha_g(x)-x\| = 0$ for all $x \in A$, where $e$ is the identity in $G$.

In the special case when $G$ is the reals, the infinitesimal generator

$$\delta(x) = \lim_{t \to 0} t^{-1}(\alpha_t(x)-x) , \qquad x \in A ,$$

defines a densely defined derivation on $A$, viz $\delta(xy) = x\delta(y)+\delta(x)y$, $x,y \in A$. The theory of densely defined derivations has been extensively developed over the last five years; for a survey on some of the developments see [22].

In order to define crossed products let $K(G,A)$ denote the set of continuous $A$-valued functions on $G$ with compact support endowed with the following structure:

(a) $\quad \|y\|_1 = \int_G \|y(t)\|\, dt$

(b) $\quad y^*(t) = \alpha_t(y(t^{-1})^*)\Delta(t)^{-1}$

(c) $\quad y \times z(t) = \int_G y(s)\, \alpha_s(z(s^{-1}t))ds$ ,

where $y,z \in K(G,A)$, $ds$ is the left Haar measure and $\Delta$ the modular function on $G$. Note that if $\alpha$ is trivial, i.e. $\alpha_s$ is the identity for all $s$, then (c) is just the convolution product of $y$ and $z$. Thus in particular, if $A$ is the complex numbers, the norm $\| \ \|_1$ is the usual $L^1$-norm and the closure of $K(G,\mathbb{C})$ is $L^1(G)$. In order to define a $C^*$-algebra we choose the largest of all possible $C^*$-norms on $K(G,A)$, so if $x \in K(G,A)$ we let

$$\|x\| = \sup\{\|\pi(x)\| : \pi \text{ representation on } K(G,A)\} .$$

The closure of $K(G,A)$ in this norm is denoted by $A \times_\alpha G$ and is called the <u>crossed product</u> of $A$ and $G$ (with respect to $\alpha$). It is a direct extension of the semi-direct product of two groups to the case when one of the factors is a $C^*$-algebra. In the case when $A = \mathbb{C}$ we get the <u>group $C^*$-algebra</u> $C^*(G)$, which has a more complicated algebraic structure than the reduced group algebra $C^*_r(G)$. Indeed, $C^*(G)$ has $C^*_r(G)$ as a quotient.

Historically it was via the crossed product construction that von Neumann in 1940 first was able to construct a factor of type III, i.e. a factor such that every nonzero projection is of the form $vv^*$ with $v$ an isometry in the factor. He used $A = L^{\infty}(\mathbb{R})$ with Lebesgue measure and $G$ the $ax+b$ group ($a,b$ rational and $a > 0$), and then took weak closure of the crossed product [9]. This is a typical example of the fact that even if $A$ is abelian, if the action $\alpha$ of $G$ on $A$ is sufficiently ergodic, then $A \times_{\alpha} G$ may be a simple C*-algebra. I think it is fair to say that probably the most interesting results in pure C*-theory over the last years are closely related to such simple C*-algebras. The main reason for this is that the techniques for handling crossed products have been improved greatly. Let me mention some problems which have been studied.

In the 1950's Kaplansky asked if every simple C*-algebra with identity contains projections different from 0 and 1. The answer is negative for abelian algebras like $C([0,1])$, and people believed it was negative in general. It is for example believed that $C_r^*(F_2)$ has no projections, but nobody has been able to prove it. Blackadar was recently able to exhibit a projectionless simple nuclear C*-algebra by considering an inductive limit of certain subalgebras of $C([0,1],B)$ with $B$ simple [1]. More recently Connes [4] took a certain diffeomorphism of the 3-sphere $S^3$ [13] and showed that the crossed product $C(S^3) \times_{\varphi} \mathbb{Z}$ of the action of the integers $\mathbb{Z}$ on $C(S^3)$ defined by $\varphi$ is a projectionless simple nuclear C*-algebra. The sophisticated proof is an example of the interplay between differential geometry, C*-algebras, and algebraic K-theory to be described later.

There is a particular simple C*-algebra which has lately attracted a great deal of attention, namely the one generated by two unitaries $u$ and $v$ satisfying the commutation relations

$$uv = e^{2\pi i\theta} vu ,$$

where $\theta$ is an irrational number, $0 < \theta < 1$. This C*-algebra, denoted by $A_\theta$, can also be described as a crossed product. Namely let $\mathbb{Z}$ act on $C(\mathbb{T})$, $\mathbb{T}$ being the circle group, by $n \to \alpha^n$, where

$$\alpha(f)(z) = f(e^{2\pi i\theta} z) , \quad f \in C(\mathbb{T}), \ z \in \mathbb{T} .$$

In a suitable representation if $v$ denotes the unitary corresponding to the function $z \to z$ in $C(\mathbb{T})$, and $u$ to the automorphism $\alpha$, i.e. $ufu^* = \alpha(f)$, then it is easy to show that

$$A_\theta = C(\mathbb{T}) \times_\alpha \mathbb{Z} .$$

From the general theory on crossed products $A_\theta$ is therefore nuclear [20], and it is also seen to be simple. The isomorphism problem for these C*-algebras was only settled last year by Pimsner and Voiculescu [17], using algebraic invariants I shall describe below.

A central problem in C*-algebra theory as in much mathematics is to find the basic building blocks for the classification theory. A nuclear algebra which is one such example is the simple C*-algebra $O_n$ of Cuntz [6], defined as the C*-algebra generated by $n$ isometries $S_1,\ldots,S_n$, $n \in \{2,3,\ldots\} \cup \{\infty\}$, on a separable Hilbert space with orthogonal ranges generating the whole Hilbert space, i.e. $\sum_{i=1}^{n} S_i S_i^* = 1$. These algebras are also obtained by crossed products construction of known algebras, indeed there is an AF-algebra $C_n$, a *-automorphism $\alpha$ of $C_n$, defining an action of $\mathbb{Z}$ on $C_n$ in the obvious way such that

$$C_n \times_\alpha \mathbb{Z} \simeq \mathcal{K} \otimes O_n .$$

Thus if $p$ is a minimal projection in $\mathcal{K}$ then $O_n \simeq (p \otimes 1)(C_n \times_\alpha \mathbb{Z})(p \otimes 1)$, so $O_n$ is nuclear by general theory [20]. The reason why I said the $O_n$'s are building blocks in the theory is that every simple C*-algebra containing an isometry which is not a unitary operator, will for each $n$ contain a C*-subalgebra a quotient of which is $O_n$.

Generalizations of the algebras $O_n$ appear naturally in some recent work of Cuntz and Krieger [8] in the classification of irreducible topological Markov chains, being defined as certain subshifts of the n-shift. They could just as for $O_n$ define an AF-algebra a crossed product of which with an action of $\mathbb{Z}$ is of the form $\mathcal{K} \otimes O_A$, where $O_A$ is a generalization of $O_n$ defined by an $n \times n$ matrix $A$ with entries $0$ or $1$. Furthermore $\mathcal{K} \otimes O_A$ is an invariant of flow equivalence of irreducible topological Markov chains. Since we can use K-theory to distinguish C*-algebras Cuntz and Krieger could obtain invariants for a problem in ergodic theory in terms of abelian groups via nontrivial constructions of simple C*-algebras.

9. K-theory. We say two projections $e$ and $f$ in a C*-algebra $B$ are equivalent, written $e \sim f$ if there exists $v \in B$ such that $v^*v = e$, $vv^* = f$. We write $e \lesssim f$ if there is a projection $f_0 \in B$ such that $e \sim f_0 \leq f$. Let now $A$ be a C*algebra and consider the tensor product $A \otimes \mathcal{K}$ of $A$ and the compacts as our $B$. If $[e]$ and $[f]$ denote the equivalence classes of $e$ and $f$ in $A \otimes \mathcal{K}$ we can add them if $e$ and $f$ are orthogonal to each other by $[e] + [f] = [e+f]$. If they are not orthogonal we may find a projection $f_0 \in A \otimes \mathcal{K}$, $f_0 \sim f$, and $f_0$ orthogonal to $e$. Then we define $[e] + [f] = [e+f_0]$. If we also define $[e] \leq [f]$ if $e \lesssim f$ we obtain an ordered abelian semi-group, and from this semi-group an abelian group $K_0(A)$, which is in many

cases an ordered group. We can also define $K_1(A)$ as the quotient group

$$K_1(A) = \mathcal{U}((A\otimes\mathcal{K})^\sim) / \mathcal{U}_0((A\otimes\mathcal{K})^\sim) ,$$

where $(A\otimes\mathcal{K})^\sim$ is $A\otimes\mathcal{K}$ with the identity adjoined, and $\mathcal{U}((A\otimes\mathcal{K})^\sim)$ and $\mathcal{U}_0((A\otimes\mathcal{K})^\sim)$ denote the unitary group of $(A\otimes\mathcal{K})^\sim$ and its connected component respectively.

Many of the usual results in algebraic K-theory hold for C*-algebras, see e.g. [23] where the theory is done for abelian algebras but goes through for nonabelian ones as well. For example a natural version of Bott's periodicity theorem easily extends to C*-algebras, and more difficult is the extended version of the Thom isomorphism

$$K_i(A \times_\alpha \mathbb{R}) \simeq K_j(A) \quad \text{if} \quad i \neq j \in \{0,1\} ,$$

where $\alpha$ is any continuous action of the reals on $A$ [4]. If $J$ is a closed two-sided ideal in $A$ we also have the 6-term exact sequence connecting $K_0$ and $K_1$ :

$$\begin{array}{ccccc}
K_0(J) & \longrightarrow & K_0(A) & \longrightarrow & K_0(A/J) \\
\delta_1 \uparrow & & & & \downarrow \delta_0 \\
K_1(A/J) & \longleftarrow & K_1(A) & \longleftarrow & K_1(J)
\end{array}$$

This sequence makes it easy to compute $K_0(A)$ in some cases; for example if $A$ is an AF-algebra its unitary group is connected, so the $K_1$-groups in the above diagram vanish. Thus the upper sequence for $K_0$ is short exact in this case. Since furthermore $K_0$ is continuous with respect to inductive limits, viz $K_0(\lim_{\to} A_n) = \lim_{\to} K_0(A_n)$ , we have powerful techniques for computing $K_0(A)$ for $A$ an AF-algebra. For AF-algebras it can even be shown that $K_0(A)$ is an ordered group with positive cone $K_0(A)^+$ .

The pair $(K_0(A), K_0(A)^+)$ is up to tensoring with the compacts a complete invariant for AF-algebras and is the same as the dimension groups of Elliott [12].

For some of the C*-algebras we have discussed, the K-groups have been computed. Let me give some examples:

$K_0(M_n) = \mathbb{Z}$ with positive cone $\mathbb{Z}^+$.

$K_0$(CAR-algebra) = dyadic numbers eith usual ordering inherited from $\mathbb{R}$.

$K_0(O_n) = \mathbb{Z}_{n+1}$ (so is not an ordered group),

$K_1(O_n) = 0$ [7].

$K_0(A_\theta) = \mathbb{Z} + \theta\mathbb{Z}$ with natural ordering as a subgroup of $\mathbb{R}$ [17]. Thus $A_\theta \simeq A_{\theta'}$ if and only if the sets $\mathbb{Z} + \theta\mathbb{Z}$ and $\mathbb{Z} + \theta'\mathbb{Z}$ are equal, $\theta, \theta' \in (0,1)$ both irrational.

An open and natural problem would be to compute the K-groups of $C_r^*(F_n)$ for different $n$.

## Part II. Injective von Neumann algebras

One of the main techniques in studying C*-algebras is to take certain representations and then to study the von Neumann algebras obtained as the weak closures of the images. The von Neumann algebras obtained in this way from nuclear C*-algebras are the injective ones. It is possible to give a categorical definition of this concept, but the following is more concrete.

Definition: Let $M$ be a von Neumann algebra acting on a Hilbert space $H$. We say $M$ is *injective* if there exists a linear *-preserving map of norm 1 of $B(H)$ onto $M$, which is the identity map, when restricted to $M$.

The main result connecting nuclear $C^*$-algebras and injective von Neumann algebras is due to Choi , Connes, Effros and Lance, [2], [3], [11].

<u>Theorem</u>. A $C^*$-algebra $A$ is nuclear if and only if $\pi(A)^-$ is injective for each nondegenerate representation $\pi$ of $A$.

In particular, each factor representation of a nuclear $C^*$-algebra is on an injective factor. I shall now describe the main result in operator algebras of the 1970's, namely Connes' classification of injective factors.

Let $A$ be an AF-algebra acting on a Hilbert space $H$. Then $A$ is a norm closure $\overline{\bigcup_{n=1}^{\infty} A_n}$, where each $A_n$ is a finite dimensional $C^*$-algebra, and $A_n \subset A_{n+1}$. A von Neumann algebra which is of the form $M = A^-$ (weak closure) if often called <u>hyperfinite</u>. If we average each operator $x$ in $B(H)$ over the automorphisms $x \to uxu^*$ with $u$ running through the unitaries in $A_n$, we obtain an idempotent map of norm 1 of $B(H)$ onto the commutant $A_n'$ of $A_n$; thus $A_n'$ is injective. A weak limit point of these projection maps will be a similar projection map of $B(H)$ onto the commutant $A'$ of $A$, hence $A'$ is injective. The celebrated Tomita theorem, which I won't discuss here, now shows that $A^- = (A')'$ is injective. The main result of Connes is that the converse is true [3].

<u>Theorem</u>. Every injective factor acting on a separable Hilbert space is hyperfinite.

Furthermore, except for one class, those of type $III_1$, where there is expected to be only one example, all injective factors are classified in the sense that their structure is completely understood, and the isomorphism problem is the same as that of flows in ergodic theory, see [3] and [5].

The proof of the theorem consists basically of two parts. The first is to refer to the structure theory of factors developed in the early 70's mainly by Connes and Takesaki, which says that except for those factors which are isomorphic to some $B(H)$, they are all obtained from von Neumann algebras of type II and crossed products. The building block for those of type II are factors of type $II_1$. Such a factor is one which is of infinite dimension, so not isomorphic to any $M_n$, and which possesses a trace, i.e. a linear functional $\tau$ such that $\tau(x^*x) = \tau(xx^*) \geq 0$ for all elements $x$ and which satisfies $\tau(1) = 1$.

The classical way to construct $II_1$-factors is to take the weak closure of the reduced group algebra $C_r^*(G)$ for a highly nonabelian countable group, the technical condition being that each conjugacy class is infinite except for that arising from the identity $e$. Recall that finite linear combinations $\sum_{g \in G} a_g \lambda_g$ with $a_g \in \mathbb{C}$, $g \in G$, are weakly dense in $C_r^*(G)^-$. The trace $\tau$ is defined by

$$\tau(\Sigma\, a_g \lambda_g) = a_e .$$

Murray and von Neumann showed there exists a unique hyperfinite $II_1$-factor $R$ and that $R \not\cong C_r^*(F_2)^-$, see [9]. Up to the late 1960's only a handful of nonisomorphic $II_1$-factors were known, but over the last years $II_1$-factors with lots of strange properties have been exhibited.

The second part of the proof of the main result of Connes consists of several characterizations of the hyperfinite $II_1$-factor $R$. Since a central part of the proof consists of an analysis of the automorphism group of a $II_1$-factor $M$ and in particular of $R$ I shall indicate some of the ideas. The topology on $\mathrm{Aut}\, M$ is given by the continuity of the functions $\alpha \to \omega(\alpha(x))$ where $x \in M$ and $\omega$ is a linear functional on $M$ which is weakly continuous on the unit ball of $M$, i.e. $\omega \in M_*$. We then have two natural normal subgroups of Aut M, namely Int M - the group of inner automorphisms $x \to uxu^*$, $u$ unitary in $M$ - and its closure $\overline{\text{Int}}\, M$ in Aut M. Since by uniqueness $R = (\bigcup_{n=1}^{\infty} M_{2^n})^-$ and each automorphism of $M_{2^n}$ is inner, it is easy to see that $\text{Aut}\, R = \overline{\text{Int}}\, R$. Also, if the tensor product of two von Neumann algebras is defined as the weak closure of their Hilbert space C*-tensor product (in the same concrete representation) it is clear that $R \otimes R$ is hyperfinite, so isomorphic to $R$, hence $\text{Aut}\, R \otimes R = \overline{\text{Int}}\, R \otimes R$. Sakai [18] showed that the flip $\sigma: a \otimes b \to b \otimes a$ on a von Neumann algebra tensor product $M \otimes M$ defines an automorphism which is never inner if $M$ is a $II_1$-factor. Thus in a sense $\sigma$ is the "most outer" automorphism one can find. One of Connes' surprising characterizations of $R$ is that:

A $II_1$-factor $M$ on a separable Hilbert space is hyperfinite if and only if the flip $\sigma \in \overline{\text{Int}}\, M \otimes M$.

Not as striking but perhaps more important is the analysis of automorphisms and central sequences in Connes' proof. Such a sequence is a sequence $(u_n)$ of (usually) unitary operators in the factor $M$ such that $u_n x - x u_n$ converges to zero for each $x \in M$. They form a powerful tool for studying $II_1$-factors because they measure the degree of commutativity in $M$. For example $R$ is "very abelian" from this point of view while $C^*_r(F_2)^-$ is "extremely nonabelian". These properties are reflected in the size of $\overline{\text{Int}}\, M$ as compared with Int M, the more abelianness the bigger is $\overline{\text{Int}}\, M$. In particular we have $\text{Int}\, C^*_r(F_2)^- = \overline{\text{Int}}\, C^*_r(F_2)^-$.

Applications of Connes' theorem. I want to conclude these notes with some results which follow from or use crucially Connes' theorem. If $N$ is a subset of $B(H)$ we denote by $vN(N)$ the von Neumann algebra generated by $N$.

Let $G$ be a locally compact group and $\pi$ a continuous unitary representation of $G$. Then we have [3]:

(1) If $G$ is amenable then $vN(\pi(G))$ is injective (and $C^*(\pi(G))$ is nuclear).

(2) If $G$ is separable and connected then $vN(\pi(G))$ is injective.

(3) If $G$ is solvable then $vN(\pi(G))$ is injective.

(4) If $G$ is amenable and discrete and $\alpha$ is a homomorphism, $\alpha: G \to \mathrm{Aut}\, L^\infty(X,\mu)$, then $(L^\infty(X,\mu) \times_\alpha G)^-$ is injective.

An interesting property of $R$ is that it is in a well defined sense the smallest infinite dimensional factor. If $M$ is a $II_1$-factor we can find two orthogonal projections $e$ and $f$ in $M$ with sum $1$ and a partial isometry $v \in M$ such that $vv^* = e$, $v^*v = f$. Thus $M_2$ is imbedded in $M$. If we do the same with $eMe$ and $fMf$ we can imbed $M_4$ in $M$. If we continue this process we end up imbedding $R$ in $M$. Thus $R$ is contained in every $II_1$-factor. Conversely assume $M$ is a $II_1$-subfactor of $R$. Just as in ergodic theory using the trace instead of the invariant integral we can construct a conditional expectation of $R$ onto $M$, i.e. a projection map. If we compose this map with the corresponding one from $B(H)$ onto $R$ we obtain a projection map of $B(H)$ onto $M$. Hence $M$ is injective, so by the theorem $M \simeq R$. This argument shows that $R$ is smallest among all $II_1$-factors.

To conclude I shall show that $R$ is "smallest" among all infinite dimensional factors. Let $M$ be a factor and suppose $G$ is a compact subgroup of $\mathrm{Aut}\, M$ which acts ergodically on $M$, i.e. if $\beta(x) = x$ for all $\beta \in G$ then $x = \lambda 1$, $\lambda \in \mathbb{C}$. Then either $M = M_n$ or $M = R$ [15]. In other words, $R$ is the only infinite dimensional factor which is small enough so that a compact group can act ergodically on it. In the proof it is first shown that $M$ has a trace; then an argument of Connes using crossed products shows that $M$ is injective.

## References

1. B. Blackadar, A simple C*-algebra with no nontrivial projections, Proc. Amer. Math. Soc. 78 (1980), 504-508.

2. M.D. Choi and E. Effros, Nuclear C*-algebras and injectivity: The general case. To appear.

3. A. Connes, Classification of injective factors, Ann. Math. 104 (1976), 73-115.

4. A. Connes, An analogue of the Thom isomorphism for cross products of a C*-algebra by an action of $\mathbb{R}$, Preprint IHES (1980).

5. A. Connes and M. Takesaki, The flow of weights on factors of type III, Tôhoku Math. J. 29 (1977), 473-577.

6. J. Cuntz, Simple C*-algebras generated by isometries Commun. Math. Phys. 57 (1977), 173-185.

7. J. Cuntz, K-theory for certain C*-algebras, Preprint Heidelberg (1979).

8. J. Cuntz and W. Krieger, A class of C*-algebras and topological Markov chains, Preprint Heidelberg (1979).

9. J. Dixmier, Les algèbres d'operateurs dans l'espace hilbertien, Gauthier-Villars, Paris 1957.

10. J. Dixmier, Les C*-algèbres et leur representations, Gauthier-Villars, Paris 1964.

11. E. Effros and C. Lance, Tensor products of operator algebras, Adv. in Math. 25 (1977), 1-34.

12. G. Elliott, On the classification of inductive limits of sequences of finite dimensional algebras, J. Algebra, 38 (1976), 29-44.

13. A. Fathi and M. Herman, Existence de diffeomorphismes minimaux, Preprint Ecole Polytechnique (1976).

14. A. Guichardet, Tensor products of C*-algebras, Aarhus University Lecture Notes Series, no.12 (1969).

15. R. Høegh-Krohn, M.B. Landstad, and E. Størmer, Compact ergodic groups of automorphisms, Preprint Oslo (1980).

16. C. Lance, On nuclear C*-algebras, J. Funct. Anal. 12 (1973), 157-176.

17. M. Pimsner and D. Voiculescu, Exact sequences for K-groups and Ext groups for certain crossed products, Preprint INCREST (1979).

18. R. Powers, Representations of uniformly hyperfinite algebras and their associated von Neumann rings, Ann. Math. 86 (1967), 138-171.

19. R. Powers, Simplicity of the C*-algebra associated with the free group on two generators, Duke Math. J. 42 (1975), 151-156.

20. J. Rosenberg, Amenability of crossed products of C*-algebras, Commun. Math. Phys. 57 (1977), 187-191.

21. S. Sakai, Automorphisms and tensor products of operator algebras, Amer. J. Math. 97 (1975), 889-896.

22. S. Sakai, Recent developments in the theory of unbounded derivations in C*-algebras, C*-algebras and applications to Physics, Lecture Notes in Math. 650, Springer-Verlag (1977), 85-122.

23. J. Taylor, Banach algebras and topology, Algebras in analysis, Academic Press (1975), 118-186.

# SPECIALIZED LECTURES

# REPRESENTATIONS OF ALGEBRAIC GROUPS VIA COHOMOLOGY OF LINE BUNDLES

Henning Haahr Andersen

## 1. Cohomology of line bundles

Let $X$ be a homogeneous space, i.e a space of the form $G/P$ where $G$ is an algebraic group over a field $k$ and $P$ is a closed subgroup. For simplicity we assume that $G$ is semi-simple and simply connected. We'll also assume that $P$ is parabolic so that $X$ is a complete variety. Consider then a line bundle $L$ on $X$ and let's try to compute the cohomology of $L$.

First we'll see how the line bundles on $X$ look like. Recall that if $E$ is a P-module then we have a locally free sheaf $L(E)$ on $X$ defined as follows: The sections of $L(E)$ over an open subset $U$ of $X$ are the morphisms $f: \pi^{-1}(U) \to E$ which satisfy $f(xb) = b^{-1}f(x)$, $x \in \pi^{-1}(U)$, $b \in B$. ($\pi$ is the natural map $G \to G/P$). The rank of $L(E)$ equals the dimension of $E$. It is now a fact that the line bundles on $X$ have the form $L(\lambda)$ for some character $\lambda$ of $P$.

Next it turns out that the problem of computing the cohomology of $L(\lambda)$ reduces to the case where $P$ is a Borel subgroup. So we'll assume that $P = B$ is a Borel subgroup and we'll fix a maximal torus $T$ in $B$.

Some basic facts from algebraic geometry tell us that the cohomology groups $H^i(G/B, L(\lambda))$ (from now on denoted $H^i(\lambda)$) have the following properties

(1.1) $H^i(\lambda) = 0$ for $i > N = \dim G/B$.

(1.2) $H^i(\lambda)$ is finite dimensional for all $i$.

(1.3) $H^i(\lambda) \simeq H^{N-i}(-\lambda-2\rho)^*$ for all $i$, where $\rho$ is half the sum of the roots of $B$ and $*$ denotes dual ((1.3) is Serre-duality).

(1.4) If $\lambda$ is a character of $B$ with $H^0(\lambda) \neq 0$ and $F$ is any coherent sheaf on $G/B$ then $H^i(F \otimes L(\lambda+\rho)^n) = 0$ for $i > 0$ and $n$ large.

To obtain a more detailed description of the $H^i(\lambda)$'s is a hard problem which in the case of positive characteristic seems far from a complete solution. Before we report on some recent progress we shall look at the connection to representation theory.

## 2. Representations of algebraic groups

In any representation theory one of the very basic problems is the problem of describing the irreducible representations. In our case it is classical that the irreducible modules for $G$ are parametrized by the dominant characters of $B$.

A very useful tool in representation theory is the process of induction. For algebraic groups the proper definition is:

If $H$ is a closed subgroup of $G$ and if $E$ is an $H$-module then the $G$-module induced by $E$ is

$$E|_H^G = \{f: G \to E \mid f(gh) = h^{-1}f(g), g \in G, h \in H\}.$$

We note that by definition $\lambda|_B^G = H^o(\lambda)$ (Hence by (1.2) $\lambda|_B^G$ is finite dimensional).

It is easy to show that for $\lambda$ dominant $\lambda|_B^G$ contains a unique $B$-stable line and hence also a unique irreducible $G$-submodule which we denote $M(\lambda)$. In characteristic zero $M(\lambda) = H^o(\lambda)$.

Assume from now on that the characteristic $p$ is positive. The single most important unsolved problem in representation theory to day is the problem of finding the characters of the $M(\lambda)$'s. It is not hard to see that this problem is equivalent to the problem of finding the composition factors of the $H^i(\lambda)$'s. (It is a rather deep result (compare 3.4 below) that it is actually enough if one can handle the $H^o(\lambda)$'s. However, for some aspects of the theory, notably the strong linkage principle (see 3.1 below), one benefits from including in the consideration also the higher cohomology groups of non-ample line bundles).

## 3. Some recent results

The following theorem is called the strong linkage principle

(3.1) Let $\lambda$ and $\chi+\rho$ be dominant characters of $B$ and suppose $\chi' = w(\chi+\rho)-\rho$ for some element $w$ in the Weyl group. If $M(\lambda)$ is a composition factor of $H^i(\chi')$ for some $i$ then $\lambda$ is strongly linked to $\chi$.

For explanation of the term "strongly linked" and for the proof of (3.1) we refer to [1]. Let's just point out here that (1.3) is crucial for one of the key arguments in the proof.

In May 1979 the following theorem was obtained independently by W. Haboush and the author.

(3.2) Let $q = p^n$ and set $St_q = M((q-1)\rho)$. If $E$ is a B-module then there are natural G-module isomorphisms $H^i(E)^{(q)} \otimes St_q \to H^i(E^{(q)} \otimes (q-1)\rho)$, $i = 0,1,\cdots,N$. (Here $(q)$ denotes Frobenius twist).

Our proof of this result consists of 3 steps: $1^o$. Reduction to the case $i = 0$, $2^o$. For $i = 0$ reduction to the case where $E$ is an injective B-module and $3^o$. Proof for $i = 0$ and $E$ injective. It is crucial for the arguments that the injective objects in the category of rational B-modules have a nice description which allows us to employ (1.4). We also use the fact that $St_p^{(q)} \otimes St_p \simeq St_{pq}$ (a special case of Steinberg's tensor product theorem). For details we refer to [2].

The following two results are immediate corollaries of (3.2).

(3.3) For any B-module $E$ and any $i \geqq 0$ the Frobenius homomorphism $H^i(E)^{(q)} \to H^i(E^{(q)})$ is injective.

(3.4) If $\lambda$ is a character of $B$ such that $H^o(\lambda) \neq 0$ then $H^i(\lambda) = 0$ for $i > 0$.

(3.3) was conjectured by E. Cline, B. Parshall and L. Scott.

(3.4) was proved by G. Kempf (by completely different methods).

For a more detailed account of the recent developments in the area and for proper references we refer to [3].

## REFERENCES


1. H. H. Andersen, The strong linkage principle, J. Reine Ang. Math. 315 (1980), 53-59.

2. H. H. Andersen, The Frobenius morphism on the cohomology of homogeneous vector bundles on G/B, Ann. of Math. 112 (1980), to appear.

3. H. H. Andersen, Line bundles on flag manifolds, Proc. of International Conference "Young tableaux and Schur Functors in Algebra and Geometry 1980, to appear in Astérisque.

# A generic description of the evolution of a disequilibrium economy by exchange of stability

Michael C. Blad

## Introduction.

In order to appreciate the characteristic features of an economic disequilibrium model as opposed to a general equilibrium model, one must first know the basic qualities of the latter.

I shall therefore first shortly outline a basic general equilibrium model, and then explain the differences between this type of model and a disequilibrium model.

Secondly, both types of models are basically static models. However, a dynamic formulation is necessary in the following analysis, and I shall therefore shortly discuss how dynamics can be introduced into these models.

I then go on to present a specific dynamic model, which uses elements from both an equilibrium and a disequilibrium model.

## A general equilibrium model.

A general equilibrium model is often referred to as an Arrow-Debreu model due to the two economists who made the first formal presentation of this type of model. A standard reference is Debreu's book "Theory of Value", [1959].

In the simplest version of this model (which is all we need here) there are a finite number of consumers, $i=1,\ldots,N$, a finite number of goods, $h=1,\ldots,\ell$, and total initial

$\omega=(\omega_1,\ldots,\omega_\ell)$. Consumer i's consumption $x_i$ is a point in a given non-empty subset $X_i$ of $\mathbb{R}^\ell$, his consumption set. Each consumer is endowed with a preference preordering $\precsim_i$ (reflexive, transitive), defined on $X_i$, and the wealth $w_i \in \mathbb{R}$. A price system p is a set of real numbers $p=(p_1,\ldots,p_\ell)$ associated with the $\ell$ goods. Each consumer chooses his consumption $x_i \in X_i$ as a greatest element for his preference preordering such that the expenditure $p\cdot x_i$ does not exceed his wealth.

An equilibrium $((x_i),p)$ is defined as a set of demand vectors and a price vector such that

1) $\sum_i x_{ih}=\omega_h$, $h=1,\ldots,\ell$
2) $x_i$ maximizes $\precsim_i$ subject to the budget constraint $p\cdot x_i \leq w_i$.

It follows that equilibrium prices clear the markets. Describing an economy solely by the equilibrium states carries implicitly the assumption that prices are able to adjust to make demand equal to supply. Only when these adjustments have been completed, actual exchanges take place, so in order to make sense these price adjustments must be extremely fast.

A disequilibrium model.

The above approach has obvious weaknesses. One major objection is the exclusion of situations, where one or several markets do not clear. Consider e.g. the labour market. Supposing the wage rate adjust such that demand is equal to supply, implies that there is always full employment. It therefore seems obvious to try to model situations where prices do not necessarily adjust to equal demand and supply. This is exactly what happens in disequilibrium theory. The term disequilibrium is unfortunate, because the theory is still concerned with equilibrium situations, only do these differ from above by not assuming infinitely fast adjustment in prices. Actually the assumption is the opposite: prices are assumed fixed and quantities adjust fast. This may lead to rationing of consumers, and a better name is therefore rationed equilibrium. But the former term has become standard, so I shall use it too.

Slightly different disequilibrium models have been introduced during the last five years, the different approaches being due to varying assumptions on the way the quantity rationing enters the description. Main contributions are presented by Benassy, [1975], Drèze,[1975], Malinvaud,[1977], and Younès,[1975].

The main idea in these models is as follows. The economy consists of N consumers and $\ell$ goods with prices $p=(p_1,\ldots,p_\ell)$, money being good no.1. Each of the $\ell$-1 goods other than money is traded on a separate market against money. The i'th consumer's total actual transaction $z_i$ satisfies the budget constraint $p\cdot z_i=0$. Besides the price system the consumers are given upper and lower limits on their actual transactions. These quantity constraints enter the description of the consumer's decision making process. The consumer then makes simultaneously effective (i.e. constrained) demands for each good (other than money), given the rationing on all markets. An equilibrium is defined as a set of prices and quantity constraints such that the sum of net trades is zero, and for each good only traders on one side of the market for that good perceive binding constraints.

Dynamic models.

The models mentioned above are purely static models. The theory developed around these models is devoted to discussing existence of equilibrium and the equilibrium's dependance on the various parameters determining the equilibrium. Nothing is said about how or if an equilibrium value will actually be obtained, i.e. the questions of adjustment processes and stability are completely ignored. In order to investigate these problems a theory describing situations out of equilibrium is called for.

In the general equilibrium model one answer has been given by the so called "tâtonnement" processes. The idea is here that "an invisible hand" adjusts prices according to excess demands on the separate markets, and no trade is allowed to take place until the equilibrium prices have been reached.

Dynamic disequilibrium models are still few in number. The main concern here has been centered around a description of the evolution of equilibrium prices over time, assuming the initial quantity adjustments to be infinitely fast.

The specific model I present uses elements from both theories by considering simultaneously a relatively fast adjustment process in the quantities and a relatively slow adjustment process in prices. The evolution over time is analyzed by taking the fast adjustment process directly into account.

The model.

We consider a rationed equilibrium economy of the Malinvaud type. There are a finite number of identical consumers, three goods (labour, a consumption good, and money) and a production sector given by an aggregated production function. The price of the good and labour are denoted by p and w, and the price of money is the unit.

In the short run prices and wages are fixed; therefore equilibrium is obtained by adjustments in the quantities of the goods only; with the different combinations of rationing on the labour market and the good market we can devide the (p,w) plane into three regions of Keynesian, Classical and Repressed inflation equilibrium as described by Malinvaud, [1977].

However, in the long run we suppose that prices and wages react to excess demands on the different markets. "Slow" adjustment processes in prices and wages over time can be defined, and they give rise to the following flow diagram, Blad & Kirman, [1978].

Fig.1.

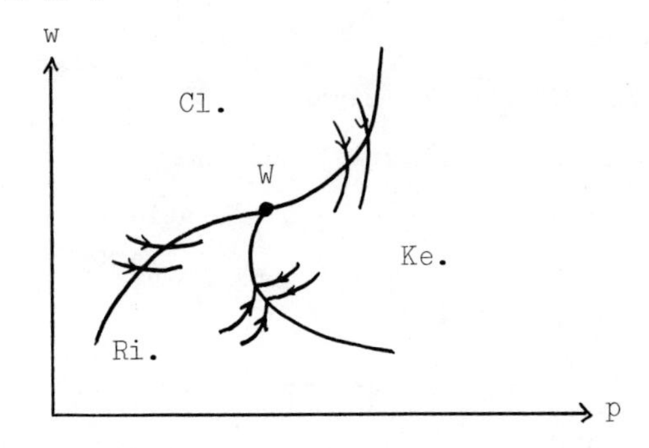


Here W denotes the Walrasian equilibrium with no rationing at all. In the following we shall only use the qualitative picture of the flow lines.

For each fixed value of p and w the consumer faces the following decision problem. How much to demand of the good, how much to supply of labour, and how much money to carry forward to the next period. In the same way the production sector must decide how much to produce and how much labour to demand for that. It turnsout that the result of these decisions conveniently can be represented by the following 3 variables:

x: the excess demand of the good (i.e. effective demand minus actual supply)

y: the extent to which notional output (i.e. profit maximizing) exceeds actual output

z: the excess supply of labour (i.e. effective supply minus actual demand.)

With this notation we have the following.

A Keynesian equilibrium is a point of the form $(p,w,0,y,z)$ with $y > 0, z > 0$. A Classical equilibrium is a point of the form $(p,w,x,0,z)$ with $x > 0, z > 0$. A Repressed inflation equilibrium is a point of the form $(p,w,x,y,0)$ with $x > 0, y > 0$.

By introducing the coordinates $(x,y,z)$ we get immediately a simple geometric picture of the equilibrium states. The $(p,w)$ plane is considered as parameter space C. Price and wage rate are supposed to remain fixed in the short run, while the quantities adjust relatively fast to an equilibrium state, determined by $(p,w)$. The 3 dimensional $(x,y,z)$ space X is denoted the state space, because for fixed $(p,w)$ the state of the economy is represented by a point in this space.

In this setting the projection of the equilibrium points in the state space is contained in the 3 faces of the positive quadrant with vertex $(0,0,0)$, the Walrasian equilibrium.

Corresponding to the partitioning of the parameter plane into a Keynesian, a Classical and a Repressed inflation equilibrium region, we get a Keynesian face, a Classical face and a Repressed inflation face.

Exchange of stability.

The analysis uses the assumption that price and wage rate do change through time. When $(p,w)$ belongs to the Classical region, the equilibrium lies in the plane $y=0$. Similarly, when the parameter $(p,w)$ belongs to the Keynesian region, the equilibrium belongs to the plane $x=0$. At first sight this description implies that if the parameter passes smoothly from the Classical to the Keynesian region, the path traced out in the state space by the equilibrium point would have to have a sharp corner, as it moved from the plane $y=0$ to the plane $x=0$. However, the model is based on the fundamental principle that the differential equation that describes the fast adjustment in X depends smoothly on C. This implies that the equilibrium varies smoothly and hence cannot turn sharp corners. Therefore there must be an "exchange of stability" between equilibria: as the parameter moves from the Classical to the Keynesian region, the Classical equilibrium $E_1$ persists in its smooth path, but changes from being stable to being unstable.

Fig.2.

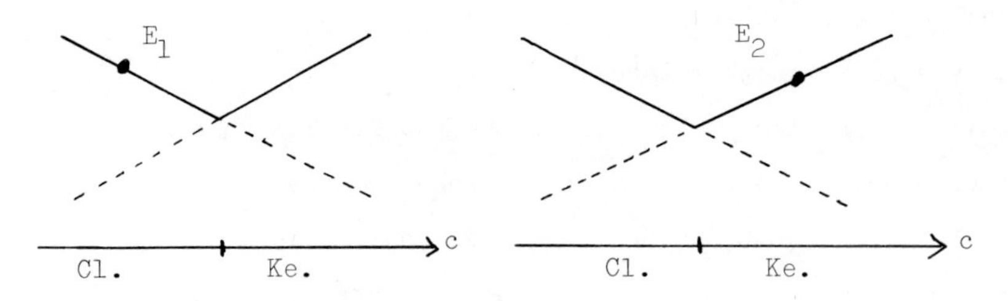


Conversely, on the return path the Keynesian equilibrium $E_2$ persists in its smooth path, but changes from being stable to unstable. In other words, over the path traced out by the parameter (p,w) lie two equilibrium points.

Fast/slow dynamics.

To get an intuitive understanding of the notion of fast/slow dynamics consider the following simple example.

$\dot{x}=-kx$ ("fast")

$\dot{\alpha}=\varepsilon$ ("slow") $\varepsilon \ll k$.

Graphically this system may be represented as follows:

Fig.3.

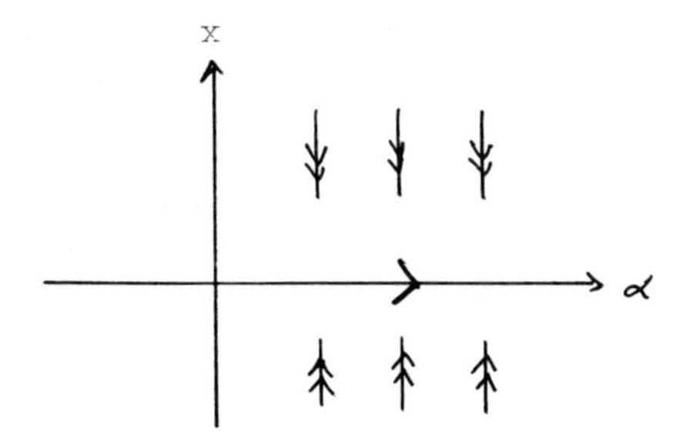


indicating that at a given point $(\alpha_0, x_0)$ the vector of changes $(\dot{\alpha}_0, \dot{x}_0)$ is almost parallel to the x-axis, as $\varepsilon \ll k$. The interpretation is as follows. The fast dynamics takes points to the line x=0 ("the slow manifold") as described by the fast equation, whereafter the slow dynamics moves the points along this line as described by the slow equation. As k increases the speed of adjustment towards the slow manifold increases, and in the limit $(k \to \infty)$ every point instantly jumps to the slow manifold.

Consider next the non linear example

$\dot{x} = -k(x^2 + \alpha)$

$\dot{\alpha} = \varepsilon \qquad\qquad \varepsilon \ll k.$

Fig.4.

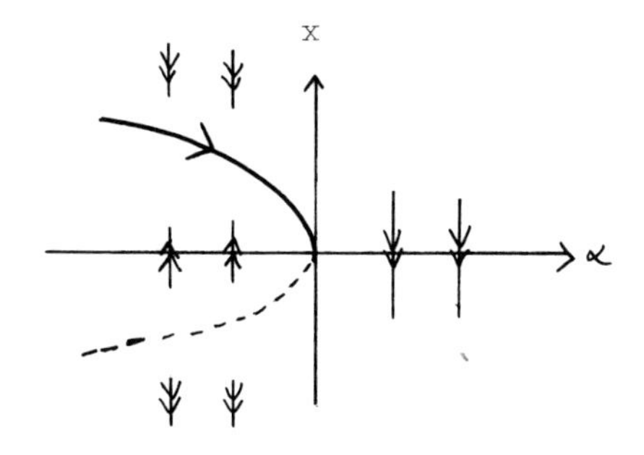


This example gives rise to the simplest catastrophe, the fold: for $\alpha < 0$ there is a fast adjustment away from the (unstable) lower branch of the parabola towards the (stable) upper branch (or infinity), followed by a slow movement along this branch.

However, when $\alpha$ reaches the critical point 0, the stable evolution breaks down and the fast dynamics takes over.

In the following we shall combine Fig.2. with this picture by considering the generic evolution of a rationed equilibrium economy for smooth variations in the parameters.

## The general long term evolution.

Above we considered a one dimensional parameter, which moved smoothly from one equilibrium region into another. The corresponding pair of smooth equilibrium paths were traced out in the two faces of the 2 dimensional equilibrium quadrant. As we now consider the general (local) situation around the Walrasian equilibrium, we must allow the parameter space to be 2 dimensional. Corresponding to paths in the (p,w) plane we shall now consider 3 smooth equilibrium paths traced out on the 3 faces of the 3 dimensional equilibrium quadrant. First we establish a simple correspondence between parameter values and equilibrium points. To simplify we assume through out that the angle between each 2 of the tangents to the boundaries at W is $120^o$. The general case can easily be reduced to this, see Blad,[1979].

Fig.5.

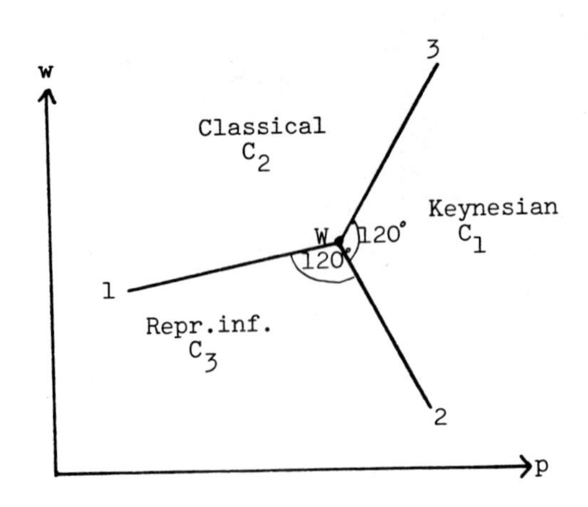

As an equilibrium in the quantities is described by 3 coordinates, we want to characterize each point in the parameter plane by 3 coordinates $(\alpha, \beta, \gamma)$. This can be done in a 1 - 1 way by imposing the restriction $\alpha+\beta+\gamma=0$. Choosing the 3 tangents as coordinate axes, the coordinates to a point $(p,w)$ are defined as follows:

$\alpha$-coordinate: the coordinate of the projection, parallel to the 1-axis, of $(p,w)$ onto the 2-axis.

$\beta$-coordinate: the coordinate of the projection, parallel to the 2-axis, of $(p,w)$ onto the 3-axis.

$\gamma$-coordinate: the coordinate of the projection, parallel to the 3-axis, of $(p,w)$ onto the 1-axis.

Fig.6.

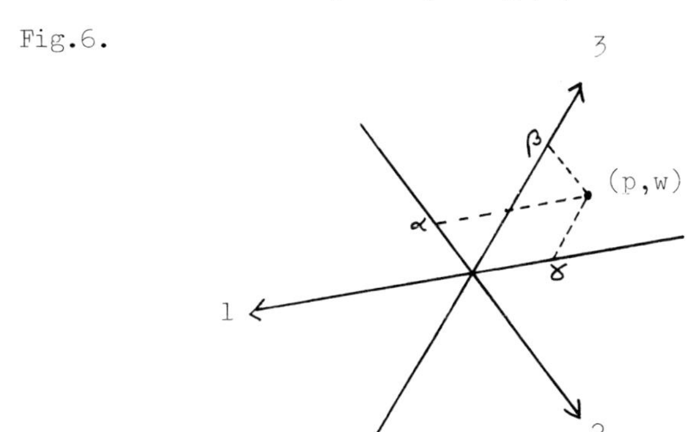


Remark.

$(p,w) \in C_1 \Rightarrow \beta > 0, \quad \gamma < 0$

$(p,w) \in C_2 \Rightarrow \gamma > 0, \quad \alpha < 0$

$(p,w) \in C_3 \Rightarrow \alpha > 0, \quad \beta < 0.$

Corresponding to each point $(\alpha, \beta, \gamma)$ in the parameter plane we now define 3 equilibrium points in the state space $X(x,y,z)$ as described in the following scheme. Note that we do not introduce any a priori restriction on the sign of $\alpha, \beta$ and $\gamma$.

| | Keynesian | Classical | Repr.inf. |
|---|---|---|---|
| x | 0 | $\gamma$ | $-\beta$ |
| y | $-\gamma$ | 0 | $\alpha$ |
| z | $\beta$ | $-\alpha$ | 0 |

We notice that due to the remark, when (p,w) belongs to the Keynesian region, the corresponding Keynesian equilibrium is characterized by positive y,z - coordinates. Therefore it is a point in the face x=0 of the equilibrium quadrant. Similarly, when (p,w) belongs to one of the other two equilibrium regions the corresponding equilibrium in the state space is a point on the corresponding face of the equilibrium quadrant.

Furthermore on each of the three coordinate axes two of the equilibrium points coalesce, e.g. if $(\alpha,\beta,\gamma)$ belongs to the boundary between two regions $C_1$ and $C_2$, then $(\alpha,\beta,\gamma)$ belongs to the 3-axis and so the point is of the form $(\alpha,\beta,0), \alpha=-\beta$. Therefore the three corresponding equilibrium points are $(0,0,\beta),(0,0,-\alpha)$ and $(-\beta,\alpha,0)$ and so the Keynesian and the Classical equilibria coincide at a point on the edge x=y=0 of the equilibrium quadrant.

Next we notice the following geometrical description of the equilibrium points.

Lemma.

For a fixed parameter value $(\alpha,\beta,\gamma)$ the corresponding 3 equilibrium points are situated on a line in the X space. The line is determined by one of the points and the vector (1,1,1).

Proof.

The coordinates of the 3 equilibrium points are

(1): $(0,-\gamma,\beta)$; (2): $(\gamma,0,-\alpha)$; (3): $(-\beta,\alpha,0)$,

so the vector determined by two of the equilibrium points are:

(2) - (1): $(\gamma,\gamma,-\alpha-\beta)=\gamma(1,1,1)$

(3) - (1): $(-\beta,\alpha+\beta,-\beta)=-\beta(1,1,1)$

(3) - (2): $(-\beta-\gamma,\alpha,\alpha)=\alpha(1,1,1)$.

Fig.7.

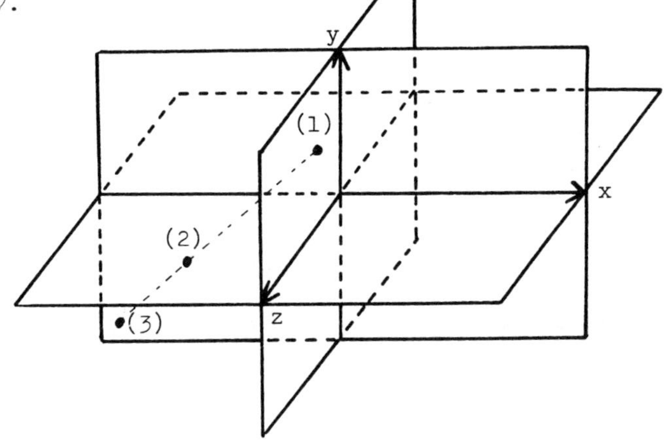


We now want to introduce the fast dynamics, i.e. the dynamics w.r.t. which the above introduced points are equilibrium (i.e. stationary) points. The fast dynamics describes the adjustment towards these equilibrium points. We are looking for a simple description of the dynamics in x,y,and z for given fixed parameters $(\alpha,\beta,\gamma)$. The above lemma indicates that a simple formulation may be obtained, if one of the coordinate axes in the state space is chosen parallel to (1,1,1). We therefore make the following coordinate change.

$$\xi = x+y+z$$

$$\eta = \alpha x+\beta y+\gamma z$$

$$\zeta = (\beta-\gamma)x+(\gamma-\alpha)y+(\alpha-\beta)z.$$

Modulo a scalar expansion this is an orthogonal change, since $\alpha+\beta+\gamma=0$ and the transformation matrix has determinant

$-3(\alpha^2+\beta^2+\gamma^2)$, which is different from zero, except at the Walrasian equilibrium.

In the new coordinate system the equilibrium points have the coordinates

| | Keynesian | Classical | Repr.inf. |
|---|---|---|---|
| $\xi$ | $\beta - \gamma$ | $\gamma - \alpha$ | $\alpha - \beta$ |
| $\eta$ | 0 | 0 | 0 |
| $\zeta$ | $-\rho^2$ | $-\rho^2$ | $-\rho^2$ |

where $\rho^2 = \alpha^2 + \beta^2 + \gamma^2$.

To describe the simplest dynamics having the required form we utilize the potential function

$$V(\xi,\eta,\zeta)=k[(\xi^4/4 - \tfrac{3}{4}\rho^2\xi^2 - \Delta\xi)+\eta^2/2 + (\zeta^2/2 + \rho^2\zeta)], k > 0,$$

where $\Delta = (\alpha-\beta)(\beta-\gamma)(\gamma-\alpha)$.

We suppose that the fast dynamics is given by the gradient of V, in other words

$$\dot{\xi} = -V_\xi = -\partial V/\partial\xi = -k(\xi-\beta+\gamma)(\xi-\gamma+\alpha)(\xi-\alpha+\beta)$$

$$\dot{\eta} = -V_\eta = -\partial V/\partial\eta = -k\eta$$

$$\dot{\zeta} = -V_\zeta = -\partial V/\partial\zeta = -k(\zeta+\rho^2) \qquad k > 0.$$

Note that

$$\dot{\xi} = -k(\xi^3 - 3/2\rho^2\xi-\Delta), \text{ where}$$

$$\Delta = (\alpha-\beta)(\beta-\gamma)(\gamma-\alpha).$$

With the lack of a standard and useful economic theory of fast quantity adjustments in disequilibrium, we have introduced a mathematically simple formulation of the fast dynamics,having the desired property of three stationary points, two stable and one unstable.

The following theorem shows that with this fast dynamics, we have a description of the long term evolution with (p,w) moving around in the parameter plane, which exhibits exchange of stability, each time one of the coordinates $\alpha$, $\beta$, and $\gamma$ changes sign.

Theorem.

If the fast dynamics is given by the gradient of V, then the following is true.

For each value of $(\rho^2,\Delta)$ there are three equilibrium points, a Keynesian, a Classical, and a Repressed inflation equilibrium, of which two are stable and one is unstable.

The Keynesian equilibrium is stable in the two regions $C_1$ and its reflection.

Similarly is true for the other two equilibrium points.

The equilibria are unstable anywhere else. (They are in fact saddle points).

Proof.

See Blad,[1979].

In the above analysis we have named as Keynesian equilibrium every point belonging to the plane x=0. However, as we initially only assigned this name to the equilibria of the form (0,y,z) with $y>0, z>0$, we shall from now on return to this terminology. The above theorem justifies the distinction in the following definition.

Definition.

Let $(p,w)=(\alpha,\beta,\gamma)$ be a point in the parameter space, and (0,y,z) an equilibrium in the state space.

If (p,w) belongs to the Keynesian region $C_1$, then (0,y,z) is called a Keynesian equilibrium.

If (p,w) belongs to the reflection of $C_1$, then (0,y,z) is called

a dual Keynesian equilibrium.
Otherwise (0,y,z) is called an unstable Keynesian equilibrium.
The analogous terminology is applied to the Classical and
Repressed inflation equilibria.

In order to analyze these new dual equilibria we now modify the potential $V(\xi,\eta,\zeta)$ such that it is a universal unfolding of the critical point (Zeeman,[1977],Chapter 18). W.l.o.g. we can assume the constant k equal to 1. Next we notice that in order to get the universal unfolding in this case, two "hidden" parameters h and k have to be added to the coefficients of $\xi^2$ and $\xi$ in the first bracket, because $\rho$ is quadratic and $\Delta$ is cubic in $\alpha,\beta,\gamma$. In order to discuss the introduction of these two new parameters, we shall specifically look at the case with $\eta=0, \zeta=-\rho^2$, since these are the equilibrium values, such that the unfolding is of the form

$$V(\xi,0,-\rho^2)=\xi^4/4 - \tfrac{3}{4}(\rho^2+h)\xi^2 - (\Delta+k)\xi + \text{constant}$$

with

$$V_\xi = \xi^3 - 3/2(\rho^2+h)\xi - (\Delta+k).$$

This is the formula for a cusp with h as splitting factor and k as normal factor. This means that for fixed parameter values $(\alpha,\beta,\gamma)$, the shape of the equilibrium surface is a cusp with the cusp point translated to the point $(h,k)=(-\rho^2, -\Delta)$ as shown below.

Fig.8. $(\alpha,\beta,\gamma)$ fixed:

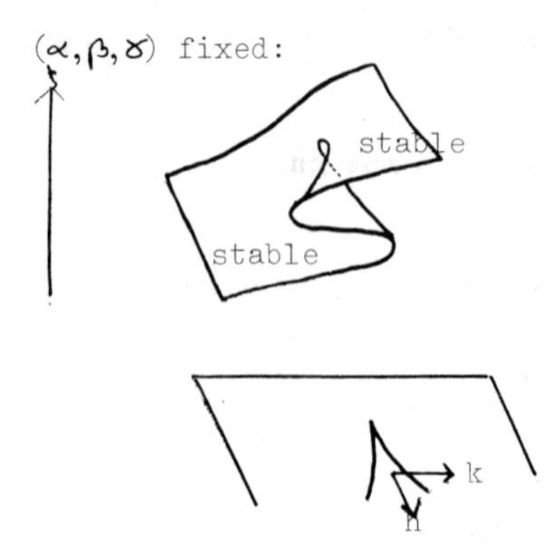

Assume that (p,w) is off the coordinate axes. By the theorem above with h=k=0, there are three equilibrium points, a stable, an unstable, and a dual stable. The same remains true for (h,k) in a neighbourhood of (0,0). Depending on the history of the economy the actual equilibrium will be on one of the two stable sheets. Let us suppose it is a stable equilibrium on the upper sheet. If the value of the parameter k is now reduced, then when the "left" part of the cusp curve in the plane is reached, the stability of the equilibrium breaks down, and there is a sudden jump to the dual equilibrium point in the lower sheet.

So for each fixed value of (p,w) the equilibrium surface forms a cusp, i.e. it is a family of cusps, parametrized by (p,w). We now do the opposite compared to above by fixing the values of h and k and allow the parameters $(p,w)=(\alpha,\beta,\gamma)$ to vary.

In particular we are interested in flowing along the flow lines shown in Fig.1., and deducing how the equilibrium states will change. We shall approach this problem by examining a section of the equilibrium surface over a circle, centre the Walrasian equilibrium in the parameter plane. For example, Fig.9. shows such a section for fixed values of (h,k) with $h<0$.k=0. As above $\xi$ is the quantity variable, as $\eta$ and $\zeta$ have been fixed at their stable equilibrium values.

We will now show how to deduce the surprising shape of Fig.9. In the discussion that follows we take a standard model of the cusp catastrophe given by the potential function

$V(\xi)=\xi^4/4 - \frac{3}{4}u\xi^2 - v\xi$

with standard 2 dimensional control space $C^2(u,v)$, splitting factor u, normal factor v and bifurcation set $u^3=2v^2$ with cusp point at the origin (see Fig.10 A).

Fig.9.

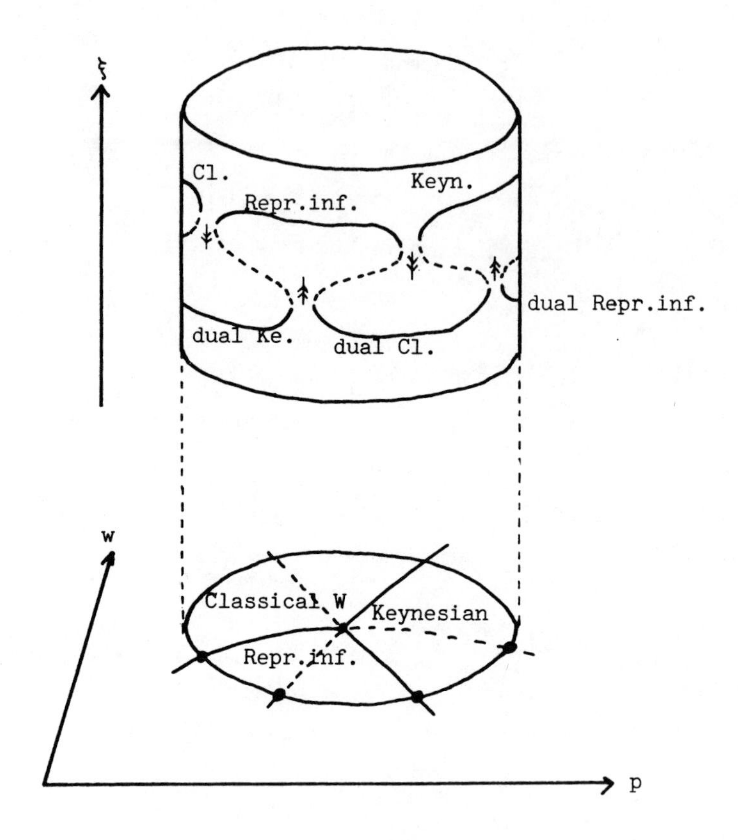


Now in our problem we have a 4 dimensional control space $C^4(p,w,h,k)$ and a potential given by

$V(\xi)=\xi^4/4 - \frac{3}{4}(h+\varsigma^2)\xi^2 - (k+\Delta)\xi$,

where $(p,w)=(\alpha,\beta,\gamma)$, $\varsigma^2=\alpha^2+\beta^2+\gamma^2$ and $\Delta=(\alpha-\beta)(\beta-\gamma)(\gamma-\alpha)$.

Therefore we can realize our problem by mapping the given control space into the standard control space

$m: C^4(p,w,h,k)\rightarrow C^2(u,v)$

by the map m given by

$(u,v)=(h+\varsigma^2,k+\Delta)$.

In particular, when $(h,k)=(0,0)$ m maps the $(p,w)$ plane by folding it into six along the three $\alpha,\beta,\gamma$- axes and mapping the resulting wedge-shape into the interior of the cusp in the $(u,v)$ plane. To see this, introduce polar coordinates in the $(\alpha,\beta,\gamma)$ plane. A straightforward computation shows that

$$\rho^2 \equiv \alpha^2+\beta^2+\gamma^2=2r^2$$

$$\Delta \equiv (\alpha-\beta)(\beta-\gamma)(\gamma-\alpha)=2r^3\cos 3\theta.$$

Therefore the circle with radius r, centre the Walrasian equilibrium is folded into six and mapped by m into the interval

$$u=2r^2 \quad , \quad |v| \leq 2r^3,$$

which is the intersection of the line $u=2r^2$ with the interior of the cusp (see Fig.10 A)

Therefore as r varies, the image of m fills out the interior of the cusp.

Fig.10 A. h=k=0

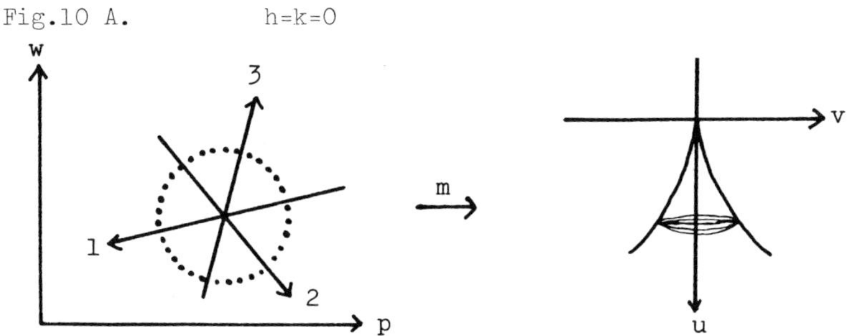


Now suppose $(h,k)=(0,0)$. The map m again folds the $(p,w)$ plane into six along the same fold lines as before, but this time the image is translated by the vector $(h,k)$ as shown in Fig.10 B,C,D,E.

Fig.10 B. $h<0, k=0$.

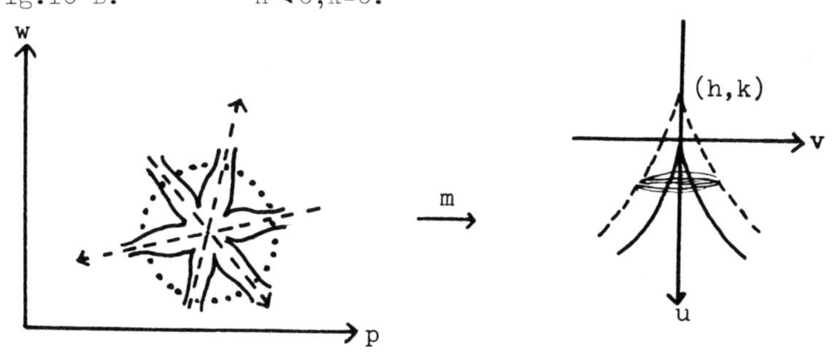

Fig. 10C  $h > 0, k=0$

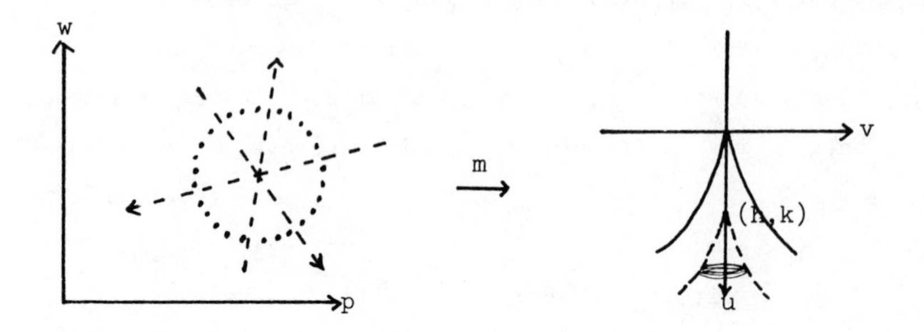


Fig. 10D  $h=o, k > 0$

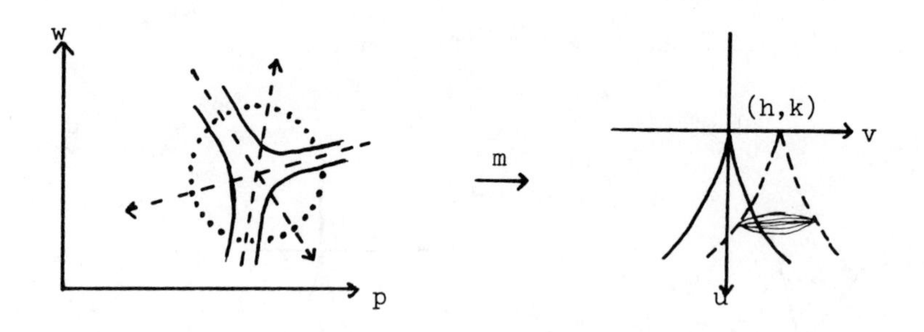


Fig. 10E  $h=o, k < 0$

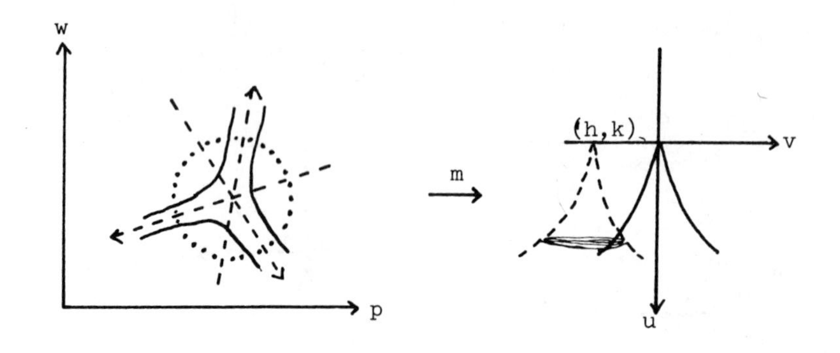

In each case the dashed lines show the $\alpha, \beta, \gamma$ axes and their image cusps in the (u,v) plane. The firm lines show the bifurcation sets in both the (p,w) plane and the (u,v) plane; in the (u,v) plane the bifurcation set is always the standard cusp, but in the (p,w) plane it depends upon (h,k), and in each case is the inverse image of the cusp under m. In each of the (p,w) planes we have drawn the dotted circle with radius r and centre the Walrasian equilibrium, and in each (u,v) plane we show its image as a horizontal interval covered six times.

We want to explain how to obtain Fig.9 from Fig.10 B. As we move round the dotted circle in the (p,w) plane in Fig.10 B, we cross the bifurcation set 12 times, corresponding to 12 fold catastrophes. The corresponding equilibrium set in the $(p,w,\xi)$ space is shown in the cylinder in Fig.9. The firm lines represent stable equilibria and the dotted lines unstable equilibria. Thus, for example if we were in the Repressed inflation equilibrium region, and proceeding anticlockwise round the circle, then just before we reach the Repressed inflation - Keynesian boundary, the equilibrium state would suffer a catastrophic jump into a dual Classical equilibrium state.

Now we have explained where Fig.9 came from and how to interpret it, we can simplify it by squashing the cylinder into a flat annulus with polar coordinates $(\xi,\theta)$ as shown in Fig.11 B Corresponding to the other combinations of values of (h,k) we get the pictures Fig.11 A,C,D,E.

Here Fig 11 A, which at first sight might seem to be the simplest case because h=k=0, in fact turns out to be the most degenerate case, because in Fig.10 A the two cusps coincide. (The dashed cusp coincides with the firm cusp). Thus instead of smooth curves of equilibria we obtain six crossing points, at each of which there is an exchange of stability. Thus Fig.11 A represents a return to Fig.2 at the beginning of the paper where we first started.

Fig.11 B,C,D,E are obtained by small pertubations of this highly degenerate situation.

Fig.11 A.

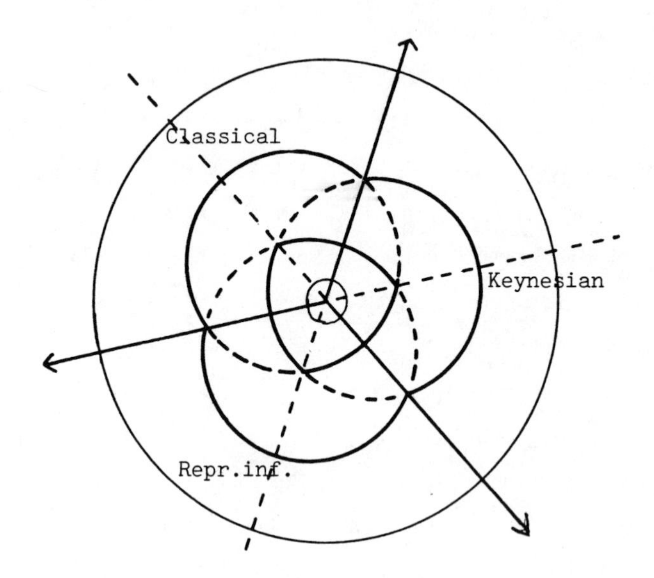

Fig. 11 B

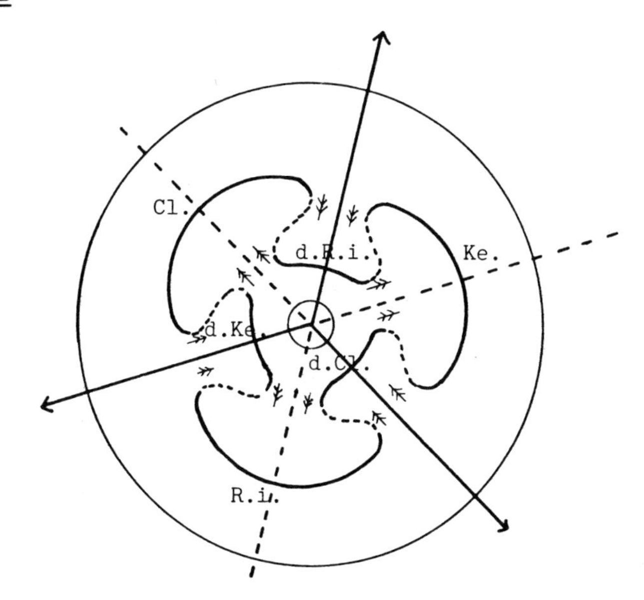


Fig. 11 C

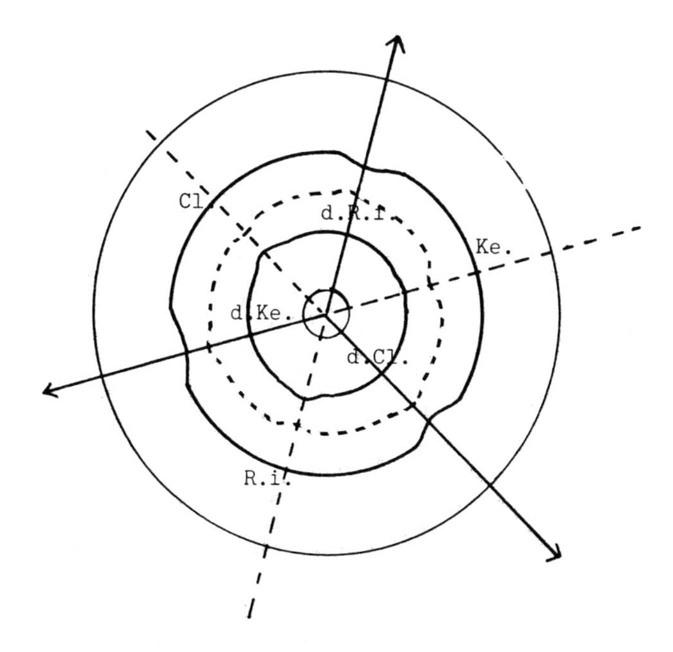

Fig. 11 D

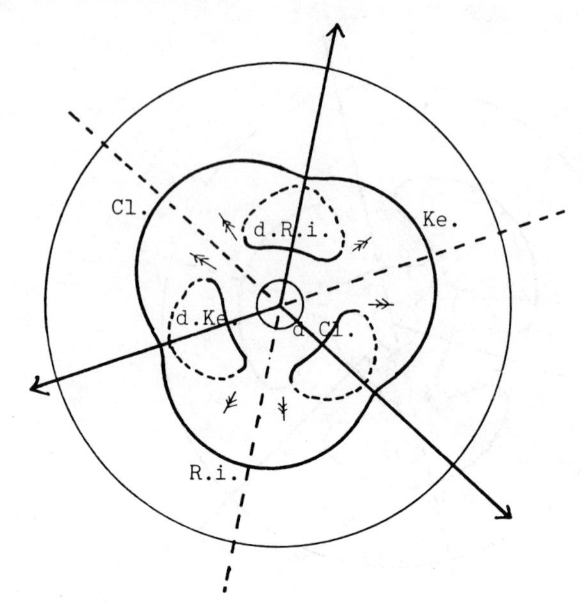


Fig. 11 E

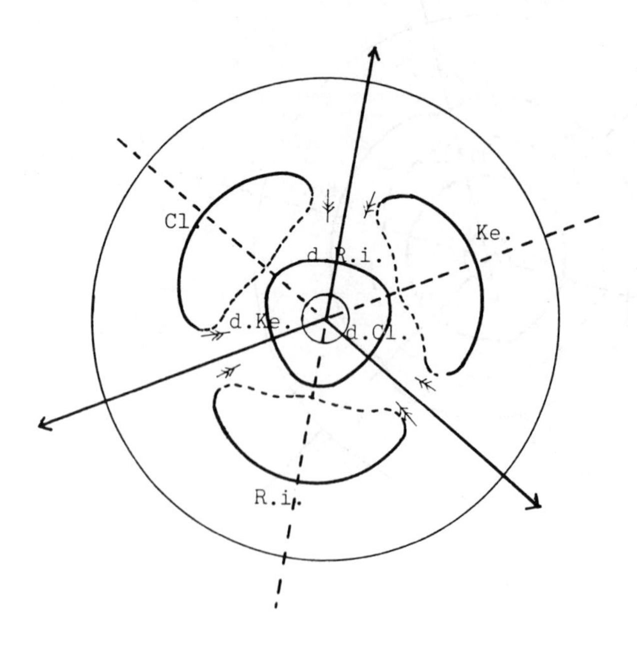

Fig.11 B represents the case with both the original and the dual equilibria present in the long term evolution. In this case the changes between types of equilibria are usually not smooth, but takes the form of sudden jumps between the equilibrium and the dual equilibrium surfaces. If, however, the radius of the circle is sufficiently small, the interval in Fig. 10 B will never cross the firm drawn cusp and so the changes between original and dual equilibria will be smooth. In this case the equilibrium set in Fig. 11 B consists of one cirle. As the radius of the circle decreases the curve of equilibria shown in Fig. 11 B with 12 folds pulls out into a convex curve, causing the folds and dotted parts to disappear.

Fig.11 C represents the situation with smooth changes between states of the original types of equilibria and smooth changes between dual types of equilibria. Depending on the history of the economy the actual state will either be one of the original types or one of the dual types, but no switches between the two types will take place.

Finally Fig.11D (E) represents the situation with smooth changes between the original equilibria (dual equilibria). In this case the economy may start out in a dual equilibrium (original equilibrium) situation, but as (p,w) reaches a boundary, there is a catastrophic switch to an original type of equilibrium (dual equilibrium), and the dual equilibria (original equilibria) will not be observed in the long run.

Altogether this description shows that the long run evolution heavily depends on the values of the hidden parameters, as different combinations of (h,k) lead to either smooth changes or catastrophic jumps.

Concluding remarks.

The analysis in this paper indicates that in order to obtain a generic description of the dynamics in a long term evolution of a disequilibrium economy, it is necessary specifically to introduce the fast adjustments in the quantities into the description. Assuming the differential equation that describes the fast adjustment to depend smoothly on the price-wage parameter leads to the introduction of "new" equilibria and a characterization of the long term evolution, based on the concept of exchange of stability around the Walrasian equilibrium.

Due to the definition of the long term adjustment process in (p,w), however, the above analysis only covers the crossing of the boundaries Classical - Keynesian and Classical - Repressed inflation. In another paper, Blad & Zeeman, [1980], we specifically analyze the behaviour around the Repressed inflation - Keynesian boundary.

References.

J.P.Benassy,[1975]: "Neo-Keynesian Disequilibrium Theory in a Monetary Economy". Review of Economic Studies, vol.42.

M.C.Blad,[1979]: "Dynamical Models in Disequilibrium Theory". Ph.D. Thesis, University of Warwick, England.

M.C.Blad & A.P.Kirman,[1978]: "The long run Evolution of a Rationed Equilibrium Model". Warwick Economic Research Papers, no.128.

M.C. Blad & E.C.Zeeman,[1980]: "Oscillations between Repressed inflation and Keynesian Equilibria due to Inertia in Decision Making". Unpublished.

G.Debreu,[1959]: "Theory of Value" John Wiley & Sons, New York.

J.H.Drèze,[1975]: "Existence of an Exchange Equilibrium under Price Rigidities". International Economic Review, vol.16,no.2.

E. Malinvaud,[1977]: "The Theory of Unemployment Reconsidered". Basil, Blackwell, Oxford.

Y.Younès,[1975]: "On the Role of Money in the Process of Exchange and the Existence of a non-Walrasian Equilibrium". Review of Economic Studies, vol.42.

E.C.Zeeman,[1977]: "Catastrophe Theory. Selected Papers 1972 - 1977". Addison Wesley.

# Rings of differential operators

by Jan-Erik Björk

Introduction During the last ten years a quite far-reaching theory has been developed in order to study systems of differential equations with algebraic or analytic coefficients. The first systematic account of this theory was given by the Japanese mathematicians M. KASHIWARA, M. SATO and T. KAWAI in [10]. Other important discoveries, restricted to the algebraic case, were done by I. N. Bernstein in [7] and [8]. There are also many interesting results due to French mathematicians such as BONY, MALGRANGE and SHAPIRA.

At present the subject is very active and the reader is referred to [11] and various publications from Kyoto University for the most recent results. In my book [9] most of the results- up to 1978 -are presented. Its aim was to supply a detailed exposition of this new branch of mathematics where algebraic and differential-geometric methods are used in order to understand the basic facts about over-determined systems of differential equations with variable coefficients.

In this talk I can only give a brief survey of course and there is no time for supplying any details of proofs. Instead I hope that the presentation at least gives some flavour of the subject.

## 1. The Bernstein class of distributions

The usual theme is to fix a differential operator $P(x,d/dx)$ and to consider homogenous (or inhomogenous) equations $P(x,d/dx)\mu = 0$ (or $P(x,d/dx)\mu = f$). More generally, we can replace the single operator $P$ by a matrix of differential operators and arrive at general over-determined systems. In the case when the differential operators have constant coefficients a very satisfying theory has been obtained, see [1] and [2]. The study of over-determined systems of differential equations with variable coefficients is much more difficult. For example, there is no analogue of the famous "Fundamental Principle" by L. EHRENPREIS when the operators have variable coefficients.

However, using algebraic results about some rings of differential operators it is possible to understand quite a lot. In particular we are able to clarify the meaning of "maximally over-determined systems", i.e. systems where it is impossible to introduce more equations if we assume that non-zero solutions exist.

So now I will explain how this is done in the algebraic case and to make a concrete approach we start as follows: Let $\mu$ be a distribution defined in $R^n$ where we use $x = (x_1 \ldots x_n)$ as coordinates. Now we can consider the set of differential operators $P(x,d/dx)$ for which $P(x,d/dx)\mu = 0$. Of course, we must specify the coefficients of these differential operators and for the moment we restrict the attention to operators with polynomial coefficients. In other words, we consider operators $P = \Sigma\, p_\alpha(x) d^\alpha/dx^\alpha$ where $\Sigma$ extends over finitely many multi-indices $\alpha$ and the coefficients $p_\alpha$ are polynomials of the n-tuple x. Remark we allow that the polynomials $p_\alpha$ have complex coefficients, i.e. that they belong to the polynomial ring $C[x_1 \ldots x_n]$.

When $\mu$ is given we denote the set of all these differential operators $P(x,d/dx)$ (with polynomial coefficients) by $L\mu$. Of course, it can happen that $L\mu$ only consists of the trivial zero-operator, i.e. that $\mu$ is not a solution to any non-trivial equation with polynomial coefficients. But for many $\mu$'s the set $L\mu$ is non-zero and it is this case we want to discuss.

By definition $L\mu$ is a subset of the family of all differential operators with polynomial coefficients. This family is denoted by $A_n$ and $A_n$ is called the Weyl algebra (in $n$ variables).

Remark $A_n$ is an associative ring where the ring structure arises from the usual composition of differential operators. For example, $A_n$ contains the zero-order operators $x_1,\ldots,x_n$ and first-order derivation operators $d/dx_1,\ldots,d/dx_n$. For convenience we denote $d/dx_j$ by $\delta_j$. Using Leibniz's rule it follows that the composed operator $\delta_j x_j = x_j\delta_j$ + the identity operator. In other words, the commutator $\delta_j x_j - x_j\delta_j =$ the identity element of the ring $A_n$, for each $1 \leq j \leq n$.

This shows that $A_n$ is a non-commutative ring and it turns out that this non-commutativity leads to very interesting results. They will be discussed later on in more detail.

Returning to the family $L\mu$ it is obvious that $L\mu$ is a left ideal of the ring $A_n$ and of course $L\mu$ is $\neq A_n$ (i.e. not the whole ring) unless $\mu \equiv 0$. To say that the distribution $\mu$ satisfies "many equations with polynomial coefficients" means that $L\mu$ is "large". So we should now try to measure the size of a left ideal in the Weyl algebra $A_n$. For this purpose we are going to associate an algebraic variety to each left ideal $L$

of $A_n$.

1.1. The definition of $\sigma(L)^{-1}(0)$. Let us first recall that if $P = \Sigma\, p_\alpha(x) d^\alpha/dx^\alpha$ is an element of $A_n$ and if $m$ is the order of this differential operator, then its principal symbol $\sigma(P) = \Sigma_m\, p_\alpha(x)\zeta^\alpha$ where $\Sigma_m$ indicates that the summation is taken over $\alpha$'s for which $|\alpha| = m$ and $\zeta$ is a new complex n-vector.

We shall also introduce a complexification of the real x-space, i.e. put $z = x + iy$ and consider $\sigma(P) = \Sigma_m\, p_\alpha(z)\zeta^\alpha$ as a polynomial in the 2n-dimensional $(z,\zeta)$-space. At this stage we can arrive at

$$\sigma(L)^{-1}(0) = \{(z,\zeta) \in C^{2n} : \sigma(P)(z,\zeta) = 0 \text{ for every } P \in L\}.$$

In other words, $\sigma(L)^{-1}(0)$ is the set of common zeros of the principal symbols of the operators in the left ideal L.

So by definition $\sigma(L)^{-1}(0)$ is an algebraic variety of the complex $(z,\zeta)$-space and observe that it is conic with respect to the $\zeta$-variables since principal symbols are $\zeta$-homogeneous polynomials. Using the Nullstellen Satz it is easily verified that $\sigma(L)^{-1}(0)$ is non-empty- unless L is the whole ring $A_n$. The algebraic variety $\sigma(L)^{-1}(0)$ has a dimension (i.e. it is the complex dimension of $\sigma(L)^{-1}(0)$ considered as a complex analytic variety) and this dimension is denoted by $d(A_n/L)$ and it is called the dimension of the left $A_n$ - module $A_n/L$.

Some examples If P is an element of $A_n$ then P generates the left principal ideal $A_nP = L$ and here $\sigma(L)^{-1}(0)$ is the algebraic hypersurface $\sigma(P)^{-1}(0)$ so that $d(A_n/L) = 2n-1$.

If $L$ is the left ideal generated by the $n$ derivations $d/dx_1, \dots, d/dx_n$ then $\sigma(L)^{-1}(0) = \{(z,\zeta) : \zeta = 0\}$ and this is an n-dimensional variety so that $d(A_n/L) = n$.

<u>Another and more difficult example goes as follows:</u> Let $p(x)$ be a given polynomial and consider its inverse $p^{-1} = 1/p(x)$. The usual rules for differentiation gives $d/dx_j(p^{-1}) = -dp/dx_j/p^2$ for each $1 \le j \le n$, and so on. So more generally, if $Q(x,d/dx) \in A_n$ then this differential operator acts on $p^{-1}$ and $Q(x,d/dx)(p^{-1})$ is some rational function of the form $r(x)/p^{m+1}(x)$ where $r(x)$ is a polynomial and $m$ is the order of the operator $Q$.

Let us now consider the set $L = \{Q \in A_n : Q(x,d/dx)(p^{-1}) = 0\}$. Then $L$ is a left ideal of $A_n$ and we would like to compute $d(A_n/L)$. Using the general theory about the Weyl algebra it can be proved that $d(A_n/L) = n$ holds for every polynomial $p$. This is a non-trivial result where the reader may try to prove it directly. Actually I do not know if there exists any simple and direct proof. The proof which works and is exposed in my book contains several steps, based upon homological methods so it is quite involved.

1.2. <u>Bernstein's inequality</u> When $L$ is a left ideal of $A_n$ (different from zero and $A_n$) then we have defined $d(A_n/L)$ and a priori this is some integer between $0$ and $2n-1$ since $\sigma(L)^{-1}(0)$ is some $\zeta$-conic algebraic variety of $C^{2n}$. However, it turns out that this dimension cannot be too small and the result is

Theorem $d(A_n/L) \geq n$ hold for every left ideal $L$ in $A_n$.

This inequality is quite remarkable. There exist several proofs of this result. The most important is due to O. Gabber. The result in [6] is that $\sigma(L)^{-1}(0)$ is an involutive variety for every left ideal $L$ of $A_n$. (This means that if two polynomials $p$ and $q(z,\zeta)$ both vanish on $\sigma(L)^{-1}(0)$ then their Poisson product also vanishes on $\sigma(L)^{-1}(0)$). Finally, the theorem above easily follows because involutive subvarieties of the $(z,\zeta)$-space are at least n-dimensional.

Remark There are also other proofs and they are given in [9].

1.3. The Bernstein class of distributions Let us now return to a distribution $\mu$ defined in $R^n$ and consider the left ideal $L\mu$. Using the theorem above we arrive at the follow ing

Definition We say that the distribution $\mu$ belongs to the Bernstein class if $d(A_n/L\mu) = n$.

The Bernstein class of distributions in $R^n$ is denoted by $\mathcal{B}(R^n)$. They form an interesting class of distributions and in [9,Chapter 7] there are several results about $\mathcal{B}(R^n)$. For example, every differential operator with constant coefficients has a fundamental solution which belongs to $\mathcal{B}(R^n)$. Here is another

Example Let $\xi = (\xi_1 \ldots \xi_n)$ be the real coordinates of another copy of $R^n$. If $f(x)$ is a test function then we get the Fourier transform $\hat{f}(\xi) = \int e^{-i(x,\xi)} f(x)dx$. Let now $G(\xi)$ be a real-valued polynomial and define a distribution $\mu$ in the x-space by the rule $\mu(f) = \int \hat{f}(\xi)\, e^{iG(\xi)} d\xi$. Then it can be proved that $\mu$ belongs to the Bernstein class and a non-trivial consequence of this is that the analytic wave front set of $\mu$ is always contained in some Lagrangian subvariety of $T^*(R^n)$. So in particular $\mu$ is a real-analytic density outside some algebraic hypersurface. In fact, this follows because every distribution in $\mathcal{B}(R^n)$ has such regularity properties.

2. Functional equations A straight forward generalization of so called $\Gamma$-functions arises as follows: Let $P(x)$ be a real-valued polynomial in $R^n$ and assume that $P$ is non-negative (but we allow zeros!). If $\lambda$ is a complex scalar whose real part is $> 0$ then $P^\lambda$ is a continuous function in $R^n$. Keeping $P$ fixed while $\lambda$ moves in the right half-plane we get holomorphic functions of the form

$$\Gamma_f(\lambda) = \int P^\lambda(x) f(x) dx \quad \text{for each} \quad f \in C_o^\infty$$

The question arises if these $\Gamma$-functions can be extended to meromorphic functions in the whole complex $\lambda$-space. This was originally asked by I.M. Gelfand at the International Congress in Amsterdam 1954. The affirmative answer was found independently by Atiyah and I.N. Bernstein and I.S. Gelfand in 1968. See [3] and [4]. The result was that if $P$ is given then there exists a positive integer $N$ such that the $\Gamma$-functions $\Gamma_f$ extend to meromorphic functions with poles contained in the negative rational numbers $\{-1/N, -2/N, \ldots\}$ for all $f \in C_o^\infty$.

Their proofs used Hironaka's theorem on the resolution of singularities. In 1972 I.N. Bernstein found a much simpler proof of the existence of meromorphic extensions and at the same time they were exhibited through functional equations. His result was: Given $P$ there exists a polynomial $b(\lambda) = \lambda^s + .. + c_{s-1}\lambda + c_s$ with complex coefficients and a finite set of differential operators $\{ Q_j \} \in A_n$ such that

$$b(\lambda)\ \Gamma_f(\lambda) = \int P^{\lambda+1}(x)\ [\Sigma\lambda^j Q_j(x,d/dx)f(x)]\ dx \quad \text{for all} \quad f \in C_o^\infty$$

Choosing $b(\lambda)$ with the smallest possible degree makes $b(\lambda)$ unique and $b(\lambda)$ is called the Bernstein-Sato polynomial attached to $P(x)$. The existence of $b(\lambda)$ and the functional equation above is proved by classical algebraic methods and is therefore quite elementary (though rather ingenious). More recently M. Kashiwara proved that the zeros of $b(\lambda)$ are strictly negative rational numbers (See [5]). This result is deep and its proof requires again the Resolution of Singularities. In addition it involves a quite involved machinery which roughly speaking requires an extension of sheaf-theoretic constructions in algebraic geometry (such as the formation of direct images) where the usual sheaves of holomorphic (or affine algebraic functions) are replaced by sheaves of differential operators with holomorphic (or affine algebraic) coefficients. A complete proof, including all the necessary constructions, is given in [9, chapter 6].

2.1. Some other functional equations. We have announced the existence of functional equations for polynomials in $R^n$. Actually they exist for other types of functions also, provided that we allow more general coefficients in the differential operators $\{Q_j\}$. The most general result applies to formal power series and the result is the following:

Let $K$ be a commutative field of characteristic zero and let $f(x) = \sum c_\alpha x^\alpha$ be an element of the local ring $\hat{\mathcal{O}}_n(K) = K[[x_1 \ldots x_n]]$.

Then there exists a polynomial $b(\lambda)$ and a finite set of differential operators $\{Q_j\}$ (where each $Q_j$ is of the form $\sum f_\alpha(x) d^\alpha/dx^\alpha$ where the coefficients $f_\alpha \in \hat{\mathcal{O}}_n(K)$, in other words each $Q_j$ is a K-linear differential operator with coefficients in the local ring $\hat{\mathcal{O}}_n(K)$)
such that the following algebraic identity holds:

$$b(\lambda) f^\lambda(x) = \sum \lambda^j Q_j(x, d/dx) f^{\lambda+1}(x)$$

Remark In this formula $\lambda$ is considered as a new parameter. The rules for the actions by the differential operators $Q_j$ on $f^{\lambda+1}$ are the familiar rules. For example, $d/dx_j(f^{\lambda+1}) = (\lambda+1)\partial f/\partial x_j f^\lambda$, and so on.

The existence of this functional equation is considerably more difficult than in the case when $f$ is a polynomial. For the proof we need much insight about the ring $\hat{\mathcal{D}}_n(K) = \hat{\mathcal{O}}_n(K) < d/dx_1 \ldots d/dx_n >$ of K-linear differential operators with coefficients in $\hat{\mathcal{O}}_n(K)$.

The ring $\hat{\mathcal{D}}_n(K)$ is isomorphic to the tensor product $\hat{\mathcal{O}}_n(K) \otimes A_n(K)$ (with $\otimes$ taken over the polynomial ring $K[x_1 \ldots x_n]$) and one of the necessary steps towards the proof of the existence of functional equations is to compute the global homological dimension of $\hat{\mathcal{D}}_n(K)$. The result is that gl.dim $(\hat{\mathcal{D}}_n(K)) = n$ which turns out to be closely related to the theorem in Section 1.2. and it leads to a similar result for finitely generated left $\hat{\mathcal{D}}_n(K)$-modules and a theory of so called "maximally over-determined" left $\hat{\mathcal{D}}_n(K)$-modules of the form $\hat{\mathcal{D}}_n(K)/L$ where $L$ is a left ideal of $\hat{\mathcal{D}}_n(K)$ .

The discussion above indicates that sophisticated algebraic methods form an essential part of the theory. All this has lead to a nice theory about filtered (non-commutative) rings whose associated graded rings are regular commutative noetherian rings and it is exposed in [9, chapter 2].

An example Returning to the Weyl algebra $A_n$ and some left ideal $L$ we can compute the integer $d(A_n/L)$ by homological methods. Here is the result: $d(A_n/L) = 2n - j(A_n/L)$ where $j(A_n/L)$ is the smallest non-negative integer for which the Ext-group $\mathrm{Ext}_{A_n}^{j(A_n/L)}(A_n/L, A_n)$ is $\neq 0$.

Using this formula and a result due to J-E ROOS which asserts that $\mathrm{gl.dim}(A_n) = n$ it follows that $0 \leq j(A_n/L) \leq n$ for all left ideals $L$ and then we obtain a new (purely homological proof) of the theorem in Section 1.2.

3. The sheaf $\mathcal{D}$. Another natural class of differential operators are those with analytic coefficients. So when $n$ is a non-negative integer we first consider the local ring $\mathcal{O}_n = \mathbb{C}\{x_1, \dots x_n\}$ of convergent power series and then the ring

$\mathcal{D}_n = \mathcal{O}_n < d/dx_1 \dots d/dx_n >$ of C-linear differential operators with coefficients in $\mathcal{O}_n$. So an element of $\mathcal{D}_n$ is written as a finite sum $\Sigma q_\alpha(x) d^\alpha/dx^\alpha$ where $\alpha$ are multi-indices and $q_\alpha$ belong to $\mathcal{O}_n$.

If we now pass to an n-dimensional complex analytic manifold $X$ then we get a sheaf $\mathcal{D}$ where the stalks $\mathcal{D}_x = \mathcal{O}_x < d/dx_1 \dots d/dx_n >$ for each point $x \in X$. (Here $\mathcal{O}_x$ denotes the germs of holomorphic functions at the point $x$).

In other words, the ring $\mathcal{D}_n$ is isomorphic to the stalks of the sheaf $\mathcal{D}$ for all points $x$ on the manifold $X$. Now we can build up a nice theory based upon the sheaf $\mathcal{D}$ and the category of coherent sheaves of left (or right) $\mathcal{D}$-modules. First it is easy to prove that $\mathcal{D}$ is a coherent sheaf of rings so the category above behaves well. Then sheaf-theoretic methods are combined with ring-theoretic results about $\mathcal{D}_n$ and this leads to several important results. All this is exposed in [9, chapter 5] and here I will just mention one typical result:

A functional equation on real-analytic manifolds: Let $X$ be a compact real-analytic manifold (without boundary) and let $p(x)$ be a real-valued and non-negativ real-analytic function on $X$ and let $dm(x)$ be some real-analytic density on $X$. Then there exists a polynomial $b(\lambda)$ and a finite set of (globally defined) differential operators $\{Q_j\}$ with real-analytic coefficients such that

$$b(\lambda)\int_X p^{\lambda}(x)f(x)dm(x) = \sum\lambda^j\int_X p^{\lambda+1}(x)Q_j(x,d/dx)f(x)\ dm(x)$$

hold for all $f \in C^{\infty}(X)$.

3. The sheaf $\mathcal{E}$. This is the sheaf of micro-local differential operators (with analytic coefficients) which can be defined outside the zero-section of the cotangent bundle of a complex analytic manifold. As long as we only study local properties of the sheaf $\mathcal{E}$ it is sufficient to consider the case when the manifold simply is some open subset of an affine space $C^n$. Here we use the following notations: Let $z = (z_1 \ldots z_n)$ be the coordinates in $C^n$ and let $\zeta = (\zeta_1 \ldots \zeta_n)$ be the covector coordinates so that the cotangent bundle $T^*(C^n)$ is the 2n-dimensional $(z,\zeta)$-space. Its zero section is the n-dimensional subspace where $\zeta = 0$ and its complement (where $\zeta \neq 0$) is denoted by $T^*_o(C^n)$.

Now we are going to define the stalks of the sheaf $\mathcal{E}$ at points $p \in T_o^*(C^n)$ and for this purpose we first introduce the notation

$\mathcal{O}(m)(U)$ which denotes the set of holomorphic functions in an open subset $U$ of $T_o^*(C^n)$ which are $\zeta$-homogenous of order $m$, i.e. if $f(z,\zeta) \in \mathcal{O}(m)(U)$ then $f(z,\lambda\zeta) = \lambda^m f(z,\zeta)$ for all complex scalars $\lambda$.

<u>The definition of the stalk</u> $\mathcal{E}_p$. Let $p \in T_o^*(C^n)$ be a given point. Then the stalk $\mathcal{E}_p$ consists of all infinite sums

$F = \Sigma f_v(z,\zeta)$ where the following hold

1) There exists a neighborhood $U$ of $p$ such that $f_v \in \mathcal{O}(v)(U)$ for each integer $v$

2) $f_v = 0$ when $v >> 0$, i.e. there exists some $w$ so that $f_v \equiv 0$ when $v > w$

3) There exist constants $A$ and $K$ such that the sup-norm $|f_v|_U \leq A\,|v|!\,K^{|v|}$ for all integers $v$.

<u>Remark</u> Condition 3 above means that we put a restriction on the infinite sum and the growth conditions which enter there reflect the fact that the stalks $\mathcal{E}_p$ are micro-local (or "pseudo-differential) operators with <u>analytic symbols</u>.

Condition 2) means that if $F = \Sigma f_v$ belongs to $\mathcal{E}_p$ (and is non-zero) then there exists an integer $m$ so that $f_m$ is not $\equiv 0$ while $f_v \equiv 0$ for $v > m$. Then $m$ is called the <u>order of F</u> and $f_m(z,\zeta)$ is called the <u>principal symbol</u> of $F$.

<u>The ring structure</u> It turns out that the stalks $\mathcal{E}_p$ are associative rings. The ring multiplication is defined

as follows:

If $F = \Sigma f_\nu$ and if $G = \Sigma g_\mu$ belong to $\mathcal{E}_p$

,then the product

$FG = \Sigma r_j$ where $r_j = \Sigma (\alpha!)^{-1}(d^\alpha f_\nu/d\zeta^\alpha)(d^\alpha g_\mu/dz^\alpha)$

with $\Sigma$ extended over triples $\alpha,\nu$ and $\mu$ for

which $\nu + \mu - |\alpha| = j$)

Of course, it requires a proof to show that this indeed is a welldefined product. It is easily seen that each $r_j \in \mathcal{O}(j)(U)$ and using Cauchy's inequalities it is not difficult to prove that $\Sigma r_j$ also satisfies condition 3) above for a sufficiently small neighborhood of p.

Some results about the rings $\mathcal{E}_p$. Keeping n fixed it is easily seen that the stalks $\mathcal{E}_p$ are isomorphic as rings for all points p in $T_o^*(C^n)$ and this ring is called the ring of germs of micro-local differential operators in n variables. Here are some results which can be proved about this ring which we denote by $E_n$:

Theorem $E_n$ is a (non-commutative) noetherian ring (i.e. every left or right ideal is finitely generated) and the global homological dimension of $E_n$ is equal to n.

The proof of this is not at all easy. To prove that $E_n$ is noetherian involves a hard analysis, where versions of the Weierstrass Preparation Theorem for germs of micro-local differential operators are used. The computation of $gl.dim(E_n)$ was originally done in the thesis by M. Kashiwara and all the details can be found in [9, Chapter 4].

Some results about the sheaf $\mathcal{E}$ . Let us now consider the sheaf $\mathcal{E}$ on $T_o^*(C^n)$ whose stalks have been defined as above. Using the noetherianess of the stalks it follows rather easily that $\mathcal{E}$ is a coherent sheaf of rings and hence we get a nice category of coherent sheaves of left (or right) $\mathcal{E}$-modules. For this we can begin to prove various results. For example, if $m$ is a coherent sheaf of left $\mathcal{E}$ -modules then the support of the sheaf $m$ is a complex analytic subvariety of $T_o^*(C^n)$ which in addition is $\zeta$-conic.

There is also the following fundamental result

Theorem supp ($m$) is an involutive subvariety of $T_o^*(C^n)$ for every coherent sheaf of $\mathcal{E}$ -modules.

This is the micro-local version of the result we have already mentioned in section 1.2. and it was actually proved in | |. The proof in | | was completely different from the proof in and employed the sheaf of micro-local differential operators of infinite order (See also |, Ch.4 par.11) for an exposition about this and a proof of the involutivness of supp ( $m$ )).

The theorem above implies in particular that the dimension of supp($m$) is at least n and this leads to

Definition A coherent sheaf $m$ of $\mathcal{E}$-modules is said to be maximally over-determined if dim(supp($m$)) = n

Remark In the literature "maximally over-determined systems" are often called holonomic

Maximally over-determined (or holonomic) systems of $\mathcal{E}$-modules (always assumed to be coherent of course) have been studied in many papers by the Japanese school. In particular the aim is to find good structure theorems of these. The case when $\mathrm{supp}(m)$ is smooth is the simplest and here quite satisfactory structure theorems are available. In order to reach these one employs the existence of contact transformations (which resemble the transformations used in the theory of Fourier integral operators). So let me now describe how this goes:

Contact transformations attached to canonical transformations

First we must recall some geometric facts. Fix a point $p$ in $T^*_o(C^n)$. Let us now consider a locally defined biholomorphic mapping $\phi$ which sends an open neighborhood $U$ of $p$ onto another neighborhood $U'$ of $p$ (here $\phi(p) = p$ is also assumed).

Recall also that $T^*(C^n)$ is a symplectic manifold where the fundamental 2-form is given by $\Omega = \sum d\zeta_v \; dz_v$. We say that $\phi$ is a canonical transformation at $p$ if $\phi$ preserves the symplectic structure, in other words if the pullback $\phi^*(\Omega) = \Omega$. In addition we say that $\phi$ is homogenous if $\phi$ is homogenous of order 1 with respect to the $\zeta$-variables.

So if we write $\phi = (\varphi, \Psi)$ where $\varphi_j(z,\zeta) = z_j$-coordinate of the image point $\phi(z,\zeta)$ while $\Psi_j(z,\zeta) =$ the $\zeta_j$-coordinate of $\phi(z,\zeta)$, then the homogenity of $\phi$ means that $\varphi_j$ are $\zeta$-homogenous of order zero while the $\Psi$-functions are $\zeta$-homogenous of order 1.

It can also be proved that the condition that $\phi$ is canonical is equivalent to the following identities of Poisson products:

$\{\Psi_j,\varphi_j\} = 1$ for each $1 \leq j \leq n$ while all other Poisson products $\{\Psi_j,\varphi_v\} = 0$ when $v \neq j$ and similarly $\{\varphi_j,\varphi_v\} = \{\Psi_j,\Psi_v\} = 0$ for all pairs $j$ and $v$.

Summing up, we have now introduced the class of <u>locally defined and homogenous canonical transformations at the point</u> p.

Let us now return to the ring $E_p$. Here we shall consider the class of $\Sigma$-preserving ring automorphisms, i.e., we consider a ring automorphism $\chi$ on $E_p$ which preserves the order of the micro-local differential operators. So if $F \in E_p$ has some order $m$, then $\chi(F)$ has the same order $m$.

Let $\chi$ be one such ring automorphism. Then we can associate a canonical transformation to $\chi$ as follows: Namely, in $E_p$ we first have the elements $Z_1,\ldots,Z_n$ where the sum which defines the micro-local operator $Z_j$ is given by the single element $z_j \in \mathcal{O}(0)$. We also have the elements $D_1,\ldots,D_n$ where $D_j$ is represented by the element $\zeta_j \in \mathcal{O}(1)$.

Now $\chi(Z_j)$ have order zero and we denote the principal symbol $\sigma(\chi(X_j))$ by $\varphi_j(z,\zeta)$. So here $\varphi_j(z,\zeta) \in \mathcal{O}(0)$.

In the same way we get the principal symbols $\psi_j(z,\zeta) = \sigma(\chi(D_j)) \in \mathcal{O}(1)$.

Using the definition of the ring product in $E_p$ it is easily seen that the commutators $D_jZ_j - Z_jD_j$ = the identity element for each $j$, and this implies that the Poisson products $\{\psi_j,\varphi_j\} = 1$ for each $j$.

In the same way we can verify that the other poisson products $\{\varphi_j,\varphi_v\}$ and $\{\Psi_j,\Psi_v\}$ and $\{\Psi_j,\varphi_v\}$ (with $j \neq v$) all vanish.

This implies that the mapping $\phi$ defined by $\phi(z,\zeta) = (\varphi,\Psi)$ is canonical.

Summing up, to a given $\sum$-preserving ring automorphism $\chi$ on the ring $E_p$ we have attached a locally defined homogenous canonical transformation which we denote by $\tilde{\chi}$.

So far we have only performed some straight forward computations. The deep fact is the converse of the discovery above. Namely, if we instead start with some locally defined and homogenous canonical transformation $\phi$ then there exists a $\sum$-preserving ring automorphism $\chi$ such that $\phi = \tilde{\chi}$.

This conclusion is called the Existence Theorem of contact transformations attached to a canonical transformation. (Of course, when $\phi = \tilde{\chi}$ then $\chi$ is called a contact transformation attached to $\phi$).

The uniqueness of a contact transformation First, if we fix an element A in the ring $E_p$ which is invertible, i.e., the inverse $A^{-1}$ also exists then we obtain a $\sum$-preserving automorphism by the rule:

$P \longrightarrow APA^{-1}$, i.e. a so called inner automorphism. For an inner automorphism it is easily seen that the attached canonical transformation is the identity. It turns out that the converse is also true. In other words if $\chi$ is a $\sum$-preserving ring automorphism on $E_p$ whose attached canonical transformation is the identity then $\chi$ is inner.

It follows of course that if $\phi$ is given and if $\phi = \tilde{\chi}$ then $\chi$ is unique (up to an inner automorphism).

Some consequences Let us now return to a coherent sheaf $m$ of left $E$-modules whose support is smooth and of minimal dimension, i.e. we assume that $V = \text{supp}(m)$ is a non-singular n-dimensional subvariety of $T_o^*(C^n)$. Since this is half of the dimension of $T_o^*(C^n)$ and since $V$ is known to be involutive it follows that $V$ is Lagrangian and then differential geometry shows that after a suitable canonical transformation we can assume that the coordinates at the point p have been given so that the equations which define $V$ are given by: $z_1 = \zeta_2 = \ldots = \zeta_n = 0$ where the point p can be assumed to be

$(0..0:1,0...0)$ (i.e. $z = 0$ while the $\zeta$-vector $= (1,0,\ldots,0)$).

Applying a contact transformation it is therefore sufficient to study the structure of the sheaf $m$ under the assumption that $V = \text{supp}(m)$ is the Lagrangian manifold above (at least when we are only interested in local properties of $m$ in small neighborhoods of the given point p).

This discussion indicates how one can arrive at reasonable structure theorems for holonomic systems whose support are smooth Lagrangian submanifolds of $T_o^*(C^n)$. Of course, there is still very much to say about this and the whole theory is not yet clarified. For example, once we have made the "reductions" above we must begin to distinguish between systems with simple and multiple characteristics and it turns out that the micro-local differential operators of infinite order play a crucial role when we want to analyze structure theorems.

The most important cases occur for holonomic systems with so called <u>regular singularities</u>. This theory has recently been developed quite succesfully in the impressive work [12] and at this stage I must refer the reader to this because of lack of time.

# R E F E R E N C E S

Included are some recent work which has not been directly refered to in the previous text.


1. L.Ehrenpreis, Fourier analysis of several variables. Pure and Appl. Math. Wiley-Intersci. 17(1970).

2. V.P.Palamodov, Linear differential operators with constant coefficients. Grundlehren 168 (Springer-Verlag).

3. M.F.Atiyah, Resolution of singularities and division of distributions. Comm. Pure and appl. Math. 23(1970).

4. I.N.Bernstein and I.S.Gelfand, Meromorphy of the function $P^{\lambda}$. Funz. Analysis Akad. Nauk. CCCR 3(1) (1969).

5. M.Kashiwara, B-functions and holonomic systems. Inventiones Math. 38(1976).

6. O.Gabber, The integrability of the characteristic variety. Preprint. Math. Dep. Tel Aviv University (1980)

7. I.N.Bernstein, Modules over the ring of differential operators. A study of the fundamental solutions of equations with constant coefficients. Funz Anal. Akad, Nauk 5(2) (1971).

8. I.N.Bernstein, The analytic continuation of generalized functions with respect to a parameter. Ibid. 6(4) (1972).

9. J-E.Björk, Rings of differential operators. North-Holland Math. Labrary Series. Vol. 21 (1979).

10. M.Sato, M.Kashiwara and T.Kawai, Hyperfunctions and pseudo-differential operators. Springer Lecture Notes Vol. 287.

11. Seminar on Micro-local analysis. Ann. of Math. Studies. Princeton University Press Vol. 93 (1979).

12. M.Kashiwara and T.Kawai, On holonomic systems of micro-local differential equations. III-Systems with regular singularities. Preprint from res. Inst. of Math. at Kyoto University (June 1979).

Some additional references:

J.M.Bony and P.Shapira, Propagation des singularités analytiques pour les solutions des équations aux dérivées partielles. Ann. L'inst. Fourier Vol.26 (1976).

M.Kashiwara and P.Shapira, Micro-hyperbolic systems. Acta Math. 142 (1979).

P.Shapira, Conditions de positivité dans une variété symplectique complexe application a l'etude des microfonctions. Preprint (March 1980) to be publ, in Ann. L'école Norm. Supérieur.

M.Kashiwara and T.Kawai, Second-microlocalisation and asymptotic expansions. Preprint from res. Inst. of Math. Kyoto University, January 1980.

# ON C* - DYNAMICAL SYSTEMS AND EQUILIBRIUM STATES

Ola Bratteli

We explain why C*-algebras in some cases provide a better framework for the study of equilibrium states in quantum statistical mechanics than the usual quantum mechanical formalism. The main benefit is that one can consider the states of an infinitely extended system directly, without taking the thermodynamic limit of systems in finite regions. In this framework the observables are represented as the self-adjoint elements of a C*-algebra $\mathfrak{A}$, and the time development is given by a strongly continous one-parameter group $t \in \mathbb{R} \longrightarrow \tau_t$ of *-automorphisms of $\mathfrak{A}$. The states are represented by positive, normalized functionals on $\mathfrak{A}$, and these latter functionals are accordingly called states. If $\omega$ is a state and $\beta \in \mathbb{R}$, then $\omega$ is called a $(\tau,\beta)$-KMS state if

$$\omega(A\tau_{i\beta}(B)) = \omega(BA)$$

for all $A,B \in \mathfrak{A}$, where $B$ is such that the function $t \longrightarrow \tau_t(B)$ has an entire analytic extension, and $\tau_{i\beta}(B)$ is the value of this extension at $i\beta$. The state $\omega$ is called a $(\tau,+\infty)$-KMS state, or a ground state, if

$$-i\omega(A^*\delta(A)) \geq 0$$

for all $A$ in the domain of the infinitesimal generator

$\delta$ of $\tau$ . Ceiling states, or $(\tau,-\infty)$ - K M S states, are characterized by the converse inequality.

If $(\mathfrak{A},\tau)$ represents a physical system, the $(\tau,\beta)$ - K M S states can be interpreted as the equilibrium states of the system at temperature $\beta^{-1}$ . We give three justifications of this statement.

1. The K M S states are characterized by a certain stability property under small perturbations of the dynamics $\tau$ , for systems with a strict ergodicity property (of $\tau$ ) called $L^1$ - asymptotic abelianess.

The stability property is that there exists invariant states for the perturbed dynamics which are close to the original state, and which evolves back to the original state under the unperturbed dynamics (Araki, Haag - Kastler - Trych - Pohlmeyer with later refinements by Hoekman, Bratteli - Kishimoto - Robinson).

2. The K M S states are also characterized by the second law of thermodynamics, i.e. that the system cannot perform work during a temporary alteration of the dynamics. This can be proved under more general circumstances than 1. (Pusz - Woronowicz).

3. For quantum lattice systems the K M S-condition can also be characterized by local, or global, maximum entropy principles, for non-translationally or translationally invariant states, respectively (Araki, Lanford, Robinson, Ruelle, Sewell and others. See Reference [1]).

We empasize that the principles above are both necessary and sufficient for a state to be a K M S state at some temperature $\beta$, where $\beta \in [-\infty,+\infty]$ in case 1, $\beta \in [o, +\infty]$ in case 2, while $\beta$ occurs as a parameter in the characterization in case 3.

For a detailed account of these principles, as well as a complete bibliography and a history of the subject, the reader is referred to Volume II of [1].

[1] O. Bratteli and D.W.Robinson,
Operator Algebras and Quantum Statistical Mechanics;

Volume I: C* and W* - algebras, Symmetry Groups,
Decomposition of States;

Volume II: Equilibrium States, Models;
Springer Verlag, New York - Heidelberg - Berlin
(1979, and to appear near end of 1980).

Matematisk Institutt
Universitetet i Trondheim
N-7034 Trondheim, NTH
Norway

# ON THE EXISTENCE OF NON-TRIVIAL OR NON-STANDARD TRANSLATION INVARIANT SUBSPACES IN WEIGHTED $\ell^p$ and $L^p$

Yngve Domar

We shall consider six different situations. It is always assumed that $1 \leq p < \infty$. Subspace means closed subspace.

1. $w = \{w_n\}_0^\infty$ denotes a positive sequence. $\ell^p(w)$ is the Banach space of all complex $c = \{c_n\}_0^\infty$, for which $cw = \{c_n w_n\} \in \ell^p$, and such that the mapping $c \mapsto cw$ is an isometric isomorphism from $\ell^p(w)$ to $\ell^p$. Translation $T$ of $c$ is defined by

$$(Tc)_n = \begin{cases} 0, & n = 0, \\ c_{n-1}, & n \geq 1. \end{cases}$$

Assuming that $\{w_{n+1}/w_n\}$ is bounded, $T$ is an operator on $\ell^p(w)$. We shall consider subspaces, invariant under $T$.

The isomorphism of $\ell^p(w)$ to $\ell^p$, defined above, maps $T$ to a so called unilateral weighted shift on $\ell^p$, and our discussion can therefore be interpreted in terms of such operators ([10], [11]). If $w = \{w'_n\} = \{w_n e^{an}\}$, where $a \in \mathbb{R}$, then the families of invariant subspaces of $\ell^p(w)$ and $\ell^p(w')$ have the same structure. Thus we can assume, without loss of generality, that $w$ decreases.

For every $m \in \{0\} \cup \mathbb{Z}^+ \cup \{\infty\}$,

$$\ell^p(w,m) = \{c \in \ell^p(w): c_n = 0,\ n < m\}$$

is an invariant subspace. These subspaces are called <u>standard invariant subspaces</u>. Does it exist non-standard invariant subspaces?

The answer is trivially <u>yes</u>, if $\{n^{-1}\log w_n\}_1^\infty$ is a bounded sequence, i.e. if there is a $d > 0$ such that

$$w_n > d^n , \quad n > 0 .$$

For then, given any $z \in \mathbb{C}$ with $|z| < d$, $z \neq 0$, the set of all $c$ with $\sum_0^\infty c_n z^n = 0$, is a non-standard invariant subspace.

So let us consider the case when $\{n^{-1}\log w_n\}$ is unbounded. Then the answer is known to be <u>no</u>, if certain regularity conditions on $w$ are fulfilled. An easy way to prove theorems of this kind is by using the elementary theory of Banach algebras, as in [4] and [8, § 3.2], with assumptions which imply that $\ell^p(w)$ is a Banach algebra under convolution (known sufficient conditions for this property are essentially due to J. Wermer [13]). A simple condition for having the answer no, for $p = 1$, is that $\{\log w_n\}$ is concave. For $p > 1$, it suffices that $\{\log w_n - \alpha \log n\}_1^\infty$ is concave for some $\alpha > \frac{p-1}{p}$. B. Styf [12] has proved, for $p = 1$, that the answer is no also in certain cases when $\ell^1(w)$ is not a Banach algebra. N.K. Nikolskii [8, § 3.2] has proved that the answer is <u>yes</u> for certain $w$, of irregular behavior, but it is not quite clear whether his proof can be used to give an example, where $\ell^p(w)$ is a Banach algebra. It would be of interest to prove or disprove the existence of such a space, since the notions of closed ideal and invariant subspace coincide, in case $\ell^p(w)$ is a Banach algebra, and hence such a result would give information on the ideal structure of $\ell^p(w)$.

2. Now let $w = \{w_n\}_{-\infty}^{\infty}$ where $w_n$ are positive numbers. $\ell^p(w)$ is defined in analogy to the previous case. Translation $T$ is defined by

$$Tc = \{c_{n-1}\}_{-\infty}^{\infty}$$

for $c = \{c_n\}$. Assuming that $\{w_{n+1}/w_n\}$ is bounded, $T$ is an operator on $\ell^p(w)$. What can be said about its invariant subspaces?

The isometric isomorphism $\ell^p(w) \to \ell^p$ maps $T$ to a so called bilateral weighted shift on $\ell^p$ ([7], [10], [11]), and thus our discussion can as well be formulated using that concept. As in Section 1, we can assume that $w$ is decreasing, without loss of generality.

The invariant subspaces

$$\ell^p(w,m) = \{c \in \ell^p(w): c_n = 0, n < m\},$$

$m \in \{-\infty\} \cup \mathbb{Z} \cup \{\infty\}$, are the standard invariant subspaces. Do we have non-standard invariant subspaces?

As in Section 1 one can prove that non-standard invariant subspaces exist if $\{n^{-1}\log w_n\}$ is bounded, either at $-\infty$ or $\infty$, or, equivalently, that there is a $d > 0$ such that

$$\text{either} \quad w_n < d^n, \quad \text{for } n < 0,$$

$$\text{or} \quad w_n > d^n, \quad \text{for } n > 0.$$

It is, however, of interest that this sufficient condition can be weakened, if the restrictions hold simultaneously at $-\infty$ and $\infty$:

<u>Theorem 1 [6]</u>. Non-standard invariant subspaces exist,

if $n^{-1}\log w_n + \log|n|$ is bounded below both at $-\infty$ and $\infty$, or, equivalently, that there is a $d > 0$ such that

$$w_n < d^n|n|!, \quad n < 0,$$

and

$$w_n > \frac{d^n}{n!}, \quad n > 0.$$

In the opposite direction, we have the following theorem.

Theorem 2 [6]. Let us assume that $\{\log w_n\}_{-\infty}^{0}$ is convex, and $\{\log w_n\}_{0}^{\infty}$ is concave, and

$$\sum_{-\infty}^{\infty} \frac{w_{n+1}}{w_n} < \infty .$$

If either

$$\overline{\lim_{n\to -\infty}} \frac{\log w_n}{n^2} > \log 3 ,$$

or

$$\underline{\lim_{n\to \infty}} \frac{\log w_n}{n^2} < -\log 3 ,$$

all invariant subspaces are standard.

Theorem 2 is in particular valid, if $w_n = (|n|!)^a$, for $n \leq 0$, $w_n = 3^{-bn^2}$, for $n > 0$, where $a > 1$, $b > 1$. Reformulated in terms of bilateral shifts, Theorem 2 shows in the case $p = 2$, that there is an operator on a Hilbert space, for which the lattice of invariant subspaces is equivalent to the lattice $\{-\infty\} \cup \mathbb{Z} \cup \{\infty\}$. This gives an affirmative answer to a problem, raised by Halmos ([10, p. 193] and [11, Question 22]).

It would be desirable to narrow the considerable gap between Theorem 1 and Theorem 2. For instance, which is the correct order of magnitude at $\infty$ in Theorem 2, if $w_n = 1/w_{-n}$, $n \in \mathbb{Z}$ ?

3. With $\ell^p(w)$ and $T$ defined as in Section 2 we assume that $\{w_{n+1}/w_n\}$ is bounded and, as well, bounded from below by a positive constant. Then $T$ and $T^{-1}$ are operators, and we ask about subspaces which are invariant under both $T$ and $T^{-1}$, or, equivalently, hyperinvariant under $T$. As before, we have an interpretation in terms of bilateral weighted shifts.

$\{0\}$ and $\ell^p(w)$ are the trivial invariant subspaces of $\ell^p(w)$. Are there any non-trivial?

Here are some cases when non-trivial invariant subspaces are known to exist:

$1^o$. if constant sequences belong to $\ell^p(w)$ or to its dual,

$2^o$. if

$$\sum_{-\infty}^{\infty} \frac{|\log w_n|}{1+n^2} < \infty ,$$

$3^o$. if $\log w_n - \log w_{n-1} = O(|n|^{-1/2})$, as $n \to \infty$, and

$$\sum_{-\infty}^{0} \frac{|\log w_n|}{1+|n|^{3/2}} < \infty ,$$

and $n^{-1/2}\log w_n$, $n > 0$, is bounded below by some positive constant.

$4^o$. if $\{w_{-n}\}$, $\{1/w_n\}$, or $\{1/w_{-n}\}$ fulfils $3^o$.

$5^o$. if $w' = \{e^{an} w_n\}$ fulfils one of the conditions $1^o - 4^o$, for some $a \in \mathbb{R}$:

Condition $2^o$ is from A. Atzmon [2]. Here is a sketch of the proof of $3^o$. Consider the set of all $c \in \ell^p(w)$, which are restrictions to $\mathbb{Z}$ of entire functions of order $1/2$, type $\leq \varepsilon$. If $\varepsilon$ is small enough, it follows from Theorem 1 in Beurling and Malliavin [3], that this is a proper subset of $\ell^p(w), \neq \{0\}$. Using a Phragmén-Lindelöf argument, it can be proved that it is a subspace. Conditions $1^o$, $4^o$ and $5^o$ are easy to prove.

An example of a space $\ell^2(w)$ with only trivial invariant subspaces would show that there is an operator on an infinite dimensional Hilbert space with only trivial hyperinvariant subspaces, and hence it would be of great interest. A particular space $\ell^2(w)$ worth investigating is, for instance, the one obtained from

$$w_n = n/\log(|n|+2) , \quad n \in \mathbb{Z} .$$

4. Now we shall treat the continuous correspondence to the problem in Section 1. Let $w$ be a positive Lebesgue measurable function on $\mathbb{R}^+$, bounded on bounded intervals. $L^p(w)$ is the Banach space of complex-valued $f$ with $fw \in L^p(\mathbb{R}^+)$, so that $f \mapsto fw$ is an isometric isomorphism. We assume that

$$\sup_{y \in \mathbb{R}^+} w(x+y)/w(y)$$

is bounded on bounded subintervals of $\mathbb{R}^+$. Then, for $a \in \mathbb{R}^+$, translation $T_a$, defined by

$$T_a f(x) = \begin{cases} 0, & x \leq a \\ f(x-a), & x > a, \end{cases}$$

is an operator on $L^p(w)$. What subspaces are invariant under $\{T_a, a > 0\}$?

As in Section 1, we can assume that $w$ is decreasing, without loss of generality. For $a \in \{0\} \cup \mathbb{R}^+ \cup \{\infty\}$,

$$\{f \in L^p(w) : f = 0 \text{ a.e. on } ]0,a]\}$$

is an invariant subspace. These subspaces are called the standard invariant subspaces. Do non-standard invariant subspaces exist?

Corresponding to the situation in the discrete case, one finds non-standard invariant subspaces, if $w^{-1}\log w(x)$ is bounded at $\infty$. Let us assume in what follows that $w^{-1}\log w(x)$ is unbounded.

Then the answer is more difficult to give than in the discrete case, due to the fact that whenever $L^p(w)$ can be interpreted as a convolution Banach algebra, it has no unit. It is true that the results in Section 1 can be used to construct non-standard invariant subspaces for certain irregular $w$ (cf. the concluding remark of this paper). But if $w$ has sufficient regularity, for instance if $\log w$ is concave, nothing is known: we do not know if there is an $L^p(w)$ with non-standard invariant subspaces, and we do not know if there is an $L^p(w)$ with only standard invariant subspaces. A comparison with the corresponding problem, if we allow $w$ to vanish for large values of the variable, lies near at hand: then Titchmarsh's convolution theorem shows that only standard invariant subspaces exist (cf. [10, Ch. 4]).

There are, however, some approaches to the problem, giving sufficient conditions for an element $f \in L^1(w)$ to be cyclic, in the sense that the invariant subspace, generated by it, is $L^p(w)$. Here are some alternative conditions:

$1^o$. $\int_0^\infty |f(x)|e^{-bx}dx < \infty$, for some $b \in \mathbb{R}$.

$2^o$. $\log w$ concave, $(x \log x)^{-1}\log w(x) \to -\infty$, as $x \to \infty$, $f$ is of bounded variation at $0$, $f(+0) \neq 0$.

$3^o$. $\log w$ concave, $x^{-q}\log w(x) \to -\infty$, as $x \to \infty$, for some $q > 2$, there exists $\xi \in \mathbb{R}$ and $\varepsilon > 0$ such that

$$\left|\int_0^\varepsilon f(x)e^{i(\xi+i\eta)x}dx\right| \geq \exp\{-\eta^{\frac{q-2}{q-1}}\},$$

for sufficiently large positive values of $\eta$.

$1^o$ follows directly from results of B. Nyman [9] (cf. [1] and [5]). $2^o$ was proved by G.R. Allan [1] (a different proof was given independently in [5]). $3^o$ can be derived from a more general result in [5].

If $\log w$ is concave (and assuming $x^{-1} \log w(x) \to -\infty$, as $x \to \infty$), $L^1(w)$ is a radical Banach algebra under convolution. Invariant subspace is then the same as closed ideal. Thus the question on the existence of non-standard invariant subspaces has interest from Banach algebra point of view.

5. Next in turn is the continuous version of the problem in Section 2. $w$ is a positive Lebesque measurable function on $\mathbb{R}$, bounded on bounded intervals. $L^p(w)$ is the Banach space of complex-valued $f$ with $fw \in L^p(\mathbb{R})$, and with the norm defined in an obvious way. Assuming that

$$\sup_{y \in \mathbb{R}^+} w(x+y)/w(y)$$

is bounded on bounded intervals, translation $T_a$, defined by

$$T_a f(x) = f(x-a), \quad x \in \mathbb{R},$$

is, for $a \in \mathbb{R}^+$, an operator. What subspaces are invariant under $\{T_a, a > 0\}$?

We can assume that $w$ is decreasing. We have, for $a \in \{-\infty\} \cup \cup \mathbb{R} \cup \{\infty\}$, the standard invariant subspaces

$$L^p(w,a) = \{f \in L^p(w) : f(x) = 0 \text{ a.e. on } ]-\infty, a[\}.$$

Are there any more invariant subspaces?

In analogy to Section 2, the answer is yes if $x^{-1}\log w(x)$ is bounded at $-\infty$ or $\infty$. As for two-sided conditions, the following holds:

<u>Theorem 3 [6]</u>. $L^p(w)$ has non-standard invariant subspaces if

$$\int_{-\infty}^{\infty} \frac{d(\log w(x))}{1+x^2} > -\infty .$$

This result is less restrictive than Theorem 1 as for the order of magnitude of $w$ at infinity, since it holds if $|\log w(x)| = = \mathcal{O}(|x|^\alpha)$, for some $\alpha < 2$. On the other hand, we have no correspondence to Theorem 2.

6. We have, finally, the continuous correspondence to Section 3. $w$ is a positive Lebesgue measurable function, such that

$$\sup_{y \in \mathbb{R}} w(x+y)/w(y)$$

is bounded on bounded intervals of $\mathbb{R}$. $L^p(w)$ and $T_a$ are defined as in Section 5, now with $a \in \mathbb{R}$. $T_a$ are operators on $L^p(w)$. A subspace is invariant, if it is invariant under $\{T_a\}$. In contrast to Section 3, we have here a general result:

Theorem 4 [6]. $L^p(w)$ has always non-trivial invariant subspaces.

Concluding remark. We shall give a comment on the relation between the discrete and continuous variants of these problems. To every sequence $w$ we can in a natural way associate a step-function $w$, defined by $w(x) = w_n$, $x \in ]n, n+1]$. Starting with one of the cases 1, 2, 3, we can, for instance by using Hahn-Banach´s theorem, map the family of invariant subspaces injectively into the family of invariant subspaces of $L^p(w)$. This shows that existence of a non-standard (non-trivial) invariant subspace in the discrete case implies existence of a non-standard (non-trivial) invariant subspace in the corresponding continuous case. It would be desirable to have an example showing that the converse is not generally true. An inspection of Theorems 2 and 3 reveals that a modest strengthening of either theorem could provide such an example.

References

[1] G.R. Allan, Ideals of rapidly growing functions. Manuscript.

[2] A. Atzmon, On the existence of hyperinvariant subspaces. Manuscript 1980.

[3] A. Beurling and P. Malliavin, The Fourier transforms of measures with compact support. Acta Math. 107 (1962), 291-309.

[4] Y. Domar, On the ideal structure of certain Banach algebras. Math. Scand. 14 (1964), 197-212.

[5] Y. Domar, Cyclic elements under translation in weighted $L^1$-spaces on $\mathbb{R}^+$. To appear in Ark. Mat.

[6] Y. Domar, Translation invariant subspaces of weighted $\ell^p$ and $L^p$ spaces. Report 1980:3, Mathematics, Uppsala University.

[7] R. Gellar and D.A. Herrero, Hyperinvariant subspaces of bilateral weighted shifts. Indiana Univ. Math. J. 23 (1974), 771-790.

[8] N.K. Nikolskiĭ, Selected problems on weighted approximation and spectral analysis. Proceedings of the Steklov Inst. of Math. 120 (1974).

[9] B. Nyman, On the one-dimensional translation group and semi-group in certain function spaces. Uppsala 1950.

[10] H. Radjavi and P. Rosenthal, Invariant subspaces. Berlin 1973.

[11] A.L. Shields, Weighted shift operators and analytic function theory. AMS Mathematical Surveys 13 (1974), 49-128.

[12] B. Styf, Closed translation invariant subspaces in a Banach space of sequences summable with weights. Report 1977:3, Mathematics, Uppsala University.

[13] J. Wermer, On a class of normed rings. Ark. Mat. 2 (1953), 537-551.

# A geometric approach to nonlinear elasticity and continuum dynamics.

Halldór I. Elíasson

Introduction. The density of mass $\rho$ and the impulse $\rho u$ should be considered as basic concepts in our description of the continuum, or motion of deformable bodies. Then

$$\int_U \rho \quad , \quad \int_F \langle \rho u, \eta \rangle$$

represent the measurable mass contained in a domain U and flowing through a pice of surface F with $\eta$ as a unit normal field. The velocity u is thus indirectly defined. Then we are able to follow a mass-point, initially at a point x in space, defining its position $\phi(t,x)$ at a later time $t > 0$ as the solution of

$$\frac{\partial}{\partial t} \phi(t,x) = u(\phi(t,x)) \quad , \quad \phi(0,x) = x.$$

Of course, we could think ourselves able to identify $\phi(t,x)$ by some kind of colouring and then we would define u by the equation above as is customary in point mechanics.

The preservation of impulse of a system of perfectly elastic bodies, say occupying a domain U tied to the bodies, is a basic experimentally verified fact:

$$\frac{d}{dt} \int_U \rho u = \int_U \rho \frac{du}{dt} = 0.$$

That is, in the absence of exterior forces. This can be interpreted by requesting the existence of an operator field $\Sigma$, such that

$$\rho \frac{du}{dt} = \text{Div}\ \Sigma,$$

and $\Sigma \cdot \eta = 0$ at a free boundary (with normal $\eta$). This however, does not determine $\Sigma$. Evidently, we need a law

$\Sigma = F(\phi)$, if the equation of motion is to determine the future. The formulation of such a law would complete the precise definition of $\Sigma$ and then we could solve:

$$\rho \frac{\partial^2 \phi}{\partial t^2} = (\mathrm{Div}\, F(\phi)) \circ \phi$$

to obtain the flow $\phi$. In section 1 we shall discuss such a law, considering a deformation $\phi$ as acting on a stress tensor $\Sigma$ to give a new stress tensor $\phi_e \Sigma$ on the deformed body, the formulation being in an arbitrary Riemannian manifold as space, in order to ensure naturality of our postulates and methods.

In section 2 we consider the stress from the energy point of view, which is based on the conventional definition of the stress tensor. Then $\Sigma \cdot \eta$ is defined as an area-density of a force acting on the domain on the negative side of a surface oriented by its normal $\eta$, from the domain on the positive side. This is in accordance with our previous conception of div $\Sigma$ as a Newton-force in the equation of motion and does determine $\Sigma$, if we accept the possibility of experimental verification of $\Sigma \cdot \eta$ for all directions, $\eta$. This is however by no means easily acceptable and we might tend to think that this generally accepted definition of the stress tensor $\Sigma$ has possibly more to do with Green's formula:

$$\int_U \mathrm{Div}\Sigma = \int_{\partial U} \Sigma \cdot \eta,$$

giving two equivalent descriptons of the total force acting on a body in U, than with physical measurements. However, once we have accepted this strange picture of a force situated on a surface and acting somehow through it, we get its power:

$$\begin{aligned} &\int_{\partial U} \langle \Sigma \cdot \eta, u \rangle \\ = &\int_U \mathrm{div}(\Sigma \cdot u) \\ = &\int_U \langle \mathrm{div}\Sigma, u \rangle + \int_U \langle \Sigma, Du \rangle. \end{aligned}$$

Here we have used that $\Sigma$ is a self-adjoint operator, which is a consequence of the conservation of angular momentum. The first expression gives us the perhaps more satisfying interpretation of $-\Sigma \cdot u$ as the flux of mechanical energy, so that $<\Sigma u, \eta>$ is the area-density of the rate of mechanical energy entering the body at a point of its surface having outward normal $\eta$. The first term in the last expression shows us the part of this energy which is used to increase the kinetic energy of the body, so the second term must be the rate of work used to increase its internal energy. Here we are refering to the first law of thermodynamics:

$$\text{I} \qquad dE = \Delta W + \Delta Q,$$

which relates the differential increment of the internal energy E of a body to the work $\Delta W$ and heat $\Delta Q$ it takes from its surroundings. We have just explained that:

$$\Delta W = \left( \int_U <\Sigma, Du> \right) dt$$

and we show how this leads to a $\Sigma$ , $\phi$ relation for adiabatic processes, where $\Delta Q = 0$. In section 3 we generalize this law to the nonadiabatic case with the aid of the second law of thermodynamics:

$$\text{II} \qquad TdS \geq \Delta Q,$$

However, we must boldly turn this inequality into an equality, when written down for densities. We feel that the entropy S can not possibly be defined in full generality except this is possible by correct accounting for $\Delta Q$, also for socalled irreversible processes. This completes our picture of the perfectly elastic continua and we write down its basic properties and dynamic laws in section 4.

In section 5, we discuss possible methods for dealing with imperfect elasticity, hoping for a unified treatment of hysteresis in solids and viscosity in fluid flow. There we are lead, rather by a deliberate violation of our previously established mathematical properties of perfect elastic behavior, than by physical understanding of why there should be a transfer of mechanical energy to heat. Our results are therefore quite speculative, more so than in our previous section. However, the classical theory with the Navier-Stoke equation as the most formidable result, does leave quite a room for improvement.

1. Perfect elasticity theories.

We consider a body, represented by a bounded domain N in a Riemannian manifold M, subjected to deformations, represented by diffeomorphisms $\phi$ of N into M. We shall assume such a diffeomorphism to be estendable to a diffeomorphism of an open set containing the closure of N and denote the set of all those diffeomorphisms by Diff(N;M). We call it the state space of deformations.

Given a deformation:

$$\phi : N \to U \subset M,$$

there is a natural way to assign to each tensor field A on N a new tensor field $\phi_* A$ on U, the field adopted to A as sometimes called in differential geometry [3]. First of all if f is a function on N, then

$$\phi_* f = f \circ \phi^{-1}$$

and if X is a vector field in N, i.e. X(x) is a vector in the tengent space $T_x M$ to M at x, for each point x in N, then

$$\phi_* X = (d\phi \cdot X) \circ \phi^{-1}.$$

Finally the operation by $\phi_*$ should commute with contractions. This implies e.g. that for a 1-form $\omega$ in N, we have:

$$\phi_* \omega = (\omega (d\phi)^{-1}) \circ \phi^{-1} ,$$

so that: $(\phi_* \omega) \cdot (\phi_* X) = \phi_* (\omega \cdot X)$ .

Let us also observe that $\phi_*$ is actually a linear map from a tensor space at each point x to the corresponding tensor space at the point $\phi(x)$. Also, if the tangent of a curve $c(t)$, starting at $x = c(0)$, then

$$(\phi_* X)(\phi(x)) = \frac{d}{dt} \phi(c(t)), \text{ at } t = 0 ,$$

or $\phi_* X$ at $\phi(x)$ is the tangent to the same curve, refering to the physical body, meaning $\phi \circ c$. This discussion of the mathematical properties of the $\phi_*$ operation, together with the interpretation of $\phi(x)$ as the same masspoint as x and with the physical meaning given to tensors, based on a comparision of the values obtained by complete contraction with relevant dual tensors, with experiments, we feel right in interpreting $\phi_* A$ as physically the same tensor as A. This means that a physical property described by A is not changed by a deformation, if its values are given by $\phi_* A$ after the deformation $\phi$.

This discussion serves only to justify our first property of a perfect elasticity theory proposed in the following definition.

Definition 1. A theory of perfect elasticity for a body N in a Riemannian manifold M is given by a set $\Sigma_U$ of sections in the bundle $L_s(TU,TU)$ of self-adjoint operators on the tangent spaces of M, restricted to U, associated to any domain U in M diffeomorphic to N and by a functor e that assigns to any $\phi$ in Diff(U;M), with U diffeomorphic to N, a map:

$$\phi_e : \Sigma_U \to \Sigma_{\phi(U)}$$

such that:

(i) $\phi_e = \phi_*$ , restricted to $\Sigma_U$, if and only if $\phi$ is an isometry in M.

(ii) $(\psi \circ \phi)_e = \psi_e \circ \phi_e$ , for any $\phi$ in Diff(U;M) and $\psi$ in Diff(V;M), with $V = \phi(U)$.

Remarks. We mean to interpret $\Sigma$ in $\Sigma_U$ as a possible state of tension in the domain U. Observe that U can not be interpreted as a body except through a given diffeomorphism $\phi : N \to U$, identifying the points in U as mass points. Then a given tension $\Sigma_0$ on N gives rise to a definite tension $\Sigma = \phi_e(\Sigma_0)$ in U. On the other hand, given a tension $\Sigma$ in U, a further deformation $\psi : U \to V$ determines a new tension $\psi_e(\Sigma)$ and property (ii) garanties the independence of the tension of the way of deformation, which is the meaning of the perfectness of the elasticity. Property (i) means that the tension does not "change", in our previously discussed physical sense, by rigid body motions. Although we have not requested any continuity or diffrentiability properties of the operations in the theory, such properties of some kind must be present in any resonable example, as well as properties of locality.

We shall begin by giving an example of a perfect elasticity theory. We define for a given domain U the subset $A_U$ of $L_s(TU,TU)$ to consist of all:

$$A = (d\phi d^*\phi)\circ\phi^{-1} ,$$

with $\phi$ in Diff(N;M) such that $\phi(N) = U$ and where $d^*\phi$ denotes the adjoint of the linear map $d\phi$, with respect to the Riemannian metric on M, i.e.,

$$\langle d\phi X, Y\rangle = \langle X, d^*\phi Y\rangle ,$$

at any point x in N, with X in $T_xM$ and Y in $T_{\phi(x)}M$.

Then if $\psi$ is in Diff(U;M), we define:

$$\psi_a(A) = (d\psi A d^*\psi)\circ\psi^{-1}.$$

Obviously $(\psi\circ\phi)_a = \psi_a\circ\phi_a$, so that $\psi_a$ maps $A_U$ to $A_{\psi(U)}$ and $e = a$ satisfies (ii). Property (i) is satisfied, since an isometry $\psi$ is characterized by:

$$(d\psi)^{-1} = d^*\psi$$

and: $$\psi_*A = (d\psi A(d\psi)^{-1})\circ\psi^{-1}.$$

One must observe at this point that A in $A_U$ can take an arbitrary value in $L_s(T_xM,T_xM)$ at a given point x in U, although A, as a section on U, does have special properties.

Despite the naturality of this construction, this can not be the correct theory for most applications. In particular, the zero tensor is not contained in any $A_U$. Suppose however, that we have a perfect elasticity theory e as in Def. 1, such that say $\Sigma_N$ contains the zero section 0. Then there is a well defined map.

$$F_U : A_U \to \Sigma_U.$$

defined by:

$$F_U(A) = \phi_e(0) \text{ , if } A = \phi_a(I) .$$

In fact, suppose

$$A = \phi_a(I) = \psi_a(I)$$

Then: $$(\psi^{-1}\circ\phi)_a(I) = I ,$$

which means that $\psi^{-1}\circ\phi$ is an isometry.

Then by (ii) and (i):

$$\phi_e(0) = \psi_e((\psi^{-1}\circ\phi)_e(0)) = \psi_e((\psi^{-1}\circ\phi)_*(0)) = \psi_e(0),$$

so that $F_U$ is well defined. Moreover, if $\psi_e(0) = \phi_e(0)$,

then $(\psi^{-1}\circ\phi)_e(0) = id_e(0) = 0 = (\psi^{-1}\circ\phi)_*(0)$ ,

so that $\psi^{-1}\circ\phi$ is an isometry by the "only if" part of (i) and then

$$\phi_a(I) = \psi_a(I).$$

Thus $F_U$ is injective. This implies, that if $\Sigma$ is given in $F_U(A_U)$, then there is a unique A in $A_U$, such that

$$\Sigma = F_U(A).$$

Writing $A = \phi_a(I)$, we get for any $\psi$ in Diff(U;M):

$$\psi_e(\Sigma) = \psi_e(\phi_e(0)) = (\psi\circ\phi)_e(0)$$

$$= F_U((\psi\circ\phi)_a I) = F_U(\psi_a(A))$$

The e-theory is thus completely determined by the a-theory and the $F_U$ 's, if restricted to the section spaces $F_U(A_U)$.

Let us assume that V is isometric to U by an isometry:

$$\psi : V \to U,$$

and that $\phi : N \to U$ is a diffeomorphism. Then

$$F_U(\phi_a(I)) = \phi_e(0) = \psi_e((\psi^{-1} \circ \phi)_e(0))$$

$$= \psi_* F_V((\psi^{-1})_*(\phi_a(I)),$$

or:

$$F_U = \psi_* F_V(\psi^{-1})_* .$$

This property does give $F_U$ some flavor of universality and even more so if we introduce some localizing condition in the theory, relaxing the dependence of $F_U$ on U. Although this seems a worthwile project, we shall henceforth restrict our attention to the case, that $F_U$ is a point dependent map, defining F at each point x as a mapping of $L_s(T_xM, T_xM)$ into itself, but the commuting of F with isometries as described above does then make F quite independent of the manifold M, if M is homogeneous.

Definition 2. A stress-strain relation in dimension n is a continuously differentiable injective map F from the set of positive definite self-adjoint operators A in the n-dimensional euclidean space $\underline{R}^n$, into the set of self-adjoint operators in $\underline{R}^n$, such that:

$$F(CAC^{-1}) = CF(A)C^{-1}$$

for any linear isometry C in $\underline{R}^n$.

As consequence of the invariance of F by the adjoint of the isometry group, F is in fact defined on the set of self-adjoint positive definite operators on any n-dimensinal vector space with an inner product, extending the invariance to hold for any linear isometry C from one such space to another.

Proposition 1. Let F be a stress-strain relation in dimension n, M a Riemannian manifold of dimension n, N a domain in M and $A_0$ a given field on N of positive definite self-adjoint operators on the tangent space of M. Define:

$$A_U = \{ \phi_a(A_0) : \phi \in \mathrm{Diff}(U,M) \}$$

for any U diffeomorphic to N and with the a-action defined as before. Then:

$$\Sigma_U = F(A_U)$$

$$\phi_e(\Sigma) = F(\phi_a F^{-1}(\Sigma)), \ \Sigma \in \Sigma_U, \text{ and } \phi \in \mathrm{Diff}(U,M),$$

defines a perfect elasticity theory.

<u>Proof</u>. We have:

$$\phi_a(A) \in A_{\phi(U)}, \text{ if } A = F^{-1}(\Sigma)$$

by the action property of a, so $\phi_e$ maps $T_U$ to $T_{\phi(U)}$. If $\phi$ is an isometry, then $\phi_a = \phi_*$, so F commutes with $\phi_a$ and we get the "if" part of (i) (Def. 1). On the other hand, if

$$F(\phi_a(A)) = \phi_* F(A),$$

we can write:

$$F(\phi_a(A)) = F(\phi_*(A)),$$

by extending F to the orbit of A by the adjoint action of the full group of linear isomorphisms of the tangent space to M at the particular point in question, so that F commutes with $\phi_*$, for any $\phi$ with nonsingular differential. Obviously F is still injective, so extended, so

$$\phi_a(A) = \phi_*(A).$$

This should hold for all A in $A_U$ and since A can take an arbitrary value, e.g. I, at a particular point, we get $\phi_a = \phi_*$, so $\phi$ is an isometry, which concludes the verification of (i). Property (ii) follows from:

$$(\psi \circ \phi)_a(A)) = F(\psi_a(\phi_a(A))$$

$$= \psi_a(F(\phi_a(A))) = \psi_e(\phi_e(F(A)))$$

$$= \psi_e \circ \phi_e(\Sigma).$$

We have not yet introduced the concept of strain, although our name for F immediately suggests to those familiar with classic elasticity theory that $A = \phi_*(A_0)$ should be the strain tensor. This is in fact our intention, giving a more concrete physical meaning to A. Observing:

$$\phi_*\left(\langle A^{-1}\cdot\phi_* X, \phi_* X\rangle\right) = \langle A_0^{-1}X, X\rangle,$$

we see, introducing on $\phi(N)$ the metric:

$$g(X,Y) = \langle A^{-1}X, Y\rangle,$$

that $\phi$ is an isometry from $(N,g_0)$ to $(\phi(N),g)$, in fact $g = \phi_* g_0$. The metric g is thus an intrinsic property of the body, and the ratio of lengths measured in outer space to the corresponding intrinsic length, contains the essence of the strain-concept. It is remarkable, that we are able to assign an arbitrary state of strain $A_0$ to N, our state of reference to the body, which is an extension from our deductive approach to a stress-strain relation F, based on Def. 1, since we found it necessary there to assume N in a "neutral state", i.e. with zero tension and $A_0 = I$. It is in fact quite natural that we run into the possibly quite difficult problem of determining the initial state in a given situation.

It now becomes apparent that we can interpret two domains $U_1$, $U_2$, each with a strain tensor $A_1$ and $A_2$ respectively as representing the same body in different states of deformation, only if there is a diffemorphism $\phi: U_1 \to U_2$, such that $\phi_a A_1 = A_2$. This equation has an integrability condition, which can be expressed by $\phi_* R_1 = R_2$, where $R_1$ and $R_2$ are the curvature tensors of $U_1$ and $U_2$ with the metrics $g_1$ and $g_2$ assigned to $A_1$ and $A_2$ as before. In particular, the search for a $\phi$ such that $\phi_a(I) = A$, i.e., the search for a "neutral state", is a difficult problem.

<u>Proposition 2.</u> A stress-strain relation F can be represented by a function f of n+1 variables, such that

$$F(A)\cdot v = f(I_1,\ldots,I_n,\lambda)v, \quad \text{if } A\cdot v = \lambda v,$$

where $I_1,..,I_n$ are the invariants of $\Lambda$, determined by:

$$\det(\lambda I-A) = \sum_{k=0}^{m} (-1)^k I_n \lambda^{n-k}.$$

Proof. It is clear that F(A) must have the same eigenvectors as A and that we may write the eigenvalues $\Lambda_i$ of F(A) as functions of the eigenvalues $\lambda_i$ of A:

$$\Lambda_i = f_i(\lambda_1,\dots,\lambda_n) .$$

If we choose an isometry C so as to permute the eigenvectors of A by some permutation $\sigma$, we obtain from

$$F(CAC^{-1}) = CF(A)C^{-1}:$$

$$\Lambda_{\sigma(i)} = f_i(\lambda_1,\dots,\lambda_n).$$

As a consequence, we must have:

$$f_i(\lambda_1,\dots,\lambda_n) = f(\lambda_1,\dots,\lambda_n,\lambda_i),$$

with f invariant under permutations of the first n variables, so we may introduce the symmetric polynomials $I_i$ as new variables instead of the $\lambda_i$.

Remarks. In particular, if the function f is analytic in the last variable $\lambda$, we may write:

$$F(A) = \sum_{k=1}^{n} F_k(I_1,\dots,I_n)\Lambda^{n-k}$$

since $\lambda^n$ (or $A^n$) is a polynomial of degree n-1 in $\lambda$ (or A), due to the characteristic eqution for $\lambda$. This form of F (n=3) has been presented earlier by workers in elasticity theory as the most general nonlinear stress strain relation for a homogeneous medium. From our point of view however, in our theory of interior strain and stress, this holds also for inhomogeneous media, the inhomogeneities in directions, due e.g. to crystal structure, are accounted for by the strain tensor A.

## 2. Work and Energy.

We shall proceed independently from our concept of elasticity in section 1, except that we shall assume that the stress tensor $\Sigma$ commutes with the strain tensor A, we introduced there, which is in fact the essence of the stress strain relation. Our starting point here is the rate of work performed by the stress on a body, which is moving from an initial position N as described by a time dependent deformation $\phi:N\to M$, say $\phi$ = id. at time t = 0. The velocity field u is given by:

$$u\circ\phi = \frac{\partial\phi}{\partial t}$$

and, as described in our introduction, the rate of work is:

$$\frac{dW}{dt} = \int_{\phi(N)} \langle\Sigma, Du\rangle$$

The strain-field is given by:

$$A = \phi_a A_0 = (d\phi A_0 d^*\phi)\circ\phi^{-1}$$

Using d/dt to denote material derivation along the flow, we have

$$\frac{dA}{dt}\circ\phi = \frac{\partial A}{\partial t}\circ\phi + (DA\circ\phi)\cdot\frac{\partial\phi}{\partial t} = \frac{\partial}{\partial t}(A\circ\phi)$$

$$= \frac{\partial}{\partial t}(d\phi A_0 d^*\phi)$$

$$= D\left(\frac{\partial\phi}{\partial t}\right)A_0 d^*\phi + d\phi A_0\, D^*\left(\frac{\partial\phi}{\partial t}\right)$$

$$= (DuA + ADu)\circ\phi,$$

or: $$\frac{dA}{dt} = DuA + ADu.$$

Here, we interpret D as a covariant derivation on a Riemannian manifold, if needed. Since $\Sigma$ commutes with A, we have:

$$\langle\Sigma, Du\rangle = \frac{1}{2}\langle A^{-1}\Sigma, DuA+ADu\rangle$$

and thus:

$$\frac{dw}{dt} = \frac{1}{2}\int_{\phi(N)} \langle A^{-1}\Sigma, \frac{dA}{dt} \rangle$$

Now we request the existience of a potential energy W which has an energy density as an interior property of the body, and by that we mean $\phi_* w_0$, with $w_0$ the density at the initial state N. The operation of $\phi_*$ on densities is the same as on the absolute value of an n-form, so:

$$\phi_* w_0 = \rho w, \text{ with}$$

$$\rho = (\det A)^{-1/2} = (\det A_0)^{1/2} \circ \phi^{-1}.$$

Here we have for convenience introduced $\rho$ as a mass-density, using sometimes the specific volume $v = 1/\rho$ instead, and the function w as an energy density measured per unit mass.

Then:

$$W = \int_{\phi(N)} \rho w = \int_N \rho_0 w \circ \phi ,$$

$$\frac{dW}{dt} = \int_{\phi(N)} \rho \frac{dw}{dt} .$$

We use for the derivation the transformation formula for integration, but we could also derive the equation for the conservation of mass from the derivative for A:

$$\frac{d\rho}{dt} = -\frac{1}{2}\rho \langle A^{-1}, \frac{dA}{dt}\rangle = -\rho\langle I, Du\rangle$$

$$= -\rho \,\mathrm{div} u,$$

and use:

$$\frac{dW}{dt} = \int_{\phi(N)} \frac{\partial}{\partial t}(\rho w) + \int_{\partial\phi(N)} \rho w\langle u, \eta\rangle$$

$$= \int_{\phi(N)} \left(\frac{\partial}{\partial t}(\rho w) + \mathrm{div}(\rho w u)\right)$$

$$= \int_{\phi(N)} \rho \frac{dw}{dt} .$$

Now a comparison of the two expressions for the rate of work gives:

$$\frac{dw}{dt} = \frac{1}{2\rho} \langle A^{-1}\Sigma, \frac{dA}{dt} \rangle$$

or, as an expression in total differentials:

$$dw = \frac{1}{2\rho} \langle A^{-1}\Sigma, dA\rangle.$$

thus w is a function of A and we obtain its gradient:

$$w = f(A), \ \nabla f(A) = \frac{1}{2\rho} A^{-1}\Sigma.$$

This in turn delivers a stress-strain relation:

Proposition 3. The existence of a density w for a potential energy in a perfect elasticity theory given by a stress-strain relation F, requires that w is a function of the invariants of the strain tensor A and

$$F(A) = 2\rho A \nabla f(A) , \ w = f(A),$$

$$\rho = (\det A)^{-1/2} .$$

If $I_1, \ldots, I_n$ are the invariants of A, then

$$\nabla f(A) = \sum_{m=1}^{n} \partial_m f \ \nabla I_m , \quad \partial_m f = \partial f / \partial I_m$$

where the gradient of $I_m$ is given by:

$$\nabla I_m = \sum_{k=0}^{m-1} I_k (-A)^{m-1-k}$$

$$= - \sum_{k=m}^{n} I_k (-A)^{m-1-k} ,$$

as can be seen from the characteristic equation. Thus our request for the existence of w does not alter the form of the last expression for F(A) obtained in section 1, but it does put a differential condition on the coefficients of that expression.

We shall work out an explicit formula in dimension $n = 3$, using:

$$\theta = \frac{1}{3}(\lambda_1+\lambda_2+\lambda_3) = \frac{1}{3}\text{Trace } A$$

$$\gamma = \frac{1}{3}(\lambda_1^2+\lambda_2^2+\lambda_3^2) = \frac{1}{3}||A||^2$$

$$v = (\lambda_1\lambda_2\lambda_3)^{1/2} = 1/\rho$$

as independent variables instead of the mentioned invariants of A. Then:

$$d\theta = \frac{1}{3}<I, dA> , \; d\gamma = \frac{2}{3}<A,dA>$$

$$dv = \frac{v}{2}<A^{-1},dA> ,$$

$$\text{grad}w = \frac{v}{2}\frac{\partial w}{\partial v}A^{-1} + \frac{1}{3}\frac{\partial w}{\partial \theta}I + \frac{2}{3}\frac{\partial w}{\partial \gamma}A ,$$

$$F(A) = \frac{\partial w}{\partial \gamma}I + \frac{2}{3v}\frac{\partial w}{\partial \theta}A + \frac{4}{3v}\frac{\partial w}{\partial \gamma}A^2.$$

In the linear approximation for the deformation of a solid body from a state with $A_0 = I$ we write:

$$\phi(x) = x + \xi(x),$$

and assume all partial derivatives of $\xi$ small compared with 1, so:

$$A\circ\phi = (I+D\xi)(I+D^*\xi) \simeq I+D\xi+D^*\xi = I+2\tau ,$$

with $\tau$ the classic strain tensor, so the stress, as the change in the internal tension, is given by:

$$\sigma = F(A\circ\phi) - F(I)$$

$$\simeq DF(I)\cdot 2\tau$$

$$= 2\mu\tau + \lambda(\text{Trace } \tau)I, \text{ with:}$$

$$\mu = \frac{2}{3v}\frac{\partial w}{\partial \theta} + \frac{8}{3w}\frac{\partial w}{\partial \gamma} , \text{ in } v = \theta = \gamma = 1,$$

$$\lambda = Y(Yw) \qquad , \text{ in } v = \theta = \gamma = 1,$$

$$\text{Where:} \quad Y = \frac{\partial}{\partial u} + \frac{2}{3v}\frac{\partial}{\partial \theta} + \frac{4}{3v}\frac{\partial}{\partial \gamma} .$$

This is the classic linear stress-strain relation, with $\mu$ and $\lambda$ the usual coefficients of elasticity.

It is further of interest to consider the mechanical energy:

$$W(\phi) = \int_{\phi(N)} \rho w = \int_N \rho_0 \, f(A\circ\phi) ,$$

as a variation-integral on some appropriate space of mappings $\phi: N \to M$. With

$$\xi = \frac{d}{dt}\phi_t |_{t=0} , \quad \phi_0 = \phi$$

an infinitesinal variation $(\xi = u\circ\phi)$ ,
the variational derivative of W in the direction of $\xi$, is:

$$dW(\phi)\cdot\xi = \int_N \rho_0 \, \langle \nabla f(A\circ\Phi), \frac{dA}{dt} \circ\phi \rangle \, |t=0$$

$$= \int_{\phi(N)} \langle \Sigma, Du \rangle$$

$$= \int_N 2\rho_0 \, \langle \nabla f(A\circ\phi) d\phi A_0, D\xi \rangle .$$

Then Div $\Sigma = 0$ is the Euler-Lagrange equation for the stationary states as expected, but the last expression shows that this can also be written as:

$$\mathrm{Div}_N\big(\nabla f(d\phi A_0 d^*\phi) d\phi\big) = 0 ,$$

where $\mathrm{Div}_N$ denotes the divergence operator acting on fields along mappings from N to M, with N carrying the Riemannian metric $g_0$ associated to $A_0$. Problems of this kind have been studied in Global Analysis [1] , especially the case with $f(A) = \text{Trace } A$, but then $\nabla f = I$ and the stationary states are the harmonic maps from N to M. This energy does however not carry sufficient "weight" [2] , in order for critical point theories to be applicable in full strength. In the case of perfect elasticity we have also dependence of $|| A ||^2$, in the case of a nondegenerate stress-strain relation, and we would expect sufficient weights (strength of ellipticity) at least for A in some neighbourhood of $A_0$. Lack of weight for larger strains might be responsible for the appearance imperfect elastic behavior. Unfortunately this nonlinear and global problems are still poorly understood.

## 3. Interaction with heat flow.

We want to generalize the concept of perfect elasticity to the case of variable temperature. This can immediately be done, if we have constant emperature T along the lines of flow. We then can demand a stress-strain relation as before along each flow line, but allow it to be given by a different function F on different lines, i.e. we then have a relation:

$$\Sigma = F(T,A),$$

which can be obtained as in section 2 from an elasticity potential w, which is a function of T and A. On the other hand, if T changes along flow lines, then the deformation is a mapping between two different states of temperature and it is not clear how the change in tension should respect this difference.

We shall base our extension on the following local forms of the two fundamental laws of thermodynamics:

$$\text{I} : \quad \rho \frac{de}{dt} - \langle \Sigma, Du \rangle = \operatorname{div}(k\nabla T) ,$$

$$\text{II} : \quad T\rho \frac{ds}{dt} = \operatorname{div}(k\nabla T).$$

Here, e and s are densities (per mass) of internal energy and entropy respectively, and the right hand term represents the accumulation of heat due to heat conduction (heat radiation could also be added). We can eliminate this heat term and obtain:

$$\frac{de}{dt} - v\langle \Sigma, Du \rangle = T \frac{ds}{dt}$$

Introducing the free energy:

$$f = e - Ts,$$

the equation can be written as:

$$\frac{df}{dt} - v\langle\Sigma, Du\rangle = -s\frac{dT}{dt}.$$

Here we have the same equation as in section 2, if we put $w = f$, in case T is constant along flow lines, i.e., if $dT/dt = 0$. However, dropping this assumption and inserting the derivative of the strain A instead of Du as before, still assuming that $\Sigma$ commutes with A, we may write this as a differential equation:

$$df = \frac{v}{2}\langle A^{-1}\Sigma, dA\rangle - s\, dT$$

This means:

$$\frac{\partial f}{\partial T} = -s,$$

$$\nabla f = \frac{v}{2} A^{-1}\Sigma,$$

with $\nabla f$ to be understood as the gradient of f with T constant. The first equation is a well-known relation in thermodynamics and can also be written:

$$\frac{\partial e}{\partial T} = T\frac{\partial s}{\partial T},$$

which is the specific heat by constant volume (fixing A implies fixing volume). The second result is the stress strain relation:

$$\Sigma = F(T,A) = \frac{2}{v} A\, \nabla f,$$

which we may also call the equation of state, using the language of thermodynamics. In fact, assuming f only a function of T and v, as in classical thermodynamics, then

$$\nabla f = \frac{v}{2}\frac{\partial f}{\partial v} A^{-1}$$

and we get $\Sigma = - p I$, with a pressure:

$$p = -\frac{\partial f}{\partial v} = T\frac{\partial s}{\partial v} - \frac{\partial e}{\partial v}$$

which is a well known representation for the equation of state. This kind of degeneracy of the energy function f, which is close to being true for gases and liquids, does mean a departure from perfect elasticity since the mapping F is not injective on the whole strain space. This departure could be taken seriously as a triggering of viscous behavour, if the flow is not conformal (when A is also a multiple of I).

## 4. Dynamics.

We are now able to formulate in full generality a definition of a perfectly elastic continua and write down equations governing the flow in terms of its state variables.

The state variables are given by a temperature T, and a strain tensor A, which are functions on the domain in space occupied by the body at each time t. Characterizing the properties of the body is an energy function $w = f(T,A)$, representing the potential energy of tension and the free energy of the body. The flow is governed by the following equations:

1. The preservation of internal strain:

$$\frac{dA}{dt} = DuA + ADu$$

2. The impulse equation:

$$\rho\frac{du}{dt} - \operatorname{div}\Sigma = -\rho\nabla h,$$

with $\rho = (\det A)^{-1/2}$ and h denoting some given exterior potential field.

3. The heat equation:

$$\rho T\frac{ds}{dt} = \operatorname{div}(k\nabla T), \quad s = -\frac{\partial f}{\partial T}$$

with k a given function of the state variables T and A.

4. The equation of state:

$$\Sigma = 2\rho A \nabla f,$$

with $\nabla f$ the gradient of f as a function of A for a fixed T.

Remark. By a "function of A" we mean here more precisely a function of the invariants of the tensor A.

We must observe that this is a dynamical system which determines the future from the present and appropriate boundary conditions, reflecting exterior influence. However, we have not given a physical meaning to A, except to its determinant, so that we are not really prepared to assign the initial or boundary conditions. Even if we have a precise physical understanding of $\Sigma$, we can not hope to construct f without a priori understanding of A. Perhaps, we must be satisfied with an instrumental definition of A, leaving the precise definition of f to depend on our agreement on A. However A has a mathematical property, the relations

$$A = \phi_a(A_0) : \frac{dA}{dt} = DuA + ADu$$

to the deformation $\phi$ and velocity field u, with $A_0$ the initial value of A. This is of importance in relation to the concept of equilibrium, e.g. solution with $u = 0$ and T constant. The equations:

$$\operatorname{div} \Sigma = \rho \nabla h, \ \Sigma = 2\rho A \nabla f$$

do not suffice to determine A or $\Sigma$. The equation:

$$A = \phi_a(A_0)$$

must be added, with some understanding of $A_0$.

## 5. Viscoelasticity.

Although hysteresis is characterized by the conversion of mechanical energy into heat, classical thermodynamics does not have much to say on the matter, except to describe it as an irreversible process, which it must reckon with by placing an inequality in the second law of thermodynamics.

We shall consider the presence of a tension $\Sigma$ not commuting with the strain A as a source of conversion of mechanical energy into heat and we shall reformulate our previous presentation of the localized laws of thermodynamic. It has been recognized that the separation of the change in internal energy into mechanical work and heat, as is done in the first law, is not at all obvious. Thinking of $\langle \Sigma, Du \rangle$ as the power density of the tension, a part r of this power could be used to produce heat and the rest $\langle \Sigma, Du \rangle - r$ to perform mechanical work. Then the two laws would give:

$$\text{I} \qquad \rho \frac{de}{dt} - \left( \langle \Sigma, Du \rangle - r \right) = r + \operatorname{div}(k \nabla T)$$

$$\text{II} \qquad T\rho \frac{ds}{dt} = r + \operatorname{div}(k \nabla T)$$

We see that I is still the same equation, but a new term has entered equation II. If we combine the two by eliminating the heat terms, we obtain:

$$\frac{df}{dt} - v \langle P, Du \rangle + s \frac{dT}{dt} = v \langle R, Du \rangle - r,$$

$$\text{with } \Sigma = P + R .$$

Here we split $\Sigma$ into two terms, such that P commutes with the strain A and R is supposed to be responsible for the generation of heat. $R = 0$, $r = 0$ is the case of perfect elastic behavour and we shall define P, as in that case by

$$P = \frac{2}{v} A \nabla_A f$$

with $\nabla_A f$ the gradient of f as a function of A alone. As before f depends also on T and we have:

$$\frac{\partial f}{\partial T} = - s.$$

We shall also consider f as dependent on:

$$q = \frac{1}{2} \| R \|^2 ,$$

Furthermore, we shall assume that R is a self-adjoint and traceless operator, as

$$Q = Du + D^*u - \frac{2}{n}(\operatorname{div} u)I.$$

Then after elimination of **P**, our equation becomes:

$$\frac{\partial f}{\partial q} < R, \frac{dR}{dt} > = \frac{v}{2} < R, Q > - \Sigma$$

This equation holds if we put:

$$5. \qquad \frac{2}{v}\frac{\partial f}{\partial q}\frac{dR}{dt} = Q - \frac{1}{\mu} R$$

with $\mu$ the coefficient of viscosity, which we take as the equation defining R, and then

$$r = \frac{v}{2\mu} \| R \|^2 = \frac{v}{\mu} q$$

is the rate of heat production.

We then have to add 5 to the dynamical equations in section 4, replace $\Sigma$ in 2 by P + R, $\Sigma$ in 4 by P and add r to the right hand side of 3. Equation 2 is then a generalisation of the Navier-Stoke equation and is asymptotically equal, as $t\to\infty$, if R becomes stable so that $R = \mu Q$ is an asymptotic solution of 5. Recall that $P = -pI$, if f as a function of A depends only on v.

It must be admitted that this introduction of viscoelasticity does not look convincing. Perhaps, a better physical understanding of the strain tensor A could clarify the reason for viscous behavior.

References

[1] J. Eells, Jr., A setting for global analysis, Bull. Amer. Math. Soc. 72 (1966), 739-807.

[2] H.I. Elíasson, Variation integrals in fibre bundles, Proc. of Symp. in pure Math. Vol. 16, AMS 1970.

[3] S. Helgason, Differential Geometry and Symmetric Spaces, Acad. Press 1962.

# THE NORMAL BUNDLE OF ELLIPTIC SPACE CURVES OF DEGREE 5.

G. Ellingsrud & D. Laksov

## § 1. The main result.

Our interest in the geometry of elliptic space curves of degree 5 was initiated by A. Van de Ven, who pointed out to us that a basic assertion in F. Ghione's work [4] (more precisely assertion (ii) of the Lemma at p. 376) on the normal bundle of such curves was contradicted by the classical assertion (see e.g. [7] or [11]) that their tri-secant scroll is of degree 5 and contains the curve doubly. Considering the elliptic space curves as projections of a fixed curve in $\mathbb{P}^4$ from different centers, we were led independently and by different methods to a nearly complete classification of the splitting types of the normal bundles according to the location of the center of projection. The differences of our approaches are reflected throughout this article where we give several proofs of most of the results. We hope that this will not obscure the presentation, but instead illuminate the properties of the elliptic space curves and contribute to the understanding of their geometry.

To describe our main result we let C be an elliptic curve defined over an algebraically closed field k and L an invertible sheaf of degree 5 on C. Then the vector space $W = H^o(C, L)$ is of dimension 5 and L defines a canonical embedding of C into $\mathbb{P}^4 = \mathbb{P}(W)$. For each point P in $\mathbb{P}^4$ we denote by $V_P$ the corresponding hyperplane in W and we denote by $C_P$ the image of C in $\mathbb{P}^3 = \mathbb{P}(V_P)$ under the projection with center P. There is a commutative diagram of bundles on C with exact rows and columns:

$$
\begin{array}{ccccccccc}
 & & & & 0 & & & & \\
 & & & & \downarrow & & & & \\
 & & & & (V_P)_C & \xrightarrow{\tau_P} & P^1_C(L) & & \\
 & & & & \downarrow & & \| & & \\
(*) \quad 0 & \longrightarrow & N^* \otimes L & \longrightarrow & W_C & \xrightarrow{\tau} & P^1_C(L) & \longrightarrow & 0 \\
 & & \nu_P \downarrow & & \downarrow \kappa_P & & & & \\
 & & \mathcal{O}_C & = & \mathcal{O}_C & & & & \\
 & & & & \downarrow & & & & \\
 & & & & 0 & & & &
\end{array}
$$

where $N^*$ is the conormal bundle of $C$ in $\mathbb{P}^4$ and $P^1_C(L)$ is the bundle of first principal parts of $L$. Note that at each point $P$ on $C$ the map $\tau$ represents the tangent line to $C$ at $P$. Consequently the maps $\tau_P$ and $\nu_P$ are surjective at each point $Q$ such that the tangent to $C$ at $Q$ does not pass through $P$. Moreover we note that, for each point $P$ not on the secant variety $\mathrm{Sec}(C)$ to $C$ in $\mathbb{P}^4$, the map $\tau_P$ is the universal map on $C_P \cong C$ and consequently $\ker \tau_P = \ker \nu \cong N^*_P \otimes L$, where $N^*_P$ is the conormal bundle to $C_P$ in $\mathbb{P}^3$.

Denote by $p$ and $q$ the projections of $\mathbb{P}^4 \times C$ onto the first and second factor. Then there is a natural commutative diagram of bundles on $\mathbb{P}^4 \times C$

$$
(**) \qquad
\begin{array}{ccc}
q^*(N^* \otimes L) & \longrightarrow & q^*W_C \cong p^*W_{\mathbb{P}^4} \\
\downarrow & & \downarrow \\
p^*\mathcal{O}_{\mathbb{P}(W)}(1) & = & p^*\mathcal{O}_{\mathbb{P}(W)}(1)
\end{array}
$$

which globalizes the left bottom square of diagram (*). Denote by $F$ the kernel of $\nu$, then by the above remarks the sequence

$$0 \to F \to q^*(N^* \otimes L) \to p^*\mathcal{O}_{\mathbb{P}(W)}(1) \to 0$$

is exact at $\mathbb{P}^4 \setminus \mathrm{Tan}(C)$ and $F(P) \cong N_P^* \otimes L$ for each point $P \in \mathbb{P}^4 \setminus \mathrm{Sec}(C)$. Twisting the map $\nu$ by $q^*(L \otimes M)$ where $M$ is an $\mathcal{O}_C$-module and applying $p_*$ we obtain a map on $\mathbb{P}^4$

$$\delta_M\colon p_*q^*(N^* \otimes L^2 \otimes M) = H^0(C, N^* \otimes L^2 \otimes M)_{\mathbb{P}^4} \to \mathcal{O}_{\mathbb{P}(W)}(1) \otimes H^0(C, L \otimes M)$$

Theorem. Let $M$ be in the group $\mathrm{Pic}_0(C)$ of locally free sheaves on $C$ of degree zero and denote by $Y_M$ the subscheme of $\mathbb{P}^4$ where the above map $\delta_M$ is not of maximal rank. With the above notation the following assertions hold:

(i) The scheme $Y_M$ is a hypersurface in $\mathbb{P}^4$ of degree 5 with underlying set

$$\{P;\ H^0(C_P, F_P \otimes L \otimes M) \neq 0\}\ .$$

(ii) We have that $Y_M = \mathrm{Sec}(C)$ when $M = \mathcal{O}_C$ and $Y_M \neq \mathrm{Sec}(C)$ when $M \neq \mathcal{O}_C$.

(iii) For each point $P \notin \mathrm{Sec}(C)$ there is an $M \in \mathrm{Pic}_0(C)$, unique up to the canonical involution $M \to M^{-1}$, such that $P \in Y_M$. In particular $Y_M = Y_{M^{-1}}$ and $Y_M \cap Y_{M'} \subseteq \mathrm{Sec}(C)$ when $M' \neq M$ and $M' \neq M^{-1}$.

(iv) If $M^2 \neq \mathcal{O}_C$ and $P \in Y_M \setminus \mathrm{Sec}(C)$ then

$$N_P^* = M(-2) \oplus M^{-1}(-2)\ .$$

(v) If $M^2 = \mathcal{O}_C$, then for each $P$ in a non-empty open subset of $Y_M \setminus \mathrm{Sec}(C)$, the bundle $N_P^*$ is indecomposable, isomorphic to the only non-trivial extension in $\mathrm{Ext}_C^1(M(-2), M(-2))$

We shall now give a proof of the theorem referring to results in the later sections. The forward references will at the same time give an idea of the contents of these sections.

Except for the claim that $Y_M \neq \mathbb{P}^4$ assertion (i) of the theorem clearly follows from the following two assertions; $h^0(C, N^* \otimes L^2 \otimes M) = h^0(C, L \otimes M) = 5$ and the bundles $p_*(N^* \otimes L^2 \otimes M)$ and $p_*(L \otimes M)$ commute with base change. These assertions for the bundle $L \otimes M$ follow immediately from the Riemann-Roch theorem and for the bundle $N^* \otimes L^2 \otimes M$ they are exactly the statement of the proposition of section 6.

To see that $Y_M \neq \mathbb{P}^4$ we factor $\delta_M$ at a point $P \in \mathbb{P}^4$ via the map $\varphi_P$: $H^0(C, W_C \otimes L \otimes M)_{\mathbb{P}^4} \to H^0(C, L \otimes M)_{\mathbb{P}^4}$ obtained from the factorization of $\nu_P$ via $\kappa_P$ in diagram (*). Now choose a basis for W corresponding to independent points $P_1, P_2, \ldots, P_5$ of $\mathbb{P}^4$. We have that the map corresponding to $P_i$ is the projection of $H^0(C, W_C \otimes L \otimes M)_{\mathbb{P}^4}$ onto the i'th factor in the decomposition induced by the $p_i$'s. Hence $\bigcap_P \ker \varphi_P = 0$ and the map $\delta_M$ is generically injective.

To prove assertion (ii) of the theorem we fix a secant S to C which is not a tangent. Then for every point $P \in S$ we prove in section 7 (Proposition 2) that $C_P$ lies on a unique quadric surface and that except for exactly one point on S the quadric is non-singular. Moreover we prove in section 7 (Proposition 1) that when the quadric is non-singular it gives rise to an exact sequence

$$0 \to \mathcal{O}_C \to F_P \otimes L \to \mathcal{O}_C \to 0$$

and that, when projecting from the exceptional point, the quadric cone gives rise to a splitting $F_P \otimes L \cong \mathcal{O}(Q) \oplus \mathcal{O}(-Q)$ where Q is the vertex of the cone. Hence at the exceptional point we have that $h^o(C, F_P \otimes L \otimes M) = h^o(C, \mathcal{O}(Q) \otimes M) = 1$ for all $M \in \mathrm{Pic}_o(C)$ and at all other points $H^o(C, F_P \otimes L \otimes M) = 0$ when $M \neq \mathcal{O}_C$ and $H^o(C, F_P \otimes L) \neq 0$. Consequently, when $M \neq \mathcal{O}_C$, all but one point on S is in $\mathrm{Sec}(C) \setminus Y_M$ and $\mathrm{Sec}(C) \subseteq Y_M$ when $M = \mathcal{O}_C$. Note that we have given a second proof of the assertion that $Y_M \neq \mathbb{P}^4$.

The only remaining part of assertion (ii) is that $\mathrm{Sec}(C) = Y_M$ when $M = \mathcal{O}_C$. However Sec(C) is of degree 5 because the image of C in $\mathbb{P}^2$ under the projection with center on a general line in $\mathbb{P}^4$ has 5 double points by the genus formula. Hence the inclusion $\mathrm{Sec}(C) \subseteq Y_M$ is an equality.

We collect some elementary observations that we shall need in the remaining part of the proof, in the following lemma which we prove in § 2 (see also M.F. Atiyah [1] in particular Theorem 5 p. 432).

Lemma. Let E be a rank 2 vector bundle on C such that $H^o(C, E) = 0$ and $\overset{2}{\Lambda} E = \mathcal{O}_C$.

(i) There exists an $M \in \mathrm{Pic}_o(C)$, unique up to the canonical involution $M \to M^{-1}$, such that $H^o(C, E \otimes M) \neq 0$.

(ii) If $M^2 \neq \mathcal{O}_C$, then $E = M \oplus M^{-1}$.

(iii) If $M^2 = \mathcal{O}_C$ and $h^o(C, E \otimes M) = 2$, then $E = M \oplus M$.

(iv) If $M^2 = \mathcal{O}_C$ and $h^o(C, E \otimes M) = 1$, then E is indecomposable, isomorphic to the only non-trivial extension in $\mathrm{Ext}^1_C(M, M)$.

To prove the remaining parts of the theorem we fix a point $P \in \mathbb{P}^4 \setminus \mathrm{Sec}(C)$ and apply the lemma to the bundle $E = N_P^*(2) = F_P \otimes L$ on $C \cong C_P$. The property $\overset{2}{\Lambda}(F_P \otimes L) \cong \mathcal{O}_C$ of the lemma is immediate from the sequence

$$0 \to F_P \to V_P \to P_C^1(L) \to 0$$

of diagram (*) and the property $H^0(C_P, F_P \otimes L) = 0$ is the assertion of the fundamental proposition of section 8 below. We see that assertions (iii) and (iv) of the theorem are restatements of the assertions (i) and (ii) of the lemma. Moreover, to prove assertion (v) of the theorem we observe that, by assertion (iv) of the lemma and the semi-continuity of $H^0(C_P, F_P \otimes L \otimes M)$ in P, it suffices to show the existence of a point $P \in \mathrm{Sec}(C)$ such that $h^0(C_P, F_P \otimes L \otimes M) = 1$. However, such a point was found in the proof of part (ii) of the theorem as the exceptional point on the secant line S where $C_P$ was lying on a quadric cone.

<u>Remark 1</u>. We shall prove in section 10 that the hypersurfaces $Y_M$ are indeed irreducible. One natural question that the theorem leaves unsettled is therefore wether $N_P$ is indecomposable for every point in $Y_M \setminus \mathrm{Sec}(C)$ when $M^2 = \mathcal{O}_C$. Another interesting problem is to find a natural geometric interpretation of the hypersurfaces $Y_M$. In order to shed some light on these problems we have included some related material in sections 9 and 10 that is not necessary for the proof of the theorem.

<u>Remark 2</u>. The above proof that the hypersurfaces $Y_M$ are locally the zero scheme of a determinant of a $5 \times 5$-matrix with linear entries, and consequently either a hypersurface of degree 5 or $\mathbb{P}^4$, was built upon the vanishing result $H^1(C, N^* \otimes L^2 \otimes M) = 0$ of section 6. The proof of the latter result given in section 6 is rather involved. It requires the

knowledge of the syzygies of the ideal defining C in $\mathbb{P}^4$ and these are obtained in sections 4 and 5 by explicit computations.

There is, however, an alternative description of the hypersurfaces as locally the zero scheme of a determinant of a $10 \times 10$-matrix which depends on an easier vanishing result. We give this description and a proof of the corresponding vanishing result in section 2.

A slightly different, but equally computational, approach to finding the syzygies of the ideal defining C in $\mathbb{P}^4$ was found by D. Eisenbud and A. Van de Ven (private communication). Instead of starting with explicit equations for C in $\mathbb{P}^4$ like we do, they use the Riemann-Roch theorem to show that C is arithmetically normal in $\mathbb{P}^4$. It follows that C is arithmetically Gorenstein. Then reducing the homogenous coordinate ring of C modulo two general linear forms one obtains an artinian Gorenstein ring with Hilbert function 1, 3, 1. It is easy to find a normal form for such a ring and to find its syzygies as a quotient of a polynomial ring in 5 variables and consequently also the syzygies for C in $\mathbb{P}^4$.

The first of the two proofs given above of the inequality $Y_M \neq \mathbb{P}^4$ was also communicated to us by Eisenbud and Van de Ven.

Remark 3. As a consequence of the proposition of section 8 we obtain that every elliptic space curve is contained doubly in a quintic surface. We prove that such a surface is necessarily equal to the tri-secant scroll of the curve, hence affirming the classical assertion mentioned above.

In section 9 we write down explicitly the equation for the tri-secant scroll of a certain projection of the elliptic curve C.

Remark 4. In the proof of assertion (ii) of the theorem we used projections with center P lying on a proper secant to C and used the splitting of $F_P$ on the projected curve $C_P$ which in this case has degree 5 and arithmetic genus 2. The properties of $F_P$ in this case and the treatment of it given in section 7 are similar to the properties of the normal bundle of non-singular space curves of degree 5 and genus 2 and the treatment of these curves given by Van de Ven [10].

## § 2. Alternative description of $Y_M$ and a proof of property (i).

Let $U = \mathbb{P}^4 \setminus \mathrm{Tan}(C)$. Then the composite map

$$q^*(\Omega^1_{\mathbb{P}(V)}(1)|C) \to q^*V = p^*V \to p^*\mathcal{O}_{\mathbb{P}(V)}(1)$$

is surjective on $U \times C$. Denote the kernel of this map by K. Then there is a commutative diagram of bundles on $\mathbb{P}^4 \times C$;

$$\begin{array}{ccccccccc}
 & & 0 & & 0 & & & & \\
 & & \downarrow & & \downarrow & & & & \\
0 & \longrightarrow & F & \longrightarrow & K & \longrightarrow & q^*(\Omega^1_C \otimes L) & \longrightarrow & 0 \\
 & & \downarrow & & \downarrow & & \| & & \\
0 & \longrightarrow & q^*(N^* \otimes L) & \longrightarrow & q^*(\Omega^1_{\mathbb{P}(V)}(1)|C) & \longrightarrow & q^*(\Omega^1_C \otimes L) & \longrightarrow & 0 \\
 & & \downarrow^{\nu} & & \downarrow & & & & \\
 & & p^*\mathcal{O}_{\mathbb{P}(V)}(1) & = & p^*\mathcal{O}_{\mathbb{P}(V)}(1) & & & & \\
 & & \downarrow & & \downarrow & & & & \\
 & & 0 & & 0 & & & &
\end{array}$$

with exact rows and columns on $U \times C$ and where $\nu$ and consequently F are the same as those defined in section 1. As in section 1 we obtain a map

$$\varepsilon_M\colon p_*(K \otimes q^*(L \otimes M)) \to p_*q^*(\Omega^1_C \otimes L^2 \otimes M) .$$

We want to prove that the subset of U where $\varepsilon_M$ drops rank is $\{P \in U \,;\, H^0(C_P, F_P \otimes L \otimes M) \neq 0\}$ and that the corresponding scheme is a hyperplane of degree 5 or is $\mathbb{P}^4$. To this end it clearly suffices to prove the following assertions: $h^0(C, K \otimes q^*(L \otimes M)) = h^0(C, \Omega^1_C \otimes L^2 \otimes M) = 10$ and $p_*(K \otimes q^*(L \otimes M))$ and $p_*(\Omega^1_C \otimes L^2 \otimes M)$ commute with base change and $\overset{10}{\Lambda}\, p_*(K \otimes q^*(L \otimes M))|U \cong \mathcal{O}_U(-5)$. These assertions for the bundle $\Omega^1_C \otimes L^2 \otimes M$ follow immediately from the Riemann-Roch theorem and the assertions for the bundle $K \otimes q^*(L \otimes M)$ is the statement of the following proposition.

It follows that the subscheme defined by $\varepsilon_M$ is equal to $Y_M$ because both are of degree 5, the underlying sets coincide on U and $U \cap Y_M$ was proved in section 1 to be non-empty.

<u>Proposition</u>. We have that $R^1p_*(K \otimes q^*(L \otimes M)) = 0$ on U. Consequently $K \otimes q^*(L \otimes M)$ commutes with base change and $p_*(K \otimes q^*(L \otimes M))$ is locally free of rank 10. Moreover $\overset{10}{\Lambda}\, p_*(K \otimes q^*(L \otimes M))|U \cong \mathcal{O}_U(-5)$.

<u>Proof</u>. From the commutative diagram

$$\begin{array}{ccccccccc}
 & & & & 0 & & & & \\
 & & & & \downarrow & & & & \\
 & & & & p^*\Omega^1_{\mathbb{P}(V)}(1) & \xrightarrow{\alpha} & q^*L & \longrightarrow & 0 \\
 & & & & \downarrow & & \| & & \\
0 & \longrightarrow & q^*\Omega^1_{\mathbb{P}(V)}(1)|C & \longrightarrow & q^*V_C = p^*V_{\mathbb{P}(V)} & \longrightarrow & q^*L & \longrightarrow & 0 \\
 & & \downarrow & & \downarrow & & & & \\
0 & \longrightarrow & p^*\mathcal{O}_{\mathbb{P}(V)}(1) & = & p^*\mathcal{O}_{\mathbb{P}(V)}(1) & & & & \\
 & & \downarrow & & \downarrow & & & & \\
 & & 0 & & 0 & & & &
\end{array}$$

which has exact rows and columns on $U \times C$ we see that $K$ is also the kernel of the map $\alpha$. We consequently obtain a long exact sequence of $\mathcal{O}_{\mathbb{P}(V)}$-modules,

$$0 \to p_*(K \otimes q^*(L \otimes M)) \to p_*(p^*\Omega^1_{\mathbb{P}(V)}(1) \otimes q^*(L \otimes M)) \overset{\beta}{\to} p_*(q^*L^2 \otimes M) \to$$
$$R^1p_*(K \otimes q^*(L \otimes M)) \to R^1p_*(p^*\Omega^1_{\mathbb{P}(V)}(1) \otimes q^*(L \otimes M)) .$$

We easily check the following four assertions,

(i) $\operatorname{rank} p_*q^*(L^2 \otimes M) = h^0(C, L^2 \otimes M) = 10$

(ii) $\operatorname{rank} p_*(p^*\Omega^1_{\mathbb{P}(V)}(1) \otimes q^*(L \otimes M)) = \operatorname{rank} (\Omega^1_{\mathbb{P}(V)}(1) \otimes H^0(C, L \otimes M)) = 20$

(iii) $R^1p_*(p^*\Omega^1_{\mathbb{P}(V)}(1) \otimes q^*(L \otimes M)) = \Omega^1_{\mathbb{P}(V)}(1) \otimes H^1(C, L \otimes M) = 0$

(iv) $\overset{20}{\Lambda}(\Omega^1_{\mathbb{P}(V)}(1) \otimes H^0(C, L \otimes M)) \cong \overset{5}{\otimes} \overset{4}{\Lambda}\Omega^1_{\mathbb{P}(V)}(1) \cong \mathcal{O}_{\mathbb{P}(V)}(-5)$

From these assertions we conclude that all that remains in order to prove the proposition is to verify that the map $\beta$ is surjective. However, if $P \in U$, then $\beta(P)$ is the canonical map

$$V_P \otimes H^0(C, L \otimes M) \to H^0(C, L^2 \otimes M) .$$

To prove that this map is surjective we choose three linearly independent sections $\sigma_1$, $\sigma_2$ and $\sigma_3$ of $L$ without common zeroes and lying in $V_P$. Let $N$ be the kernel of the resulting surjective map $\sigma\colon L^{-1} \oplus L^{-1} \oplus L^{-1} \to \mathcal{O}_C$. We obtain after twisting by $M \otimes L^2$ an exact sequence

$$\bigoplus_{i=1}^{3} H^0(C, L \otimes M) \to H^0(C, L^2 \otimes M) \to H^1(C, N \otimes M \otimes L^2) .$$

Consequently it suffices to prove that $H^1(C, N \otimes M \otimes L^2) = 0$ or dually that $H^0(C, N^* \otimes M^{-1} \otimes L^{-2}) = 0$. From the Koszul complex of the map $\sigma$ we obtain an exact sequence

$$0 \to L^{-3} \overset{\gamma}{\to} L^{-2} \oplus L^{-2} \oplus L^{-2} \to N \to 0$$

where $\gamma$ is the dual of $\sigma$ twisted by $L^{-3}$. Dualizing the latter sequence, twisting by $M^{-1} \otimes L^{-2}$ and passing to cohomology we obtain an exact sequence

$$0 \to H^o(C, N^* \otimes M^{-1} \otimes L^{-2}) \to \bigoplus_{i=1}^{3} H^o(C, M^{-1}) \xrightarrow{\delta} H^o(C, M^{-1} \otimes L) .$$

When $M = \mathcal{O}_C$ the map $\delta$ of the latter sequence is induced by the three linearly independent sections $\sigma_1$, $\sigma_2$ and $\sigma_3$ and is consequently surjective and when $M \neq \mathcal{O}_C$ we have that $H^o(C, M^{-1}) = 0$. In both cases we conclude that $H^o(C, N^* \otimes M^{-1} \otimes L^{-2}) = 0$.

## § 3. Proof of the lemma of section 1.

We first show the existence of an M with $H^o(C, E \otimes M) \neq 0$. Fix a point $Q \in C$. Denote by $p_i$ the projection of $C \times C$ onto the i'th factor and by $\Delta$ the diagonal in $C \times C$. It is sufficient to prove that the $\mathcal{O}_C$-module $R^1 p_{2*}(p_1^*(E \otimes \mathcal{O}_C(Q)) \otimes \mathcal{O}_{C\times C}(-\Delta))$ is non-zero because this module commutes with base change and at a point $P \in C$ the fiber is $H^1(C, E \otimes \mathcal{O}_C(Q - P))$.

The short exact sequence

$$0 \to p_1^*(E \otimes \mathcal{O}_C(Q)) \otimes \mathcal{O}_{C\times C}(-\Delta) \to p_1^*(E \otimes \mathcal{O}_C(Q)) \to E \otimes \mathcal{O}_C(Q) \to 0$$

gives rise to a long exact sequence

$$p_{2*}p_1^*(E \otimes \mathcal{O}_C(Q)) \xrightarrow{\alpha} E \otimes \mathcal{O}_C(Q) \to R^1 p_{2*}(p_1^*(E \otimes \mathcal{O}_C(Q)) \otimes \mathcal{O}_{C\times C}(-\Delta)) \to$$
$$R^1 p_{2*}p_1^*(E \otimes \mathcal{O}_C(Q)) \to 0 .$$

We see that if $H^1(C, E \otimes \mathcal{O}_C(Q)) \neq 0$ then we have proved the existence of M. On the other hand we see that if $H^1(C, E \otimes \mathcal{O}_C(Q)) = 0$ then the existence of M will follow if we prove that $\alpha$ is not surjective. However, in the latter case we have that $p_{2*}p_1^*(E \otimes \mathcal{O}_C(Q)) = H^o(C, E \otimes \mathcal{O}_C(Q))_C$ is a free $\mathcal{O}_C$-module of rank 2 by the Riemann-Roch theorem and the $\mathcal{O}_C$-module $E \otimes \mathcal{O}_C(Q)$ is not free because $\overset{2}{\Lambda}(E \otimes \mathcal{O}_C(Q)) = \mathcal{O}_C(2Q)$ by the assumption of the lemma.

To prove the uniqueness of M we note that if $h^0(C, E \otimes M) \neq 0$ then since $E \cong E^\vee$ there is a map $\varphi\colon E \to M$ which is surjective. Indeed, if $\operatorname{im} \varphi \neq M$ then the image is of negative degree. Consequently E has a rank one subbundle, $\ker \varphi$, of positive degree. This is impossible because $H^0(C, E) = 0$. Hence we obtain an exact sequence

$$(*) \qquad 0 \to M^{-1} \to E \to M \to 0 .$$

Tensor this sequence by an $M' \in \operatorname{Pic}_0(C)$ and pass to cohomology. We see that $h^0(C, E \otimes M') \neq 0$ implies that either $h^0(C, M^{-1} \otimes M') \neq 0$ or $h^0(C, M \otimes M') \neq 0$, that is, either $M' = M$ or $M' = M^{-1}$ and the uniqueness is proved.

The assertions (ii), (iii) and (iv) of the lemma also follow from the exact sequence (*). Indeed, if $M^2 \neq \mathcal{O}_C$, then the sequence splits because $H^1(C, M^{-2}) = 0$ and if $M^2 = \mathcal{O}_C$ then the sequence splits if and only if $h^0(C, E \otimes M) = 2$.

## § 4. Equations for an embedding of C in $\mathbb{P}^4$.

We shall, in order to keep the expressions below as simple as possible, assume that the characteristic of the ground field is different from 2. Then the curve C is isomorphic to a plane curve $y^2 = x(x - 1)(x - \lambda)$ for some $\lambda \neq 0, 1$. The point at infinity is $P_\infty = (0; 1; 0)$ and we shall choose the embedding of C into $\mathbb{P}^4$ given by the divisor $L = \mathcal{O}_C(5P_\infty)$.

To find the equations for C in this embedding we observe that $(x) = 2(0; 0; 1) - 2P_\infty$ and $(y) = (0; 0; 1) + (1; 0; 1) + (\lambda; 0; 1) - 3P_\infty$.

Thus $\{1, x, x^2, y, xy\} \subseteq L(5P_\infty)$ gives a basis for the complete linear system and thus a morphism $f: C \to \mathbb{P}^4$ defined on $x_2 \neq 0$ by $f(a; b; 1) = (1; a; a^2; b; ab)$ and such that $f(P_\infty) = (0; 0; 0; 0; 1)$. The image $f(C)$ is clearly contained in the scheme X defined by the five polynomials

$$F_1 = y_0y_2 - y_1^2 \,,\; F_2 = y_0y_4 - y_1y_3 \,,\; F_3 = y_1y_4 - y_2y_3 \,,$$

$$F_4 = y_3^2 - y_1y_2 + (1 + \lambda)y_0y_2 - \lambda y_0y_1 \,,\; F_5 = y_3y_4 - y_2^2 + (1 + \lambda)y_1y_2 - \lambda y_0y_2 \,.$$

It is easily checked that X is non-singular in $f(P_\infty)$ and (using e.g. the hyperplane $y_0 = 0$ in the affine piece $y_4 = 1$) that X is of degree 5. Moreover, f clearly induces an isomorphism

$$C \cap (\mathbb{P}^2 \setminus V(y_2)) \to X \cap (\mathbb{P}^4 \setminus V(y_0)) \,.$$

Hence f induces an isomorphism between C and X.

## § 5. Syzygies for the cone over C in $\mathbb{P}^4$.

The purpose of the following computations was to find an explicit equation for the tri-secant scroll of a projection of C into $\mathbb{P}^3$. Interestingly enough they yield at the same time the syzygies of C in $\mathbb{P}^4$.

We project the curve C from the point (0; 1; 0; 0; 0) into $\mathbb{P}^3$. To find the equations of the projected curve we eliminate successively $y_1$ from the equations $F_i$ and $F_j$ of section 4 for all $i < j$. We obtain

$S_1$: $y_3F_1 - y_1F_2 = -y_oF_3$

$S_2$: $y_4F_1 + y_1F_3 = y_2F_2$

$S_3$: $(y_2 + \lambda y_o)F_1 - y_1F_4 = -y_oF_5 + y_3F_2$

$S_4$: $(1 + \lambda)y_2F_1 + y_1F_5 = y_2F_4 + y_3F_3$

$G_1$: $y_4F_2 + y_3F_3 = y_oy_4^2 - y_2y_3^2$

$G_2$: $(y_2 + \lambda y_o)F_2 - y_3F_4 = y_oy_2y_4 - y_3^3 + \lambda y_o^2y_4 - (1 + \lambda)y_oy_2y_3$

$G_3$: $(1 + \lambda)y_2F_2 + y_3F_5 = y_3^2y_4 - y_2^2y_3 + (1 + \lambda)y_oy_2y_4 - \lambda y_oy_2y_3$

$S_5$: $(y_2 + \lambda y_o)F_3 + y_4F_4 = G_3 = (1 + \lambda)y_2F_2 + y_3F_5$

$G_4$: $(1 + \lambda)y_2F_3 - y_4F_5 = y_2^2y_4 - y_3y_4^2 + \lambda y_oy_2y_4 - (1 + \lambda)y_2^2y_3$

$G_5$: $(1 + \lambda)y_2F_4 + (y_2 + \lambda y_o)F_5 = (1 + \lambda)y_3^2y_2 + (1 + \lambda^2)y_oy_2^2 + \lambda y_oy_3y_4 - \lambda^2y_o^2y_2$

$= y_2y_3y_4 - y_2^3 + (1 + \lambda)y_3^2y_2 + (1 + \lambda^2)y_oy_2^2 + \lambda y_oy_3y_4 - \lambda^2y_o^2y_2$ .

The equations $G_1, \ldots, G_5$ define a variety $Y$ of degree 5 (as is seen e.g. by using the hyperplane $y_o = 4$ in the affine piece $y_4 = 1$) and clearly contain the image of $C$ by the composite map $g: C \to \mathbb{P}^3$ of $f$ with the projection. We have $g(a; b; 1) = (1; a^2; b; ab)$ and it is easily checked that $g$ induces an isomorphism

$$C \cap (\mathbb{P}^2 - V(y_2)) \to Y \cap (\mathbb{P}^3 - V(y_o))_{\frac{y_4}{y_3}}$$

and that $Y$ is non-singular at $(0; 0; 0; 1)$ and at the points $(1; 0; 0; 0)$ and $(1; 1; 0; 0)$ lying in $y_4 = 0$. Hence $g$ induces an isomorphism between $C$ and $Y$.

The relations $S_1, \ldots, S_5$ give all the syzygies for the equations $F_i$ in the ring $A = k[y_o, y_1, y_2, y_3, y_4]$. Indeed let $B = A/(F_1, \ldots, F_5)$. We obtain a complex

$$(*) \qquad A^5 \xrightarrow{\gamma} A^5 \xrightarrow{\varphi} A \to B \to 0$$

where $\varphi$ sends a basis of $A^5$ to the polynomials $F_1, \ldots, F_5$ and in this basis

$$\gamma = \begin{vmatrix} 0 & , -(1+\lambda)y_2 & , y_2 + \lambda y_o & , y_4 & , -y_3 \\ (1+\lambda)y_2 & , 0 & , -y_3 & , -y_2 & , y_1 \\ -(y_2 + \lambda y_o) & , y_3 & , 0 & , y_1 & , -y_o \\ -y_4 & , y_2 & , -y_1 & , 0 & , 0 \\ y_3 & , -y_1 & , y_o & , 0 & , 0 \end{vmatrix} .$$

It is easily checked that the $4 \times 4$-pfaffians of $\gamma$ are the polynomials $F_1, \ldots, F_5$ and since these define a subvariety of $\mathbb{P}^4$ of codimension 3 it follows that the ring B that they define is Gorenstein of codimension 3 and that the sequence (*) is exact ([3] Theorem 2.1 p. 456 and the proof of that theorem p. 464, [6] Corollary 16 p. 178 and the proof of the corollary).

## § 6. Proof of (i).

Proposition. We have that $H^1(C, N^* \otimes L^2 \otimes M) = 0$. Consequently $p_*(N^* \otimes L^2 \otimes M)$ commutes with base change and $h^o(C, N^* \otimes L^2 \otimes M) = 5$.

Proof. From the sequence (*) of section 5 we obtain a partial resolution

$$W \otimes L^{-3} \to W \otimes L^{-2} \to N^* \to 0$$

of $N^*$. Twisting this sequence by $L^2 \otimes M$, dualizing and passing to cohomology we obtain an exact sequence

$$0 \to H^0(C, N^* \otimes L^{-2} \otimes M^{-1}) \to H^0(C, W \otimes M^{-1}) \to H^0(C, W \otimes L \otimes M^{-1}) .$$

We conclude that $H^0(C, N^* \otimes L^{-2} \otimes M^{-1}) = 0$. Indeed, if $M \neq \mathcal{O}_C$ we have that $H^0(C, W \otimes M^{-1}) = 0$ and if $M = \mathcal{O}_C$ the map $H^0(C, W_C) \to H^0(C, W \otimes L)$ of the sequence is the transpose of the matrix $\gamma$ of section 5 and by inspection the two first rows of $\gamma$ are linearly independent, so $\gamma$ is injective. By duality $H^0(C, N \otimes L^{-2} \otimes M^{-1}) = H^1(C, N^* \otimes L^2 \otimes M)$ and we have proved the proposition.

## § 7. Normal bundles of space curves of arithmetic genus 2 and degree 5 with an ordinary double point.

In the proof of assertion (ii) of the theorem of section 1 we needed to consider the image of the curve C under projections with center at points in $\mathrm{Sec}(C) \setminus \mathrm{Tan}(C)$. This leads to the study of curves of arithmetic genus 2 and degree 5 with an ordinary double point. Non-singular curves of this genus and degree were treated by Van de Ven [10] who proved that their normal bundle decomposes or is indecomposable according to wether the curve is on a singular or non-singular quadric surface. Our treatment and results below in the case of curves with an ordinary double point are similar to those of Van de Ven.

Proposition 1. Assume that $C_P$ has an ordinary double point.

(i) If $C_P$ is on a smooth quadric, then the quadric induces an exact sequence

$$0 \to \mathcal{O}_C \to F_P \otimes L \to \mathcal{O}_C \to 0$$

(ii) If $C_P$ lies on a cone over a plane conic with vertex Q, then Q is a smooth point of $C_P$ and

$$F_P \otimes L \cong \mathcal{O}_C(Q) \oplus \mathcal{O}_C(-Q) .$$

Proof. Let I denote the ideal defining $C_P$ in $\mathbb{P}^3$. The quadric that contains $C_P$ induces a section of $I(2)/I^2(2)$ and hence a section of $F_P \otimes L$. The composite of the resulting map $\sigma: L^{-1} \to F_P$ with the inclusion $F_P \subseteq V_C$ has zeroes exactly on the singularities of the quadric and thus the same is true for $\sigma$.

If the quadric is non-singular, then $\sigma$ is nowhere zero and assertion (i) follows from the isomorphism $\overset{2}{\Lambda}(F_P \otimes L) \cong \mathcal{O}_C$.

If the quadric is singular, then the vertex lies on a non-singular point of $C_P$. Indeed, in this case the curve projects from the vertex of the cone onto a plane conic and if we let d = (the multiplicity at Q of the intersection of $C_P$ with a hyperplane in $\mathbb{P}^3$ in general position) and e = (the degree of the fibers of the projection of C onto the plane conic), then the following formula holds

(degree of $C_P$) = (degree of conic)e + d, that is $5 = 2e + d$.

This formula rules out the possibility $d = 0$, that is Q is not on $C_P$, and the possibility $d = 2$, that is Q is a singular point of $C_P$. Hence, when $C_P$ is on a quadric cone $\sigma$ induces an exact sequence

$$0 \to \mathcal{O}_C(D) \to F_P \otimes L \to \mathcal{O}_C(-D) \to 0$$

where D is an effective divisor with support on the non-singular point Q of $C_P$. This sequence splits because $H^1(C, \mathcal{O}_C(2D)) = 0$. It remains to prove that D has multiplicity 1. Choose parameters $t_1$, $t_2$ and $t_3$ for $\mathbb{P}^3$ at Q in such a way that I is the ideal $(t_1, t_2)$ in the completion $k[[t_1, t_2, t_3]]$ of the local ring of $\mathbb{P}^3$ at Q. Let $q = f_1t_1 + f_2t_2$ be the equation for the quadric cone containing $C_P$. Then $f_1$ and $f_2$ are not contained in the ideal $(t_1, t_2)$ because then q would be reducible. Consequently at least one of the series $f_i$ contains a term $ut_3$ with u a unit in $k[[t_1, t_2, t_3]]$ and the section of $I_Q/I_Q^2$ induced by q at the point Q, has a simple zero.

We shall now prove that the two cases of the above proposition materialize.

<u>Proposition 2</u>. Choose two different points $P_1$ and $P_2$ on C such that

$$(*) \qquad 3P_1 + 2P_2 \not\equiv H \quad \text{and} \quad 2P_1 + 3P_2 \not\equiv H$$

where H is a hyperplane section of C. Then the image $C_P$ of C under a projection with center at a point P on the line joining $P_1$ and $P_2$ lies on a unique irreducible quadric surface and this surface is non-singular except for exactly one point of the line.

More precisely, define points Q and R on C by $Q \equiv H - 2P_1 - 2P_2$ and $R \equiv 3P_1 + 3P_2 - H$. Then the exceptional point P on the line joining $P_1$ and $P_2$ is given by the hyperplane

$$H^o(C, L(-P_1 - P_2))t + kt_Q^3t_R^2 \subseteq W ,$$

where we denote by $t_P$ a non-zero section in $H^o(C, \mathcal{O}(P))$ and by t the product of non-zero sections in $H^o(C, \mathcal{O}(P_i))$ for $i = 1, 2$.

<u>Remark</u>. It is clear that the condition (*) of the above proposition is satisfied by nearly all choices of points $P_1$ and $P_2$. Moreover, the existence and uniqueness of a quadric containing $C_P$ is immediate because $h^o(\mathbb{P}^3, \mathcal{O}(2)) = 10$ and by the Riemann-Roch theorem $h^o(C_P, \mathcal{O}(2)) = 9$ such that $h^o(C_P, I(2)) \geqq 1$ and the intersection of two different quadrics is of degree 4 and can not contain a curve of degree 5. Finally the quadric is irreducible because $C_P$ is not plane.

The only object of the following proof is therefore the construction of the exceptional point on the line joining $P_1$ and $P_2$.

<u>Proof</u>. Assume first that $2Q \not\equiv P_1 + P_2$. Then $Q \neq P_i$ because of the assumption (*) of the proposition. Similarly $R \neq P_i$. It follows that $H^o(C, L(-P_1 - P_2))t \cap H^o(C, L(-2R - Q))t_R^2 t_Q = 0$, because if $P_1 + P_2 + P_3 + P_4 + P_5 = 2R + Q + Q_1 + Q_2$, then $P_1 = Q_1$ and $P_2 = Q_2$, and consequently $H \equiv 2R + Q + P_1 + P_2$ which as a consequence of the equivalence $3Q + 2R \equiv H$ implies that $2Q \equiv P_1 + P_2$.

Let P be a point on the line joining $P_1$ and $P_2$, that is $H^o(C, L(-P_1 - P_2))t \subseteq V_P$. Then, by the above observation, $H^o(C, L(-2R - Q))t_R^2 t_Q \subsetneqq V_P$ and since $h^o(C, L(-2R - Q)) = 2$ we have that $\dim(H^o(C, L(-2R - Q)t_R^2 t_Q \cap V_P) = 1$. Consequently there are unique points $R_1$ and $R_2$ on C such that $st_R^2 t_Q \subseteq V_P$ where s is the product of non-zero sections in $H^o(C, \mathcal{O}(R_i))$ for i = 1, 2. As above one checks that $R_i \neq P_j$.

The following four sections of L

$$a_1 = t_Q t^2 \ , \ a_2 = t_Q^2 t_R t \ , \ a_3 = t_Q t_R^2 s \text{ and } a_4 = stt_R$$

clearly satisfy the equation $a_1 a_3 - a_2 a_4 = 0$ and thus define a quadric containing $C_P$. This quadric is non-singular if and only if the $a_i$'s are linearly independent. However $a_1$, $a_2$ and $a_3$ are linearly independent because we have shown that $R \neq P_i$ and we have $Q \neq R$ because equality would mean that $P_1 + P_2 = 3P_1 + 3P_2 - 2P_1 - 2P_2 \equiv H + R + Q - H = 2Q$. The only possible linear relation between the $a_i$'s is therefore of the form $a_4 = \sum_{i=1}^{3} \alpha_i a_i$. However, from $Q \neq P_i$ and $Q \neq R$ we conclude that a linear relation implies that Q is equal to say $R_1$, and then it follows from $2R + Q + R_1 + R_2 \equiv H \equiv 2R + 3Q$ that $Q = R_2$. Hence the quadric is singular exactly when $Q = R_1 = R_2$, that is, P is defined by the hyperplane

$$H^0(C, L(-P_1 - P_2))t + kt_Q^3 t_R^2 \subseteqq W .$$

Secondly we assume that $2Q \equiv P_1 + P_2$. Then $H \equiv 5Q$, $R = Q$ and $Q \neq P_i$. In this case $\dim(H^0(C, L(-2R-Q))t_R^2 t_Q \cap H^0(C, L(-P_1-P_2))t) = 1$ because $t_Q^3 t$, but not $t_Q^5$, is in the intersection. We have $\dim(V_P \cap H^0(C, L(-2Q))t_Q^2) \geqq 2$ so there are points $R_1$, $R_2$ and $R_3$ on C such that if r is the product of non-zero sections of $H^0(C, \mathcal{O}(R_i))$ for $i = 1, 2, 3$ then $rt_Q^2$ is in $V_P$ but is linearly independent from $t_Q^3 t$. The four sections

$$a_1 = t_Q^2 r \ , \quad a_2 = t_Q^3 t \ , \quad a_3 = t_Q t^2 \quad \text{and} \quad a_4 = tu$$

of L satisfy $a_1 a_3 - a_2 a_4 = 0$ and thus define a quadric containing C, which is singular if and only if the $a_i$'s are linearly dependent. The sections $a_1$, $a_2$ and $a_3$ are linearly independent. Indeed $Q \neq P_i$ so a linear dependence would imply say $R_1 = P_1$, $R_2 = P_2$ and $R_3 = Q$ which is impossible because $a_1$ and $a_2$ are chosen linearly independent. The only possible linear relation between the $a_i$'s is therefore of the form $a_4 = \sum_{i=1}^{3} \alpha_i a_i$. Such a relation clearly implies say $Q = R_3$ and since we can not have

$R_1 = P_1$ and $R_2 = P_2$ it follows that $\alpha_1 = 0$. We obtain a relation

$$r = \alpha_2 t_Q^3 + \alpha_3 t_Q t$$

and since $rt_Q^2$ and $t_Q^3 t$ are in $V_P$ it follows that $t_Q^5 \in V_P$. Consequently we have that

$$V_P = H^0(C, L(-P_1 - P_2))t + kt_Q^5 .$$

## § 8. A fundamental vanishing result.

Proposition. For each point $P \notin Sec(C)$ we have that

$$H^i(C_P, N_P^* \otimes L^2) = 0 \quad \text{for} \quad i = 0, 1 .$$

Remark 1. Since $N_P^* \otimes L^2$ is of degree zero it follows from the characterization of the underlying set of the hypersurfaces $Y_M$ given in part (i) of the theorem of section 1, that the above proposition is equivalent to the assertion that $Y_M \subseteqq Sec(C)$ when $M = \mathcal{O}_C$. However, we have already shown the equality $Y_M = Sec(C)$ for $M = \mathcal{O}_C$, as a part of (ii) of the theorem of section 1. Consequently we have already proved the above proposition. Because of the fundamental character of the proposition we offer however below two more proofs of it. They can, as observed above, also be considered as alternative proofs of the result that $Y_M \subseteqq Sec(C)$ when $M = \mathcal{O}_C$.

Remark 2. Somewhat surprisingly we shall need both during the first proof of the proposition and in the proof of the corollary below, the observation that the tri-secant scroll of $C_P$, when $P \notin Sec(C)$, is of degree at least 4. To see this we map $C_P$ into $\mathbb{P}^2$ by a projection with center on a point Q of $C_P$.

We then obtain a plane curve of geometric genus 1 and degree 4 which by the genus formula has two different double points. Hence there are exactly two tri-secants passing through Q. Choose a secant which is not contained in the tri-secant scroll and which does not pass through the at most finite set of points through which there pass infinitely many tri-secants. Then the secant meets at most a finite number of tri-secants, but at least the two pairs of tri-secants that pass through the points of intersection of the secant with $C_P$.

Proof. Method 1.

We assume in this proof that the ground field k is of characteristic 0, and without loss of generality we assume that $k = \mathbb{C}$.

The curve $C_P \cong C$ is locally a complete intersection and has a trivial dualizing sheaf. Hence there is a rank 2 vector bundle on $\mathbb{P}^3$, which we write E(2), and a section of this bundle whose zero scheme is $C_P$ (see [9] where this connection between codimension 2 subvarieties and zeros of sections of rank 2-bundles was first described. For an explicit statement of the result used above see [8] Theorem 5.1.1 pp. 93-94). It follows (see [8] loc. cit) that $c_1(E) = 0$, $c_2(E) = 1$ and that we have an exact sequence

$$(*) \qquad 0 \to \mathcal{O}(-4) \to E(-2) \to I \to 0$$

where I is the ideal in $\mathcal{O}_{\mathbb{P}^3}$ defining $C_P$. Now $C_P$ is not contained in a quadric because every tri-secant to $C_P$ must be contained in such a quadric and as observed in Remark 2 above the tri-secant scroll is of degree at least 4. Hence $0 = H^0(\mathbb{P}^3, I(2)) = H^0(\mathbb{P}^3, E)$, that is E is stable (For the notion of stability see [8]

Chapter 1.2 especially Lemma 1.2.5). Stable bundles with first and second Chern classes 0 and 1 are called null-correlation bundles and have been classified by W. Barth [2]. They are all given by an exact sequence

$$0 \to \mathcal{O}_{\mathbb{P}^3}(-1) \to \Omega^1_{\mathbb{P}^3}(1) \to E \to 0$$

(see [8], Lemma 4.3.2 pp. 362-363). From this it follows that $H^1(\mathbb{P}^3, E(-2)) = H^1(\mathbb{P}^3, E) = 0$ and that any two null-correlation bundles are projectively equivalent. We shall also need that $H^1(\mathbb{P}^3, E \otimes E(-2)) = 0$. If E is a real instanton (see [8] pp. 371-372 for a definition) this is the Atiyah-Hitchin-Drinfeld-Manin-vanishing theorem (see [8] p. 372). Since every null-correlation bundle is projectively equivalent to one being a real instanton it follows for any null-correlation bundle.

Now from the exact sequence (*) it follows that $N^*_P \otimes L^2 \cong E|C_P$ (we have $E \cong E^*$ because E is of rank 2 and $\overset{2}{\Lambda} E = \mathcal{O}_{\mathbb{P}^3}$). Hence tensorizing the sequence (*) by E and passing to cohomology we get an exact sequence

$$H^1(\mathbb{P}^3, E \otimes E(-2)) \to H^1(\mathbb{P}^3, E \otimes I) \to H^2(\mathbb{P}^3, E(-4)) .$$

By duality $H^2(\mathbb{P}^3, E(-4)) = H^1(\mathbb{P}^3, E) = 0$, and we conclude that $H^1(\mathbb{P}^3, E \otimes I) = 0$. From the exact sequence

$$0 \to I \otimes E \to E \to N^*_P \otimes L^2 \to 0$$

we now conclude that $H^0(\mathbb{P}^3, N^*_P \otimes L^2) = 0$.

Method 2. Let J denote the ideal defining C in $\mathbb{P}^4$. Then $H^0(\mathbb{P}^3, J^2(2)) = 0$ because a quadric that contains C doubly must contain every secant to C. However, we have seen in the proof of assertion (ii) of the theorem that the secant variety is of degree 5. From the sequence

$$0 \to J^2(2) \to J(2) \to N^* \otimes L \to 0$$

we therefore conclude that the resulting map $H^o(\mathbb{P}^3, J(2)) \to H^o(C, N^* \otimes L)$ is a surjection, that is every section of $H^o(C, N^* \otimes L)$ comes from a quadric containing C. From the cohomology of the diagram

$$\begin{array}{ccccccccc} 0 & \longrightarrow & N_P^* \otimes L & \longrightarrow & V_P & \longrightarrow & P_C^1(L) & \longrightarrow & 0 \\ & & \downarrow & & \downarrow & & \| & & \\ 0 & \longrightarrow & N^* \otimes L & \longrightarrow & W & \longrightarrow & P_C^1(L) & \longrightarrow & 0 \end{array}$$

we see that it is sufficient to prove that every quadratic form F in variables $y_o, \ldots, y_4$ such that $\sum_{i=o}^{4} \frac{\partial F}{\partial y_i} p_i = 0$, with $P = (p_o; \ldots; p_4)$, is identically zero. However, since F is a quadric we have that

$$\sum_{i=o}^{4} \frac{\partial F}{\partial y_i} p_i = \sum_{i=o}^{4} \frac{\partial F}{\partial y_i} (P) y_i \ .$$

Hence $\frac{\partial F}{\partial y_i}(P) = 0$ for $i = 0, \ldots, 4$ so that C is contained in a quadric cone with vertex at P. Then $C_P$ is contained in a quadric and we have already shown in the proof of the proposition by the first method that this is impossible.

Corollary. We have $h^o(\mathbb{P}^3, I^2(5)) = 1$. That is, there is a unique degree five surface containing $C_P$ doubly. Moreover this surface is the tri-secant scroll.

Proof. From the lemma of section 1 we conclude that $h^o(C_P, N_P^*(5)) = h^o(C_P, M(3) \oplus M^{-1}(3)) = 30$, where $M \in Pic_o(C)$ is the bundle such that $h^o(C_P, N_P^*(2) \otimes M) \neq 0$. Moreover, from the sequence

$$0 \to I(5) \to \mathcal{O}_{\mathbb{P}^3}(5) \to \mathcal{O}_C(5) \to 0$$

we see that $H^i(\mathbb{P}^3, I(5)) = 0$ for $i > 1$ and that

$$h^o(\mathbb{P}^3, I(5)) = h^o(\mathbb{P}^3, \mathcal{O}_P(5)) - h^o(C, \mathcal{O}_C(5)) + h^1(\mathbb{P}^3, I(5)) \geqq 56 - 25 = 31 .$$

Consequently, we obtain, using the sequence

$$0 \to H^o(\mathbb{P}^3, I^2(5)) \to H^o(\mathbb{P}^3, I(5)) \to H^o(C_P, N_P(5)) \to$$

that $h^o(\mathbb{P}^3, I^2(5)) \geqq 1$. That is there is at least one quintic containing the curve $C_P$ doubly. However, every tri-secant to $C_P$ must be contained in such a quintic and consequently it must have the tri-secant scroll as a component. By Remark 2 above the tri-secant scroll is of degree 4 and since the curve is not plane the quintic must coincide with the tri-secant scroll.

## § 9. Equations for the tri-secant scroll of 4.

We shall give the equations for the tri-secant scroll of a particular projection of the curve defined by the equations $G_1, \ldots, G_5$ of section 5.

The open subset $\mathbb{A}^4$ of the grassmannian of lines in $\mathbb{P}^3$ that do not intersect the linear space $z_o = z_2 = 0$ has coordinates

$$\begin{vmatrix} 1, & x_{1,1}, & 0, & x_{1,2} \\ 0, & x_{2,1}, & 1, & x_{2,2} \end{vmatrix}$$

The line that intersect the curve $(1; a^2; b; ab)$ is determined by the vanishing of the maximal minors of the matrix obtained by adjoining the row $(1, a^2, b, ab)$ to this matrix, that is by $x_{2,1}b - a^2 + x_{1,1} = 0$ and $ab - bx_{2,2} - x_{1,2} = 0$. Elimination of b from these equations and the equation $b^2 = a^3 - (1 + \lambda)a^2 + \lambda a$

give two equations

$$f(a) = a^3 - x_{2,2}a^2 - x_{1,1}a + x_{2,2}x_{1,1} - x_{2,1}x_{1,2}$$

$$g(a) = a^4 - x_{2,1}^2 a^3 + (-2x_{1,1} + (1+\lambda)x_{2,1}^2)a^2 - x_{2,1}^2\lambda a + x_{1,1}^2 .$$

Division gives

$$g(a) = (a + (x_{2,2} - x_{2,1}^2))\, f(a) + (-x_{1,1} + (1+\lambda)x_{2,1}^2 + x_{2,2}(x_{2,2} - x_{2,1}^2))a^2$$

$$+ (-\lambda x_{2,1}^2 + x_{2,1}x_{1,2} - x_{2,1}^2 x_{1,1})a + x_{1,1}^2 - (x_{2,2} - x_{2,1}^2)(x_{1,1}x_{2,2} - x_{2,1}x_{1,2}) .$$

A line in $\mathbb{P}^3$ intersect the curve in three points if and only if f and g have three roots in common, that is if f divides g. After a slight rearrangement of the coefficients in the above division we obtain 3 equations

$$-x_{1,1} + (1+\lambda)x_{2,1}^2 + x_{2,2}(x_{2,2} - x_{2,1}^2) = 0 ,$$

$$-\lambda x_{2,1}^2 + x_{2,1}x_{1,2} - x_{2,1}^2 x_{1,1} = 0 \quad \text{and}$$

$$(1+\lambda)x_{1,1}x_{2,1}^2 + (x_{2,2} - x_{2,1}^2)x_{2,1}x_{1,2} = 0 .$$

The condition $x_{2,1} = 0$ implies that $x_{1,1} = x_{2,2}^2$ and thus $f(a) = (a - x_{2,2})^2(a + x_{2,2})$, $g(a) = (a - x_{2,2})^2(a + x_{2,2})^2$ corresponding to a solution with a tangent tri-secant. Disregarding this solution we get 3 equations.

$$-x_{1,1} + (1+\lambda)x_{2,1}^2 + x_{2,2}(x_{2,2} - x_{2,1}^2) = 0 ,$$

$$(*) \qquad -\lambda x_{2,1} + x_{1,2} - x_{2,1}x_{1,1} = 0 \quad \text{and}$$

$$(1+\lambda)x_{1,1}x_{2,1} + (x_{2,2} - x_{2,1}^2)x_{1,2} = 0 .$$

As above a point (1; x; y; z) in $\mathbb{P}^3$ lies on a tri-secant if

$$x_{2,1}y - x + x_{1,1} = 0$$

$$(**) \qquad z - yx_{2,2} - x_{1,2} = 0 .$$

The five equations (*) and (**) determine together the points (1; x; y; z) on the tri-secant scroll. Using (**) to eliminate $x_{1,1}$ and $x_{1,2}$ we get after a slight simplification, the following three equations,

$$x_{2,1}y - x + (1 + \lambda)x_{2,1}^2 + x_{2,2}(x_{2,2} - x_{2,1}^2) = 0$$

$$-y(x_{2,2} - x_{2,1}^2) + z - \lambda x_{2,1} - x_{2,1}x = 0$$

$$((1 + \lambda)x + y^2)x_{2,1} + z(x_{2,2} - x_{2,1}^2) - yx = 0 \ .$$

We eliminate $x_{2,2}$ using the second equation and obtain

$$(\lambda + x)yx_{2,1}^3 - ((1+\lambda)y^2 + zy + (\lambda + x)^2)x_{2,1}^2 - (y^3 - 2(\lambda + x)z)x_{2,1} - z^2 + xy^2 = 0$$

and

$$((1 + \lambda)xy + y^3 - (\lambda + x)z)x_{2,1} + z^2 - y^2x = 0 \ .$$

Elimination of $x_{2,1}$ from these equations gives(after division by $y^2$ and $xy^2 - z^2$) the inhomogenous form of the equation for the tri-secant scroll

$$(\lambda^2 + \lambda + 1)x^2y^3 - 2x^2yz^2 - (1+\lambda)x^4y + (1+\lambda)y^3z^2 + y^2z^3 + \lambda y^4z + x^4z$$
$$+ (\lambda^2 - \lambda + 1)xyz^2 + (\lambda^3 + \lambda^2 + \lambda + 1)x^3y + (1+\lambda)\lambda xy^2z - (\lambda^2 - \lambda + 1)x^3z - \lambda^2xy^3$$
$$- (1 + \lambda)\lambda^2x^2y - (1 + \lambda)\lambda z^3 - (\lambda^3 - \lambda^2 + \lambda)x^2z - \lambda^2yz^2 + \lambda^3xz \ .$$

Differentiation of this polynomial with respect to x, y and z and substitution of the coordinates of (1; $a^2$; b; ab) shows the curve is contained in the singular locus of the tri-secant scroll.

## § 10. The tangent scroll to C in $\mathbb{P}^4$.

In order to prove that the degree 5 hypersurfaces $Y_M$ are irreductible it suffices to prove that they contain a surface which is not contained in a hypersurface of degree 4. The tangent scroll Tan(C) to C in $\mathbb{P}^4$ has the latter property. Indeed, G. Horrocks and D. Mumford [5] have proved that Tan(C) is the scheme of zeroes of a section of a rank 2 vector bundle $F$ on $\mathbb{P}^4$ ([5] fact (a) p.79), that is there is an exact sequence

$$0 \to G \to F^{\vee} \to I \to 0$$

where I is the ideal defining Tan(C) in $\mathbb{P}^4$. We see that $\overset{2}{\Lambda} F^{\vee} = G$ and that $G = \mathcal{O}(-5)$ and consequently there are equalities $H^o(\mathbb{P}^4, I(4)) = H^o(\mathbb{P}^4, F^{\vee}(4)) = H^o(\mathbb{P}^4, F(-1))$ and by the calculations of Horrocks and Mumford ([5] table p.24) we have $H^o(\mathbb{P}^4, F(-1)) = 0$. Hence Tan(C) is not contained in a hypersurface of degree 4.

It remains to prove that $\text{Tan}(C) \subseteq Y_M$. To this end we let $P \in \text{Tan}(C)$. Then the curve $C_P$ has a simple cusp R and the image of the map $\Omega^1_{\mathbb{P}^3}|C_P \to \Omega^1_C \cong \mathcal{O}_C$ of differentials is $\mathcal{O}(-R)$. Consequently there is an exact sequence

$$0 \to F_P \to \Omega^1_{\mathbb{P}^3}(1)|C_P \to L(-R) \to 0$$

on C. We obtain that $\overset{2}{\Lambda}(F_P \otimes L) = \mathcal{O}(R)$.

If $C_P$ lies on a smooth quadric it gives rise to a section of $F_P \otimes L$ without zeroes and (reasoning as in the proof of Proposition 1 of section 7) we obtain an exact sequence

$$0 \to \mathcal{O}_C \to F_P \otimes L \to \mathcal{O}(R) \to 0 \ .$$

We conclude that $H^o(C, F_P \otimes L \otimes M) \neq 0$ for all $M \in \text{Pic}_o(C)$, that is, $P \in Y_M$.

If $C_P$ lies on a quadric cone with vertex Q we obtain (again reasoning as in the proof of Proposition 1 of section 7) an exact sequence

$$0 \to \mathcal{O}(Q) \to F_P \otimes L \to \mathcal{O}(R - Q) \to 0 .$$

This sequence splits because $H^1(C, \mathcal{O}(2Q - R)) = 0$. In particular we have that $H^0(C, F_P \otimes L \otimes M) \neq 0$ for all $M \in \mathrm{Pic}_0(C)$ so that also in this case $P \in Y_M$.

Note that when $M = Q - R$ then we have that $h^0(C, F_P \otimes L \otimes M) = 2$. Hence, for this choice of M, the $4 \times 4$-minors of $\delta_M$ vanishes at P.

We also note that the degree of Tan(C) is 10. Indeed, since the quotient $W_C \to P^1(L)$ represents at each point P of C the tangent to C at P, we have that Tan(C) is the projective bundle $\mathbb{P}(P^1(L))$ over C. The degree of this bundle in $\mathbb{P}(W)$ is 10 because $c_1(P^1(L)) = 10$ and $c_2(P^1(L)) = 0$.

## Bibliography.


[1] M.F. Atiyah. Vector bundles over an elliptic curve. Proc. London Math. Soc. 7 (1957), 414-452.

[2] W. Barth. Some properties of stable rank 2 bundles on $\mathbb{P}_M$. Math. Ann. 226 (1972), 125-150.

[3] D.A. Buchsbaum & D. Eisenbud. Algebra structures for finite free resolutions and some structure theorems for ideals of codimension 3. Amer. J. Math. 99 (1977), 447-485.

[4] F. Ghione. Quelques exemples de courbes de $\mathbb{P}^3$ dont le fibré normal ne se decompose pas. C.R. Acad. Sc. Paris (1977), 375-377.

[5] G. Horrocks and D. Mumford. A rank 2 vector bundle on $\mathbb{P}^4$ with 15.000 symmetries. Topology 12 (1973), 63-81.

[6] H. Kleppe & D. Laksov. The algebraic structure and deformation of pfaffian schemes. J. Algebra 64 (1980), 167-189.

[7] D. Montesano. Su la curva gobba di 5° ordine e di genere 1. Rend. Acc. Napoli 2 (1888), 181-188.

[8] C. Okonek, M. Schneider & H. Spindler. Vector bundles on complex projective spaces. Birkäuser 1980.

[9] J.P. Serre. Sur les modules projectifs. Sem. Dubreil-Pisot 1960/61, exposé 2. Paris (1963).

[10] A. Van de Ven. Le fibré normal d'une courbe dans $\mathbb{P}_3$, ne se compose pas toujours. C.R. Acad. Sc. Paris (1979), 111-113.

[11] E. Weyr. Über Raumkurven fünfter Ordnung vom Geschlecte Eins, I, II & III. Sitzungsberichte der Kaiserlichen Akademie der Wissenschaften, Wien. 90 (1884), 206-225, 92 (1885) 498-523, 97 (1888), 592-617.

# SOME RECENT RESULTS ON CONJUGATE FUNCTIONS IN THE UNIT DISK

Matts Essén
and
Daniel F.Shea

1. Introduction. This talk is a survey (essentially without proofs) of some recent results on conjugate functions in the unit disk. Many of these questions have been studied both by probability methods and by methods from classical function theory. Due to the bias of the authors, the emphasis will be on results with connections with classical function theory. For conjugate functions and probability, the main source is the work of B.Davis (further references will be given below).

Let $U$ be the unit disk in the complex plane and let $T = \partial U$ be the unit circle. For a subset $E \subset T$, $|E|$ denotes the Lebesgue measure of $E$. For functions $f$ on $T$, we write

$$||f|| = (\int_T |f(e^{i\varphi})|^p dm)^{1/p} ,$$

where $dm = d\varphi/(2\pi)$. For basic facts on $H^p$-theory, we refer the reader to Duren [7].

If $f \in L^1(T)$ is real-valued, we define an analytic function $F$ in $U$ by

$$F(z) = \int_T (e^{i\varphi}+z)(e^{i\varphi}-z)^{-1} f(e^{i\varphi}) dm .$$

A conjugate function $\tilde{f}$ is defined a.e. by

$$\tilde{f}(e^{i\varphi}) = \lim_{r \to 1-} \operatorname{Im} F(re^{i\varphi}) .$$

We note that $F(0)$ is real. Similarly, if $M(T)$ is the class of all real-valued Borel measures on $T$ and $\mu \in M(T)$, we define

$$F(z) = \int_T (e^{i\varphi}+z)(e^{i\varphi}-z)^{-1} d\mu(e^{i\varphi}) ,$$

$$\tilde{\mu}(e^{i\theta}) = \lim_{r \to 1-} F_\mu(re^{i\theta}) .$$

2. Some sharp inequalities for conjugate functions.

I. Kolmogorov's weak 1-1 inequality.

We define the set $E_t(\tilde{f}) \subset T$ by $E_t(\tilde{f}) = \{z \in T : |\tilde{f}(z)| \geq t\}$. Long ago, Kolmogorov proved that there exists an absolute constant $C$ such that

(1a) $\qquad |E_t(\tilde{f})| \leq C\, t^{-1} \|f\|_1 , \quad$ for all $t \in (0,\infty)$ .

If $E(\tilde{f}) = E_1(\tilde{f})$, we shall also consider the special case

(1b) $\qquad |E(\tilde{f})| \leq C \|f\|_1$ .

It is clear that (1a) and (1b) are equivalent.

In 1974 B.Davis found the best constant $C$ (cf. [4]): If $G(z) = (2/\pi)\log(((1+z)/(1-z))$, (1a) and (1b) are true with $C = 2\pi \|\mathrm{Re}\, G(e^{i\theta})\|_1^{-1}$. Furthermore, there is equality in (1b) with this constant $C$ if $f = \mathrm{Re}\, G$ on $T$. We note that $G$ maps $U$ conformally onto the strip $\{w \in \underline{C} : |\mathrm{Im}\, w| < 1\}$ with $G(0) = 0$.

In 1978 A.Baernstein "translated" the Brownian motion argument of Davis into a very elegant function-theoretic proof (cf. Section 2 in [2]).

II. Kolmogorov's $L^p$-norm inequality, $o < p < 1$ .

Let $\mu \in M(T)$ and let $\|\mu\|$ be the total variation of $\mu$ . In 1976, B.Davis improved another result of Kolmogorov: He proved that

$$(2) \qquad \|\tilde{\mu}\|_p \leq C_p \|\mu\| , \qquad o < p < 1 ,$$

where $C_p = (\int_T |\sin \varphi|^{-p} dm)^{1/p}$ is the best constant. There is equality in (2) for the singular measure $\nu = (\delta_1 - \delta_{-1})/2$ , where $\delta_\xi \in M(T)$ is the unit mass concentrated at $\xi \in T$ . We note that $F\ (z) = 2z(1-z^2)^{-1}$ maps $U$ conformally onto $\underline{C} \setminus \{iv : v \in \underline{R} , |v| \geq 1\}$ with $F_\nu(O) = O$ .

In 1978, A.Baernstein gave an alternative, function-theoretic proof, deducing (2) with the best constant from a more general result (cf. Section 3 in [2]). According to Baernstein, there is, in contrast to the situation in case I, no obvious connection between the two ways of proving inequality (2).

III. A rearrangement inequality.

Let $f \in L^1(T)$ be real-valued and let $F$ and $\tilde{f}$ be defined as above.

Let $g$ be the symmetric, decreasing rearrangement of $f$ on $(-\pi,\pi)$ , i.e., the function $g \in L^1(T)$ which satisfies

i) $\qquad |\{z \in T : f(z) > t\}| = |\{z \in T : g(z) > t\}|$, for all $t \in \underline{R}$ ,

ii) $\qquad g(e^{i\Theta}) = g(e^{-i\Theta})$ , $\Theta \in (O,\pi)$ .

To $g$ , we associate in the usual way a function $G$ analytic in and a conjugate function $g$. Baernstein proved (cf.[2], Theorem 5) that if $G \in H^p$ for some $p \in (O,2]$, then also $F \in H^p$ and

(3a) $$||F||_p \leq ||G||_p , \qquad 0 \leq p \leq 2 ,$$
$$||\tilde{f}||_p \leq ||\tilde{g}||_p , \qquad 1 \leq p \leq 2 .$$

Next, Essén and Shea noticed that Baernstein's arguments from [2] work also in the case $p > 2$ . However, the inequality sign is reversed (cf.Essén and Shea [8],Section 6, and Baernstein [3],p.412). If $F \in H^p$ for some $p > 2$ , then $G \in H^p$ and

$$||F||_p \geq ||G||_p , \qquad p > 2 ,$$
$$||\tilde{f}||_p \geq ||\tilde{g}||_p , \qquad p > 2 .$$

The extremal function $G$ maps $U$ conformally onto a Steiner-symmetric domain, i.e., a domain of the form $\{w = u+iv : u, v \in \underline{R} , |v| < L(u)\}$ , where $L$ is a non-negative function on $\underline{R}$ . We note that $F(0) = G(0)$ (cf. Baernstein [2],Section 6).

An important tool in the function-theoretic proofs in cases II and III is a kind of maximal function introduced by Baernstein, the *-function (cf. Baernstein [1].

Above we have described three classes of inequalities, where sharp constants are known. For what analytic functions $F$ is there equality in each of these three cases ? This problem has been treated by Essén and Shea in [8], and the answer is as follows:

Case I. Let $G$ be the extremal analytic function defined above. If $||f||_1 \neq 0$ , there is equality in (1b) if and only if $F(z) = G(\omega(z))$, $z \in U$, where $\omega$ is an inner function in $U$ such that $\omega(0) = 0$ (cf. Duren [7],p.24, for the definition of inner function).

Case III. Let $f \in L^1(T)$, $\|f\|_1 \neq 0$, be such that there is equality in one of the inequalities (3a) - (3d) for some relevant $p \neq 2$. Let $G$ be the associated extremal analytic function. This can be true if and only if $F(z) = G(\omega(z))$, $z \in U$, where $\omega$ is an inner function in $U$ such that $\omega(0) = 0$.

Case II. Assume that $\mu \in M(T)$ has $\|\mu\| = 1$ and that there is equality in (2) for some $p \in (0,1)$. Then $\mu = \mu_1 - \mu_2$, where $\mu_1$ and $\mu_2$ are positive measures singular with respect to Lebesgue measure such that $\mu_1(T) = \mu_2(T) = \frac{1}{2}$ and

(4) $$\tilde{\mu}(e^{i\theta}) = \tilde{\nu}(\omega(e^{i\theta}))$$

holds a.e. for some inner function $\omega(z)$ in $U$ with $\omega(0) = 0$. (We recall that $= (\delta_1 - \delta_{-1})\frac{1}{2}$).

Conversely, given any such $\omega(z)$, there exist positive measures $\mu_1$ and $\mu_2$, each of mass $\frac{1}{2}$, which are singular with respect to Lebesgue measure and mutually singular, such that if $\mu = \mu_1 - \mu_2$ then (4) is valid a.e., and equality holds in (2).

In Case I, the proof depends on elementary properties of subharmonic functions. In Cases II and III, we use the *-function technique of Baernstein which was mentioned above.

## 3. A Zygmund-type problem.

Zygmund has proved the following well-known results (cf. Duren [7], Section 4.3):

(5a) If $F$ is analytic in $U$ and $\operatorname{Re} F \in h \log^+ h$, then $F \in H^1$.

(5b) If $F \in H^1$ and $\operatorname{Re} F \geq 0$, then $\operatorname{Re} F \in h \log^+ h$.

(We say that $\operatorname{Re} F \in h \log^+ h$ if we have

$$\sup_{0<r<1} \int_T |\operatorname{Re} F(re^{i\varphi})| \log^+ |\operatorname{Re} F(re^{i\varphi})| \, dm < \infty \text{ .)}$$

In (5b) a condition on the value-distribution of $F$ gives a certain integrability of $\operatorname{Re} F$. We want to discuss more general results of this type. All this is joint work of M.Essén and D.F.She

We begin with some terminology. Let $F$ belong to the Nevanlinna class in $U$, i.e., $F$ is analytic in $U$ and

$$\sup_{0<r<1} \int_T \log^+ |F(re^{i\varphi})| dm < \infty .$$

Let $n(r,w)$ be the number of roots of $F(z) = w$ in $\{|z| \leq r\}$ and define $N(r,w) = \int_0^r n(t,w)\, dt/t$ , $w \notin F(0)$ .

It is well-known that $N(1,w) = \lim_{r \to > 1-} N(r,w)$ exists and is uniformly bounded except near $F(0)$ . The upper regularization $N(w)$ of $N(1,w)$ , defined by $N(w) = \limsup_{\xi \to > w} N(1,\xi)$ , is subharmonic in $\underline{C} \setminus \{F(0)\}$ . The function $N(w) + \log|w - F(0)|$ can be defined to be subharmonic in $\underline{C}$ . (For further information and references, we refer to Essén and Shea [8], Section 4.)

Using the subharmonicty of $N(w)$ , we can prove the following (unpublished) result.

Theorem 1. Let $F \in H^1$ . The following are equivalent:

(6) $$\int_{-\infty}^{\infty} N(u+iv) \log^+|v|\, dv < \infty \quad \text{for some } u \in \underline{R} .$$

(7) $\quad \mathrm{Re}\, F \in h \log^+ h$ .

Remark. If the integral in (6) converges for some $u \in \underline{R}$ , it converges for all $u \in \underline{R}$ .

It is easy to see that (5b) is a consequence of Theorem 1.

If $\mathrm{Re}\, F \geq 0$ and $u < 0$ , then $N(u+iv) = 0$ for all $v \in \underline{R}$ , and we see that (6) and thus also (7) is true.

Another consequence is the following one. If $f \in L^1(T)$ and $F, \tilde{f}, G$ and $\tilde{g}$ are defined as in Case III above, it follows from (3b) that we have

(8) $$\|\tilde{f}\|_1 \leq \|\tilde{g}\|_1 .$$

Thus (8) gives us a sufficient condition for $F$ to be in $H^1$ if $f \in L^1(T)$ is given; this will be true if $\|\tilde{g}\|_1 < \infty$. It turns out, however, that this is equivalent to Zygmund's criterion (5a), as follows from

Corollary 1. $\tilde{g} \in L^1(T) \iff g \in h \log^+ h$.

Proof. We have already mentioned that $G$ maps $U$ onto a Steiner-symmetric domain. If there exists $u_o$ such that $L(u_o) < \infty$, condition (6) will hold for $u = u_o$ and Theorem 1 implies that $g \in h \log^+ h$ which is equivalent to $f \in h \log^+ h$ ($g$ is a measurepreserving rearrangement of $f$!). But such a number $u_o$ exists, because if $L(u) = \infty$ for all $u \in \underline{R}$, $G$ would map $U$ onto the complex plane $\underline{C}$ which is known to be impossible. The converse statement is simply Zygmund's statement (5a). This completes the proof of Corollary 1.

Remark. When I talked about this result in Uppsala in September of 1979, L. Carleson showed me a simple, real-variable proof of Corollary 1.

The success of the complex-variable proof of Corollary 1 depended on the fact that $G(z)$ did not assume any value on a set of the form $\{w = u_o + iv : |v| \geq c > 0\}$. Assume that we have a function $F \in H^1$ which omits all values on two curves going out to infinity. When does it follow that $R \in h \log^+ h$ ?

Corollary 2. Let $F \in H^1$ and assume that $N(w)$ vanishes on two curves $\Gamma_1$ and $\Gamma_2$ in the upper and the lower half-plane, respectively, which have the following properties: There exists constants $c > 0$ and $\alpha > 1$, such that

$$\Gamma_1 \subset \{(u,v) : v \geq c|u|^\alpha\}, \quad \Gamma_2 \subset \{(u,v) : v \leq -c|u|^\alpha\}.$$

Both curves go out to infinity. Then $\operatorname{Re} F \in h \log^+ h$.

In the proof, we use estimates of harmonic measure to get (6). The result then follows from Theorem 1. The method breaks down if $\alpha = 1$.

We have also studied problems of the following type: Assume that $F \in H^1$ and that $N(w)$ vanishes on the imaginary axis except on a certain subset $E$. We wish to conclude that $\operatorname{Re} F \in h \log^+ h$. How large may the set $E$ be?

4. <u>More rearrangements</u>.

From Baernstein's result (8), we see that if $g \in L^1(T)$ is symmetric on $(-\pi,\pi)$ and nonincreasing on $(0,\pi)$, if $f$ is a measure-preserving rearrangement of $g$ and $G$ and $F$ are the corresponding analytic functions in $U$, then $F \in H^1$ if $g \in L^1(T)$. B.Davis has asked the following question: Let $f \in L^1(T)$ be given. When does there exist a rearrangement of $f$ in $\operatorname{Re} H^1$?
To state Davis' answer, we need the decreasing rearrangement $f_d$ of $f$ defined by

(i) $|\{z \in T : f_d(z) > t\}| = |\{z \in T : f(z) > t\}|$, for all $t \in \underline{R}$,

ii) $f_d(e^{i\theta}) \geq f_d(e^{i\varphi})$, $o \leq \theta < \varphi < 2\pi$.

Let $M(\theta) = \int_{-\theta}^{\theta} f_d(e^{i\varphi})\,d\varphi$, $\theta \in (0,\pi)$. To $f_d$, we associate in the usual way a function $F_d$ which is analytic in $U$. The following theorem characterizes the distributions of functions in $\operatorname{Re} H^1$ (cf. Davis [6], Theorem 1.1).

<u>Theorem 2</u>. <u>There is a rearrangement of $f$ in</u> $\operatorname{Re} H^1$ <u>if and only if</u> $\int_0^{\pi} |M(\theta)|\,d\theta/\theta < \infty$, <u>and in this case</u> $f_d \in \operatorname{Re} H^1$. <u>There are absolute positive constants $C$ and $c$ such that</u>

$$C\,\|F\|_1 \geq \int_0^{\pi} |M(\theta)|\,d\theta/\theta \geq c\,\|F_d\|_1 - \|f\|_1 .$$

*Remark*. In this special case when $f$ is non-negative, Theorem 2 is equivalent to Zygmund's result (5a): this is so because the integral above will be convergent if and only if $f \in L \log^+ L$. In the impressive paper [6], Davis studies also other problems on rearrangements and conjugate functions. His proofs use probability theory.

## References:

[1] Baernstein,A., Integral means, univalent functions and circular symmetrization, Acta Math. 133 (1974), 133-169.

[2] Baernstein,A., Some sharp inequalities for conjugate functions, Indiana Univ. Math. Journ. 27 (1978), 833-852.

[3] Baernstein,A., Some sharp inequalities for conjugate functions, Proc. of Symposia in Pure Mathematics XXXV, part 1, Amer.Math.Soc. 1979, 409-416.

[4] Davis,B., On the weak type (1.1) inequality for conjugate functions, Proc.Amer.Math.Soc. 44 (1974), 307-311.

[5] Davis,B., On Kolmogorov's inequalities $\|\tilde{f}\|_p < C_p \|f\|_1$, $0 < P < 1$, Trans.Amer.Math.Soc. 222 (1976), 179-192.

[6] Davis,B., Hardy spaces and rearrangements, Proc. of the Conf. on "Aspects of Contemporary Complex Analysis", Durham, England 1979. To appear.

[7] Duren,P., Theory of $H^p$ - spaces, Academic Press, New York - London, 1970.

[8] Essén,M. and Shea,D.F., On some questions of uniqueness in the theory of symmetrization, Ann.Acad.Scient.Fenn., Series A.I.Math., 4 (1978/1979), 311-340.

Matts Essén
Department of Mathematics
Royal Institute of Technology
S - 10044 Stockholm
Sweden

Daniel F. Shea
Department of Mathematics
University of Wisconsin
Madison, Wisconsin 53706
U S A

# THE GEOMETRIC AND TOPOLOGICAL SIGNIFICANCE OF CURVATURE AND DIAMETER

Karsten Grove

## Introduction.

A major part of Global Riemannian Geometry is concerned with the study of relations between geometry and topology of n-dimensional $(n \geqq 2)$ Riemannian manifolds $(M^n, g)$. Here the term "topology" refers to the algebraic - and differential topology of the underlying smooth manifold $M^n$, whereas the term "geometry" refers to the structures on $M^n$ which are derived entirely from the Riemannian metric $g$. Thus a typical result in this area will on the one hand be concerned with restrictions on say curvature, diameter, volume etc. and on the other hand with restriction on the (differential-) topological type and/or with (algebraic-) topological invariants.

So far most efforts have been confined to the study of Riemannian manifolds with non-negative or non-positive (sectional, Ricci or Scalar) curvature. In the case of strictly positive (or negative) sectional curvature the ratio between lower and upper bounds for (the absolute value of) the curvature is an important geometrical invariant called the "pinching". It has turned out, however, that the most celebrated "pinching"-theorems by Rauch, Berger and Klingenberg [B], [K] can be generalized and extended by restricting the diameter and lower bound for the curvature instead of restricting the "pinching" [GS], [GG]. Furthermore this point of view is of interest in the general case of non-signed curvature as a recent striking result of Gromov [G] shows. It is the purpose of this lecture to discuss a circle of classical and new results all of which can be interpreted in terms of one simple geometrical invariant involving only the diameter and lower bound for the sectional curvature.

Basics.

Let $M^n$ denote a smooth connected n-dimensional $(n \geq 2)$ manifold without boundary and let $g$ be a Riemannian metric on $M^n$. The length of a (piecewise) smooth curve $c: [0,1] \to M^n$ joining two points $p = c(0)$ and $q = c(1)$ on $M^n$ is given by

$$L(c) = \int_0^1 g(\dot{c}(t), \dot{c}(t))^{\frac{1}{2}} \, dt$$

and the distance between $p$ and $q$ is

$$d(p,q) = \inf\left\{L(c) \mid c(0) = p \wedge c(1) = q\right\} .$$

Then $d: M^n \times M^n \to \mathbb{R}$ is a metric on $M^n$ and the geodesics on $M^n$ are characterized locally by being distance minimizing and globally as stationary curves for the length function. The Hopf-Rinov theorem states that $M^n$ is complete as a metric space if and only if all geodesics admit an extension to the whole real line.

Let $C^\infty(TM)$ denote the smooth vectorfields on $M^n$. The Levi-Civita connection $\nabla$ on $M^n$ is determined by the requirements

$$\text{(i)} \quad \nabla_X Y - \nabla_Y X - [X,Y] \equiv 0 , \qquad X,Y \in C^\infty(TM)$$

$$\text{(ii)} \quad Zg(X,Y) = g(\nabla_Z X,Y) + g(X,\nabla_Z Y) , \quad X,Y,Z \in C^\infty(TM) ,$$

where (i) expresses that $\nabla$ is torsion free and (ii) that $\nabla$ is metric i.e. $g$ is parallel with respect to $\nabla$. A smooth vector field $X: [0,1] \to TM$ along a smooth curve $c: [0,1] \to M$ is said to be parallel if $\nabla_{\dot{c}} X \equiv 0$. Since any tangent vector $u \in TM$ at $c(0)$ determines a unique parallel field along $c$ one has the notion of parallel transport. The path dependence of parallel transport is locally governed by the curvature tensor, $R$ which is defined as

$$R(X,Y)Z = \nabla_X \nabla_Y Z - \nabla_Y \nabla_X Z - \nabla_{[X,Y]} Z , \quad X,Y,Z \in C^\infty(TM) .$$

The curvature tensor is completely determined by the sectional curvature $K$, which for an arbitrary 2-plane $\sigma$ spanned by orthonormal vectors $u,v$ is

$$K(\sigma) = g(R(u,v)v,u) .$$

An important geometrical interpretation of the sectional curvature is contained in Toponogov's triangle comparison theorem [CE]:

Let $c_i$ be the sides of a geodesic triangle in the complete Riemannian manifold $(M^n,g)$ and let $\alpha_i$ denote the angle at the vertex opposite $c_i$, $i = 0,1,2$. If the sectional curvature satisfies $K \geqq H \in \mathbb{R}$, then there exists in the canonical simply connected 2-dimensional space $M^2_H$ of constant curvature $H$ a geodesic triangle with sides $c_i^*$ and angles $\alpha_i^*$ such that $L(c_i) = L(c_i^*)$ and $\alpha_i \geqq \alpha_i^*$ for all $i$.

Significance of curvature and diameter. The point of view of considering geodesics as stationary (critical) points for the length (or energy) function leads via Morse theory to a beautiful connection between the geometry of $M^n$ and the topology of the loop space $\Omega M^n$ and hence of $M^n$. Here the index form, which is the second variation of arclength (or energy), plays a crucial role, and the formula for this involves the curvature. The following classical result of Myers [M] is a fairly simple consequence of just the second variation formula:

Theorem 1. Let $M^n$ be a complete, connected Riemannian manifold with $K \geqq \delta > 0$. Then $M^n$ is compact and $\operatorname{diam}(M^n) \leqq \pi/\sqrt{\delta}$.

The idea of the proof is to show that any geodesic $c$ with $L(c) > \pi/\sqrt{\delta}$ is not minimal. This follows by constructing a vectorfield along $c$ for which the second variation of arclength is negative. The construction of such a vector field along a corresponding geodesic $c^*$ in the 2-sphere of constant curvature $\delta$ is straightforward. Such a vector field can by means of parallel transport be transferred to a vector field along $c$ for which the second variation of arclength is (in general more) negative.

As an application of the triangle comparison theorem, Toponogov [T] obtained the following geometric uniqueness theorem:

<u>Theorem 2</u>. Let $M^n$ be a complete connected Riemannian manifold with $K \geqq \delta > 0$ and $\mathrm{diam}(M^n) = \pi/\sqrt{\delta}$. Then $M^n$ is isometric to the n-sphere $S^n(1/\sqrt{\delta})$ of constant curvature $\delta$.

Combining Morse theory and an extension of the triangle comparison theorem, Berger [CE] was then able to show that a simply-connected complete Riemannian manifold with $K \geqq \delta > 0$ and $\mathrm{diam}(M^n) > \pi/2\sqrt{\delta}$ is a homotopysphere. This result was extended and sharpened by Grove and Shiohama [GS], yielding the following <u>topological uniqueness theorem</u>:

<u>Theorem 3</u>. Let $M^n$ be a complete connected Riemannian manifold with $K \geqq \delta > 0$ and $\mathrm{diam}(M^n) > \pi/2\sqrt{\delta}$. Then $M^n$ is homeomorphic to $S^n$.

This result is optimal as e.g. the example of the real projective space shows. The idea of the proof is to exhibit $M^n$ as the union of two balls and a cylinder joined along their common boundaries. The balls in question are small metric balls centered at points of maximal distance. In order to show that the complement of such balls is a cylinder one constructs a non-singular (gradient) vector field whose integral curves define the product structure. A crucial observation for this construction is that the angle between any pair of minimal geodesics to points at maximal distance exceeds $\pi/2$. This implies that the distance functions from these points in effect behave like functions with only two critical points corresponding to minimum and maximum points.

It turns out that the conclusion of the above theorem fails only in a restricted way when the inequality $\operatorname{diam}(M^n) > \pi/2\sqrt{\delta}$ is replaced by equality. More precisely one has in particular the following <u>rigidity theorem</u> of Gromoll and Grove [GG]:

<u>Theorem 4</u>. Let $M^n$ be a complete connected Riemannian manifold with $K \geqq \delta > 0$ and $\operatorname{diam}(M^n) = \pi/2\sqrt{\delta}$. Then $M^n$ is homeomorphic to $S^n$ or the universal covering $\tilde{M}^n$ of $M^n$ is isometric to a symmetric space of rank 1.

In the proof of this result the two points at maximal distance are replaced by convex sets. If both these sets have boundary it follows very much like before that $M^n$ is homeomorphic to a sphere. In the remaining cases the complete geometric structure of $M^n$ emerges from a rather long and intriguing sequence of geometric and topological arguments.

The last result I want to discuss along these lines is a very striking <u>topological finiteness theorem</u> of Gromov [G]:

<u>Theorem 5</u>. There exist a $c_n > 0$ such that for any compact connected Riemannian manifold $M^n$

$$\sum_{i=0}^{n} \beta_i(M^n) \leqq \begin{cases} c_n & \text{if } K \geqq 0 \\ c_n^{\,1+\sqrt{|H|}\cdot \operatorname{diam}(M^n)} & \text{if } K \geqq H \in \mathbb{R}_- \end{cases}$$

where $\beta_i(M^n)$ denotes the i'th Bettinumber (any field of coefficients) of $M^n$.

The proof is based on an analysis of the crude notion of critical points for the distance function as in [GS]. The crucial observation is vaguely that the Toponogov triangle theorem gives an upper bound for the number of critical values with a given large bound for their ratios. More generally a similar bound exists for the rank of any set, which in a sense is a measure for the complexity of the set in terms of critical points away from the set. Then a suitable notion of content of metric balls is introduced, such that the content of $M^n$ is the sum of its Betti numbers and the proof is completed, by giving inductively over the rank (using the Leray spectral sequence) an a priori bound for the content of all balls, in particular of $M^n$.

This theorem gives in particular an a priori bound for the Betti numbers of any compact n-dimensional non-negatively curved manifold. By the following results of Cheeger, Gromoll and Meyer [GM], [CG] the same conclusion is true for non-compact manifolds:

Theorem 6. Let $M^n$ be a complete connected Riemannian manifold with $K > 0$. Then $M^n$ is diffeomorphic to $\mathbb{R}^n$.

Theorem 7. Let $M^n$ be a complete connected Riemannian manifold with $K \geqq 0$. Then $M^n$ contains a compact totally geodesic submanifold $S$ (the soul of $M^n$) such that $M^n$ is diffeomorphic to the normal bundle of $S$ in $M^n$.

In view of these results we consider the following geometric invariant of any complete connected Riemannian manifold $(M,g)$

$$\partial_M = \inf K \cdot \operatorname{diam}(M)^2/\pi^2 .$$

With the convention $0 \cdot \infty = 0$ we can now conclude by reformulating the theorems discussed here in a very compact way:

I. $\partial_M \in [-\infty, 1]$ for all $(M,g)$ .

II. $\partial_M = 1 \Rightarrow (M,g)$ is a sphere of constant curvature.

III. $\partial_M > \frac{1}{4} \Rightarrow M$ is a topological sphere.

IV. $\partial_M = \frac{1}{4} \Rightarrow M$ is a topological sphere or $(\tilde{M}, \tilde{g})$ is a symmetric space of rank 1.

V (III, VII). There is a $c(n) > 0$ so that $\sum_1^n \beta_i(M^n) \leqq c(n)^{1-\partial_M}$ for all $(M^n, g)$ .

## REFERENCES


[B] M. Berger, Les variétés riemanniennes (1/4)-pincées, Ann. Scuola Norm. Sup. Pisa, Ser. III, 14 (1960), 161-170.

[CE] J. Cheeger and D. G. Ebin, Comparison Theorems in Riemannian Geometry, North Holland Math. Library 9, Amsterdam-Oxford-New York, 1975.

[CG] J. Cheeger and D. Gromoll, On the structure of complete manifolds of nonnegative curvature, Annals of Math. 96 (1972), 413-443.

[GG] D. Gromoll and K. Grove, A generalization of Berger's Rigidity Theorem for positively curved manifolds.

[GM] D. Gromoll and W. Meyer, On complete open manifolds of positive curvature, Annals of Math. 90 (1969), 75-90.

[G] M. Gromov, Curvature, Diameter and Betti numbers, preprint IHES 1980.

[GS] K. Grove and K. Shiohama, A generalized sphere theorem, Annals of Math. 106 (1977), 201-211.

[K] W. Klingenberg, Über Riemannsche Mannigfaltigkeiten mit positiver Krümmung, Comment. Math. Helv. 35 (1961), 47-54.

[M] S. B. Myers, Riemannian manifolds in the large, Duke Math. J. 1 (1935), 39-49.

[T] V. A. Toponogov, Riemannian spaces having their curvature bounded below by a positive number, A.M.S. Transl. (2) 37 (1964), 291-336.

COPENHAGEN, DENMARK

# A RESOLVENT CONSTRUCTION FOR PSEUDO-DIFFERENTIAL BOUNDARY VALUE PROBLEMS, WITH APPLICATIONS

By Gerd Grubb

The classical pseudo-differential operators were introduced in the 1960s as a close generalization of differential operators, comprising the parametrices of elliptic differential operators, and other operators arising in the calculus of differential operators. We recall a few of the properties of such operators $P$ and a calculus of boundary value problems associated with them, and then go on to our principal subject, the resolvent of realizations $P_T$ determined from $P$ by homogeneous boundary conditions for bounded domains. The study of the resolvent $(P_T - \lambda I)^{-1}$ (for complex $\lambda$) is motivated by spectral theory and operational calculus. Our main result on the resolvent construction is presented with a few indications of the underlying theory; and some conclusions are drawn for the fractional powers $(P_T)^{-s}$, which are used to derive a new spectral estimate (on the asymptotic distribution of eigenvalues of $P_T$ in the non lower bounded case).

1. The pseudo-differential operators we shall be concerned with are the so-called classical ps.d.o.s, that can be described very briefly as the following generalization of differential operators: Recall that when $A$ is a differential operator on $\mathbb{R}^n$ of order $m$, of the form

$$A = \sum_{|\alpha| \leq m} a_\alpha(x) D^\alpha$$

(here $D^\alpha = D_1^{\alpha_1} \cdots D_n^{\alpha_n}$, with $D_j = \frac{1}{i} \frac{\partial}{\partial x_j}$, and $\alpha$ denotes a multi-index $(\alpha_1, \cdots, \alpha_n)$ of length $|\alpha| = \alpha_1 + \cdots + \alpha_n$), then $A$ can be written by use of the Fourier transform as

$$(1) \qquad Au = \sum_{|\alpha|\le m} a_\alpha(x)\mathcal{F}^{-1}(\xi^\alpha[\mathcal{F}u](\xi))$$
$$= (2\pi)^{-n}\int e^{ix\cdot\xi}\sum_{|\alpha|\le m} a_\alpha(x)\xi^\alpha\hat{u}(\xi)d\xi$$
$$= (2\pi)^{-n}\int e^{ix\cdot\xi}\, a(x,\xi)\hat{u}(\xi)d\xi \equiv Op(a(x,\xi))u\,,$$

where $a(x,\xi) = \sum_{|\alpha|\le m} a_\alpha(x)\xi^\alpha$. This function of $x$ and $\xi\in\mathbb{R}^n$ is called the symbol of $A$; it is polynomial in $\xi$. Now, in the last line of (1), $a$ can be replaced by other symbol functions, and the classical pseudo-differential operators arise when we insert functions of the form

$$(2) \qquad p(x,\xi) \sim p_d(x,\xi) + p_{d-1}(x,\xi) +\cdots+ p_{d-j}(x,\xi)+\cdots\,,$$

where $p_{d-j}(x,\xi)$ is homogeneous in $\xi$ of degree $d-j$, and smooth in $x$ and $\xi$, for $\xi \neq 0$. $p_d$ is called the principal symbol, often denoted $p^0$.

There are certain ambiguities in the definition of the operator $P = Op(p(x,\xi))$: For one thing, the homogeneous terms $p_{d-j}$ in general have singularities at $\xi = 0$, that are eliminated by modifying $p_{d-j}$ to a $C^\infty$ function near $\xi = 0$; for two different modified functions $\tilde{p}_{d-j}$ we get operators $Op(\tilde{p}_{d-j})$ that differ by a smoothing operator, i.e. an integral operator with $C^\infty$ kernel (an operator of order $-\infty$). Moreover, the conventions for associating a function $p(x,\xi)$ with a given series $\Sigma\, p_{d-j}(x,\xi)$ are such that $p$ is determined up to a symbol defining a smoothing operator. In fact, the whole calculus of pseudo-differential operators is most conveniently expressed modulo smoothing operators; this entails that some results have a qualitative character. In the following, the "modulo smoothing operators" is generally understood in the formulas.

The composite of two classical ps.d.o.s of orders $d$ and $d'$, respectively, is a classical ps.d.o. of order $d+d'$, its principal symbol being simply the product of the respective principal symbols.

A classical ps.d.o. $P = Op(p(x,\xi))$ of order $d$ is called elliptic, when $p_d(x,\xi) \neq 0$ for all $\xi \neq 0$; then there exists another ps.d.o. $Q$ of order $-d$ such that $PQ - I$ and $QP - I$ are smoothing operators ($Q$ is called a parametrix of $P$); here $q_{-d}(x,\xi) = p_d(x,\xi)^{-1}$. In particular, when $P$ is an elliptic differential operator, it has a classical ps.d.o. as parametrix. - A very simple example is the operator $P = -\Delta + 1$, with parametrix $Q = \mathrm{Op}((|\xi|^2 + 1)^{-1})$ (this is in fact an inverse). Note that $Q$ is a non-local operator, as are most ps.d.o.s. (However, ps.d.o.s are "pseudo-local", i.e. preserve the property of being $C^\infty$ in an open subset of $\mathbb{R}^n$.)

The theory extends easily to matrix-formed operators (also called systems); the symbols are then matrices of functions (which of course in general do not commute). Two-sided parametrices exist for elliptic square-matrix formed systems.

We content ourselves with these hints, and refer to the literature, e.g. Kohn-Nirenberg [ 9 ], Seeley [11] and Hörmander [ 8 ] for deeper and more complete presentations of the theory of ps.d.o.s; see also the lecture of L. Hörmander at this conference.

2. Let $\Omega$ be an open subset of $\mathbb{R}^n$, with smooth boundary. When $P$ is a ps.d.o. given on $\mathbb{R}^n$, we define its restriction $P_\Omega$ to functions on $\Omega$ brutally by setting

$$P_\Omega u = rPeu , \tag{3}$$

where $e$ denotes the operator that extends $u$ by zero outside $\Omega$, and $r$ is the restriction operator from $\mathbb{R}^n$ to $\Omega$. When $P$ is a differential operator, $P_\Omega$ sends $C^\infty(\bar{\Omega})$ into $C^\infty(\bar{\Omega})$, for the restriction $r$ kills off any singularities at $\partial\Omega$ arising from the jump of $eu$ at $\partial\Omega$. This is not the case for general ps.d.o.s, but there is a condition on the symbol $p(x,\xi)$ called the transmission property, which implies that $P_\Omega$ preserves $C^\infty(\bar{\Omega})$, discovered in 1965 by Boutet de Monvel [1 ]. (Also Hörmander treated the problem, and gave a more general discussion in a lecture series at Princeton 1966.) In [2 ], Boutet de Monvel developed a complete calculus of boundary value probelms for elliptic pseudo-differential operators having the transmission property (tr.p.). Since the main results there are formulated for integer degrees (where also the adjoint of $P$ has the tr.p.), we shall restrict to that case. Differential operators and parametrices of elliptic differential operators have the transmission property, as do all ps.d.o.s with rational symbols.

We assume from now on that P is an elliptic square-matrix of ps.d.o.s of integer degree $d > 0$, having the tr.p. with respect to $\bar{\Omega}$, which is bounded.

Under some circumstances, for instance when P is strongly elliptic (i.e. $p_d(x,\xi) + p_d(x,\xi)^*$ is positive for $\xi \neq 0$, d necessarily even), there exist <u>trace operators</u> T (boundary operators, sending functions on $\Omega$ into functions on $\partial\Omega$), such that the problem

$$(4) \qquad \begin{aligned} P_\Omega u &= f \quad \text{in} \quad \Omega , \\ Tu &= \varphi \quad \text{on} \quad \partial\Omega , \end{aligned}$$

is well posed (in the strongly elliptic case, T can for example consist of the normal derivatives up to order $d/2-1$, defining the Dirichlet condition). The solvability of (4) means that $\begin{pmatrix} P_\Omega \\ T \end{pmatrix}$ has a parametrix

$$(5) \qquad \begin{pmatrix} P_\Omega \\ T \end{pmatrix}^{-1} = (R \quad K) ,$$

where R operates on functions on $\Omega$, and K maps functions on $\partial\Omega$ into functions on $\Omega$.

The structure of R and K is illustrated by the example where $P = -\Delta$, and $T = \gamma_0 : u \mapsto u|_{\partial\Omega}$. Here K is the well known Poisson integral sending $\varphi$ into u when $f = 0$ in (4), and R solves the problem (4) with $\varphi = 0$; we recall that

$$(6) \qquad R = Q_\Omega + G ,$$

where Q is the convolution with $\frac{c_n}{|x|^{n-2}}$ (for $n \geq 3$) (this is in fact a pseudo-differential operator inverse of $-\Delta$), restricted to $\Omega$; and G defines the correction needed to get the solution of the particular boundary value problem. - Also for general problems (4), R is of the form (6), where Q is a ps.d.o. parametrix of P, and G is a so-called <u>singular Green operator</u> (s.g.o.), having its importance in the neighborhood of $\partial\Omega$. K in general is called a <u>Poisson operator</u>. All the operators in question can be matrix formed (as long as the dimensions fit together).

Remark 1. When $P$ is a general elliptic ps.d.o. having the tr.p., there need not exist a trace operator so that (4) is well posed; however, wellposedness can be obtained (at least micro-locally) by adjoining a suitable column $\begin{pmatrix} K' \\ S' \end{pmatrix}$ to the column matrix $\begin{pmatrix} P_\Omega \\ T \end{pmatrix}$; here $K'$ is a Poisson operator and $S'$ is a ps.d.o. in $\partial\Omega$. The general theory established by Boutet de Monvel [ 2 ] is concerned with such systems

$$\begin{pmatrix} P_\Omega + G' & K' \\ T & S' \end{pmatrix} : \begin{matrix} C^\infty(\overline{\Omega})^M \\ \times \\ C^\infty(\partial\Omega)^N \end{matrix} \to \begin{matrix} C^\infty(\overline{\Omega})^M \\ \times \\ C^\infty(\partial\Omega)^{N'} \end{matrix}$$

(where a singular Green operator $G'$ has furthermore been inserted), and it is shown that under a very natural ellipticity condition, such systems have parametrices of the same kind. ($N'$ is in general different from $N$.) Time does not allow us to elaborate on this; however, we use the theory of Boutet de Monvel in the following. (Already Vishik and Eskin [13 ] studied systems $\begin{pmatrix} P_\Omega & K' \\ T & S' \end{pmatrix}$, for more general symbols.)

When a problem (4) is given, one can define the $L^2$- realization $P_T$ of $P$ under the condition $Tu = 0$, as follows:

$$P_T\colon u \mapsto P_\Omega u, \qquad D(P_T) = \{u \in H^d(\Omega) \mid Tu = 0\}.$$

This is a closed, unbounded operator, and it is not hard to exhibit conditions under which it is selfadjoint in $L^2(\Omega)$, positive, or belongs to some other nice operator class. For instance, when $P$ is strongly elliptic and formally selfadjoint on $\mathbb{R}^n$, then $P_T + c$ is positive selfadjoint if $T$ represents the Dirichlet condition ($c$ large). Note that $R$ in (5) plays the role of an inverse to $P_T$.

3. The main subject of this talk is the resolvent of $P_T$, i.e., the operator

$$R_\lambda = (P_T - \lambda I)^{-1},$$

considered for the values of $\lambda \in \mathbb{C}$ for which it exists as a bounded operator. We assume that the system $\begin{pmatrix} P_\Omega - \lambda \\ T \end{pmatrix}$ satisfies an ellipticity condition with $\lambda$ entering in the principal symbol, for $\lambda$ on some half-ray $\{\lambda = re^{i\theta} \mid r \geq 0\}$; one can obtain by a deplacement that the ray is $\overline{\mathbb{R}}_-$, and $R_\lambda$ is defined for all $\lambda \in \overline{\mathbb{R}}_-$ (and a neighborhood thereof). In analogy with $R$ considered above, one has that

$$(7) \qquad R_\lambda = Q_{\lambda,\Omega} + G_\lambda ,$$

where $Q_\lambda$ is a parametrix of $P - \lambda$ on $\mathbb{R}^n$, and $G_\lambda$ is a family of s.g.o.s parametrized by $\lambda \in \overline{\mathbb{R}}_-$.

There are several motivations for studying $R_\lambda$.

$1^\circ$ Spectral Theory. When $P_T$ is selfadjoint positive, $P_T^{-1}$ is a compact operator, so the spectrum of $P_T$ is a sequence of eigenvalues (counted with multiplicities) going to $+\infty$, and one is interested in estimating

$$N(t\,;P_T) = \text{the number of eigenvalues in } [0,t]$$

for $t \to \infty$. By a study of the resolvent (permitting the application of a Tauberian theorem of Å. Pleijel), we showed in [5], [6]

Theorem 1. When $P_T$ is positive selfadjoint, matrix-formed, then for any $\varepsilon > 0$

$$(8) \qquad N(t\,;P_T) = c_P t^{n/d} + \mathcal{O}(t^{(n-\frac{1}{2}+\varepsilon)/d}) \quad \text{for } t \to \infty ,$$

where $c_P$ is a constant derived from $p_d(x,\xi)$.

This did not necessitate a full constructive description of $R_\lambda$, but rather a study of its integral operator kernel for $\lambda$ close to $\mathbb{R}_+$. (Results for the spectral function were also obtained.)

Remark 2. The results for pseudo-differential boundary problems have an interest also for differential problems. For example, the spectral asymptotics for realizations of the system

$$A = \begin{pmatrix} \Delta^2 & B \\ B^* & -\Delta \end{pmatrix}$$

(where $B$ is a third order operator for which $A$ is (Douglis-Nirenberg) strongly elliptic), are quite close to the spectral asymptotics for realizations of the ps.d.o.

$$P = -\Delta - B^* \Delta^{-2} B$$

(since $P^{-1}$ is the highest order block in $A^{-1}$). More generally, elliptic systems can often be reduced to a simpler form at the cost of introducing ps.d.o.s (cf. e.g. [3]).

Another motivation for studying $R_\lambda$ is:

$2^0$ Operational Calculus. When $f(\lambda)$ is a complex function that is holomorphic in the neighborhood of the spectrum of $P_T$, one may define $f(P_T)$ by the Cauchy formula

$$f(P_T) = \frac{1}{2\pi i} \int_C \frac{f(\lambda)d\lambda}{\lambda - P_T} = \frac{i}{2\pi} \int_C f(\lambda) R_\lambda \, d\lambda ,$$

where $C$ is a curve around the spectrum of $P_T$ (provided the integral converges in a suitable sense). Then the detailed knowledge of the structure of $R_\lambda$ leads to information on $f(P_T)$. In view of (7), we have in fact

$$\begin{aligned} f(P_T) &= \left[ \frac{i}{2\pi} \int_C f(\lambda) Q_\lambda \, d\lambda \right]_\Omega + \frac{i}{2\pi} \int_C f(\lambda) G_\lambda \, d\lambda \\ &= f(P)_\Omega + \frac{i}{2\pi} \int_C f(\lambda) G_\lambda \, d\lambda , \end{aligned} \tag{9}$$

where the first term comes from the function of $P$ (in $\mathbb{R}_n$); such functions have already been studied widely, and it has been proved that $f(P)$ is a ps.d.o. in many cases (in works of R. Seeley, D. Robert, R. Beals; and $C^\infty$-functions $f(P)$ have been treated by R. Strichartz, J. Dunau, H. Widom and others). In particular, the fractional powers $P^s$ have been studied (here $f(\lambda) = \lambda^s$, defined to be positive for $\lambda > 0$ and holomorphic in $\mathbb{C} \setminus \overline{\mathbb{R}}_-$, and $C$ is a curve going in $\{\text{Im}\,\lambda \geq 0\}$ from $-\infty$ to a circle around the origin (clockwise) and back to $-\infty$ in $\{\text{Im}\,\lambda \leq 0\}$). Seeley showed in [10] that $P^s$ can be defined as a classical ps.d.o. of order $ds$, when $P$ is a classical ps.d.o. of order $d > 0$ with $p_d(x,\xi) - \lambda$ invertible for $\lambda \in \overline{\mathbb{R}}_-$. There are also other functions of interest, e.g. the exponential function (solving the "heat equation"), etc.

The point of the operational calculus is of course not so much to show the existence of the operators (which is often well known) as to analyze their finer properties. In view of the mentioned studies of $f(P)$, the new object to investigate is the last term in (9), coming from the singular Green operators $G_\lambda$.

The operational calculus has consequences also for the spectral theory. For instance, there is an amusing application of the analysis of the square root:

Let $P_T$ be a selfadjoint but not lower bounded realization; then it has two infinite sequences of eigenvalues $\lambda_j^+ \to +\infty$ and $\lambda_j^- \to -\infty$, and one can ask for estimates of

$$N^{\pm}(t\,;P_T) = \text{number of evs.} \gtrless 0 \text{ in } [-t,t].$$

These are closely related to the spectral asymptotics for the two positive operators

$$A_\pm = |P_T|^{-1} \pm \frac{1}{2} P_T^{-1}, \tag{10}$$

where $|P_T| = (P_T^2)^{\frac{1}{2}}$. Defining

$$P_\pm = (|P|^{-1} \pm \frac{1}{2} P^{-1})^{-1}, \tag{11}$$

strongly elliptic ps.d.o.s of order $d$, we may write

$$A_\pm = (P_\pm{}^{-1})_\Omega + G_\pm, \tag{12}$$

where the $G_\pm$ are linear combinations of a singular Green operator and the non-pseudo-differential term in $(P_T^2)^{-\frac{1}{2}}$. Now if $G_\pm$ and $P_\pm$ have convenient properties, we can estimate the spectral behavior of $A_\pm$ by the help of Theorem 1 applied to

$P_{\pm}$ . Let us state right away what we have obtained by this method.

Theorem 2. If $P$ is a selfadjoint elliptic system of even order (not necessarily strongly elliptic), and $P_T$ is a selfadjoint realization, then for any $\varepsilon > 0$ ,

$$(13) \qquad N^{\pm}(t\,;P_T) = c_P^{\pm} t^{n/d} + \mathcal{O}(t^{(n-\frac{1}{2}+\varepsilon)/d}), \quad \text{for } t \to \infty .$$

The result is new also for general differential operators. It was shown already in [5], [6] that when $P$ is strongly elliptic with $P_T$ selfadjoint, not lower bounded, then (13) holds for $N^+$ , and $N^-$ is $\mathcal{O}(t^{(n-1)/d})$. The condition that $P$ be of even order assures that $|P|$ has the same transmission property as $P$ .

Concerning the general project of the operational calculus, our work is at the stage where we have found a satisfactory (it seems) description of $R_\lambda$ , and we have drawn some consequences for $P_T^{-s}$ for $s > 0$ . The latter analysis will hopefully be made more precise, but it was sufficient to show Theorem 2 (see also Corollary 4 and Theorem 5 below).

4. Rather than describing the full calculus leading to $R_\lambda$ (impossible in a few lines), we shall indicate some of the difficulties that enter in the description. For $\lambda \in \overline{\mathbb{R}}_-$ , we write $-\lambda = \mu^d$, $\mu \geq 0$ . In the decomposition (7), $Q_\lambda$ is a parametrix of $P - \lambda$ , with principal symbol

$$q_{-d}(x,\xi,\mu) = (p_d(x,\xi) + \mu^d I)^{-1} .$$

The principal symbol of $G_\lambda$ is harder to describe. For each $\lambda$, $G_\lambda$ is of the form (in local coordinates at a boundary point, where $\Omega$ is replaced by $\mathbb{R}^n_+ = \{x \mid x_n > 0\}$)

$$(14) \qquad G_\lambda u = (2\pi)^{-n+1} \int e^{ix'\cdot\xi'} g(x',\xi',\mu,D_n) u(\widehat{\xi'},x_n)\, d\xi' ,$$

where $g(x',\xi',\mu,D_n)$ for each $(x',\xi',\mu)$ is a Hilbert-Schmidt operator on $\mathbb{R}_+$ , the kernel $\tilde{g}(x',\xi',\mu,x_n,y_n)$ belonging to

$\mathscr{S}(\overline{\mathbb{R}}_+ \times \overline{\mathbb{R}}_+)$. The operator $g(x',\xi',\mu,D_n)$ is determined from a pseudo-differential boundary problem on $\mathbb{R}_+$, the "symbolic" problem at each $(x',\xi',\mu)$ associated with $\begin{pmatrix} P_\Omega - \lambda \\ T \end{pmatrix}$. When $P$ and $T$ are differential operators, there are explicit formulas for $g$ (involving the roots in $\xi_n$ of the polynomial $\det[p_d(x',0,\xi',\xi_n) + \mu^d I]$); but in the ps.d.o. case, $g$ is obtained more qualitatively by a meticulous use of the Boutet de Monvel theory, in a version that takes the parameter $\mu$ into account.

One difficulty is encountered already in the description of the symbol of $Q_\lambda$: Since $p_d(x,\xi)$ is generally only $d-1$ times differentiable at $\xi = 0$, we find, when considering the symbols as functions of $(\xi,\mu) \in \mathbb{R}^n \times \overline{\mathbb{R}}_+$, that <u>either</u> the symbols are homogeneous in $(\xi,\mu)$ but have singularities all along the axis $\xi = 0$, <u>or</u> if we smooth out $p_d(x,\xi)$ near $\xi = 0$, the symbols are smooth but not homogeneous in a tube around the axis $\xi = 0$. Choosing the latter possibility and denoting $(|\xi|^2 + 1)^{\frac{1}{2}} = \langle\xi\rangle$ and $(|\xi|^2 + \mu^2 + 1)^{\frac{1}{2}} = \langle\xi,\mu\rangle$, one finds, more precisely,

$$(15)\quad \begin{aligned} &\text{(i)} \quad |D_\xi^\alpha q_{-d}(x,\xi,\mu)| \leq c_\alpha \langle\xi,\mu\rangle^{-d-|\alpha|}, \quad \text{when } |\alpha| \leq d, \\ &\text{(ii)} \quad |D_\xi^\alpha q_{-d}(x,\xi,\mu)| \leq c_\alpha \langle\xi\rangle^{d-|\alpha|}\langle\xi,\mu\rangle^{-2d}, \quad \text{when } |\alpha| > d, \end{aligned}$$

so the derivatives of $q_d$ are never better than $\mathcal{O}(\mu^{-2d})$. For a systematic calculus, we introduce symbol classes $S^{m,\nu}$, where $m$ is the order (in $\xi$), and $\nu$ is a number that, when positive, measures the number of derivatives that are well behaved as in (15)(i). We have that

$$r(x,\xi,\mu) \in S^{m,\nu} \Rightarrow D_\xi^\alpha r \in S^{m-|\alpha|,\nu-|\alpha|}$$

but for compositions,

$$S^{m,\nu} \cdot S^{m',\nu'} \subset S^{m+m',\min(\nu,\nu',\nu+\nu')}$$

(nonlinear in $\nu$ and $\nu'$ unless both are $\leq 0$). However, when $r \in S^{m,\nu}$ <u>with</u> $\nu \geq 0$ and is elliptic $(|r^{-1}| \leq c\langle\xi,\mu\rangle^{-m})$, then there is a parametrix symbol in $S^{-m,\nu}$. - Such symbol classes are introduced also for the various operator classes at the boundary.

Another difficulty is the dependence on $x_n$, which requires Taylor expansions in $x_n$ at the boundary, leading to another kind of remainder terms of the same order as the given operator, that have to be handled in their dependence on $\mu$. It was this problem that made us enter into the systematic resolvent construction in the first place.

The final result is (in local coordinates):

Theorem 3. $G_\lambda$ is of the form (14), where the kernel $\tilde{g}(x',\xi',\mu,x_n,y_n)$ of $g(x',\xi',\mu,D_n)$ has an asymptotic expansion

$$\tilde{g} \sim \tilde{g}_{-d} + \tilde{g}_{-d-1} + \dots + \tilde{g}_{-d-\ell} + \dots$$

with the following properties: $\tilde{g}$ and the $\tilde{g}_{-d-\ell}$ are in $\mathcal{S}(\overline{\mathbb{R}}_+ \times \overline{\mathbb{R}}_+)$ with respect to $(x_n,y_n)$, and

$$\| D^{\beta'}_{x'} D^{\alpha'}_{\xi'} D^j_\mu x_n^{N_1} D_{x_n}^{N_1'} y_n^{N_2} D_{y_n}^{N_2'} [\tilde{g} - \sum_{\ell<M} \tilde{g}_{-d-\ell}] \|_{L^2(\mathbb{R}_+ \times \mathbb{R}_+)}$$

$$\leq c(x')(\langle\xi'\rangle^\nu + \langle\xi',\mu\rangle^\nu)\langle\xi',\mu\rangle^{-2d-j+\frac{1}{2}},$$

$$\text{with } \nu = d-M-|\alpha'|-N_1+N_1'-N_2+N_2'-\tfrac{1}{2},$$

for all $j$, $N_1 \geq N_1'$, $N_2 \geq N_2'$ and $M \geq 0$, all $\alpha'$ and $\beta'$.

The theory is developed in [7] (treating large $N_1', N_2'$ also).

For differential problems, $\langle\xi'\rangle^\nu$ can be omitted, and the estimates follow from estimates proved by Seeley in [12]. For ps.d.o. problems, we note that the decrease in $\mu$ will not get below $\mathcal{O}(\mu^{-2d-j+\frac{1}{2}})$.

Now consider

$$P_T^{-s} = (P^{-s})_\Omega + G^{(-s)}, \quad \text{where} \quad G^{(-s)} = \frac{i}{2\pi}\int_C \lambda^{-s} G_\lambda \, d\lambda,$$

cf. (9). From Theorem 3 follows:

Corollary 4. $G^{(-s)}$ is defined for $s > 0$ as an operator

$$G^{(-s)}u = (2\pi)^{-n+1}\int e^{ix'\cdot\xi'} g^{(-s)}(x',\xi',D_n)\hat{u}(\xi',x_n)\,d\xi',$$

where $g^{(-s)}$ has a kernel $\tilde{g}^{(-s)}(x',\xi',x_n,y_n)$ of the form

$$\tilde{g}^{(-s)} \sim \tilde{g}^{(-s)}_{-ds} + \tilde{g}^{(-s)}_{-ds-1} + \dots + \tilde{g}^{(-s)}_{-ds-\ell} + \dots ,$$

satisfying

$$(16)\quad \left\| D^{\beta'}_{x'} D^{\alpha'}_{\xi'} x_n^{N_1} D^{N_1'}_{x_n} y_n^{N_2} D^{N_2'}_{y_n} \left[ \tilde{g}^{(-s)} - \sum_{\ell<M} \tilde{g}^{(-s)}_{-ds-\ell} \right] \right\|_{L^2(\mathbb{R}_+ \times \mathbb{R}_+)}$$

$$\leq c(x') \langle \xi' \rangle^{-ds-M-|\alpha'|-N_1+N_1'-N_2+N_2'} ,$$

for the indices satisfying

$$(17)\qquad N_1' \leq N_1 \quad \text{and} \quad N_2' \leq N_2 .$$

In order for $G^{(-s)}$ to be a singular Green operator in the sense of Boutet de Monvel, the estimates (16) should hold for all $N_1$, $N_2$, $N_1'$ and $N_2'$, but (17) gives a serious restriction on the derivatives $D^{N_1'}_{x_n}$ and $D^{N_2'}_{y_n}$ that are allowed (note however, that $(x_n D_{x_n})^N$ and $(y_n D_{y_n})^N$ can be applied for any $N$). There is also other evidence that $G^{(-s)}$ need not be a genuine s.g.o. even when $ds$ is integer.

However, $G^{(-s)}$ does have the form of a "ps.d.o. in $x'$ with values in the class of Hilbert-Schmidt operators in $x_n$", with <u>some</u> nice properties of the kernel (so it could be viewed as a generalized s.g.o.). This enables us to prove the following result, that we have shown earlier for s.g.o.s of Boutet de Monvel type.

<u>Theorem 5</u>. The eigenvalues of $|G^{(-s)}| = (G^{(-s)*}G^{(-s)})^{\frac{1}{2}}$ satisfy an estimate

$$(18)\qquad \lambda_j(|G^{(-s)}|) \leq c\, j^{-ds/(n-1)}, \quad \text{for } j \in \mathbb{N} .$$

It is well known that the mere continuity of $G^{(-s)}$ from $L^2(\Omega)$ to $H^{sd}(\Omega)$ implies the estimate (18) with $n$ instead of $n-1$ ; but it is interesting here that $G^{(-s)}$ behaves "as if the dimension were $n-1$", the dimension of the boundary. We end by sketching how this leads to Theorem 2 above:

Theorem 5 implies that the terms $G_{\pm}$ in (12) satisfy the eigenvalue estimates

$$\lambda_j(|G_{\pm}|) \leq c\ j^{-d/(n-1)}\ , \tag{19}$$

and furthermore, the $A_{\pm}$ deviate from parametrices of the Dirichlet realization of $P_{\pm}$ (cf. (11)), respectively, by operators with the spectral behavior (19). Theorem 1 applies to $P_{\pm}$, and then perturbation and inversion arguments as in [4, Prop. 6.1 and Lemma 6.2] show that $A_{\pm}$ have the same asymptotic estimates as $(P_{\pm}^{-1})_{\Omega}$ ; this leads to (13). (The perturbation argument requires high order $d$ , which can always be obtained by iteration.)

REFERENCES


[1] L. Boutet de Monvel: Comportement d'un opérateur pseudo-différential sur une variété à bord, C. R. Acad. Sci. Paris 261 (1965), 4587-4589.

[2] L. Boutet de Monvel: Boundary problems for pseudo-differential operators, Acta Math. 126 (1971), 11-51.

[3] G. Grubb: Spectral asymptotics for Douglis-Nirenberg elliptic systems and pseudo-differential boundary problems, Comm. Part. Diff. Equ. 2 (1977), 1971-1150.

[4] G. Grubb: Remainder estimates for eigenvalues and kernels of pseudo-differential elliptic systems, Math. Scand. 43 (1978), 275-307.

[5] G. Grubb: Estimation du reste dans l'étude des valeurs propres des problèmes aux limites pseudo-différentiels auto-adjoints, C. R. Acad. Sci. Paris 287 (1978), 1017-1020.

[6] G. Grubb: On the spectral theory of pseudo-differential boundary problems. (To appear in the Proceedings of the Banach Center semester on P.D.E. 1978; available as Preprint no. 33 (1978), Copenhagen Univ. Math. Dept.)

[7] G. Grubb: On Pseudo-differential Boundary Problems. (In preparation, available in part as no.s 2(1979), 1 and 2 (1980) in Publication Series, Copenhagen Univ. Math. Dept.)

[8] L. Hörmander: Fourier integral operators I, Acta Math. 127 (1971), 79-183.

[9] J.J. Kohn and L. Nirenberg: An algebra of pseudo-differential operators, Comm. Pure Appl. Math. 18 (1965), 269-305.

[10] R. Seeley: Complex powers of an elliptic operator, Amer. Math. Soc. Proc. Symp. Pure Math. 10 (1967), 288-307.

[11] R. Seeley: Topics in pseudo-differential operators, C.I.M.E. Conference on Pseudo-differential operators, Ed. Cremonese, Roma 1969, 169-305.

[12] R. Seeley: The resolvent of an elliptic boundary problem, Amer. J. Math. 91 (1969), 889-920.

[13] M.I. Vishik and G.I. Eskin: Elliptic equations in convolution in a bounded domain and their applications, Uspehi Mat. Nauk 22 (1967), 15-76 = Russian Math. Surveys 22 , 13-75.

MATEMATISK INSTITUT
UNIVERSITETSPARKEN 5
DK-2100 COPENHAGEN, DENMARK

Added October 1980: It has come to our attention that a result overlapping with our Theorem 2 (at least in the case of differential operators) has been announced by V.A. Kozlov in Vestnik Leningrad No. 19(1979), pp. 11-113.

# The reduced C*-algebra of the free group on two generators

Uffe Haagerup

## Introduction.

To any locally compact group $G$ are associated two C*-algebras, namely the (full) group C*-algebra $C^*(G)$ and the reduced group C*-algebra $C^*_\lambda(G)$. The two algebras coincide if and only if $G$ is amenable. The reduced C*-algebras associated with connected, locally compact groups and amenable discrete groups share many "nice" properties. For instance they are all nuclear (see § 2). The free group on two generators $\mathbb{F}_2$ is the simplest known example of a non-amenable discrete group, and the algebra $C^*_\lambda(\mathbb{F}_2)$ is therefore a suitable starting point for studying non-nuclear C*-algebras.

These notes contain a report on two recent papers [9],[1] concerning $C^*_\lambda(\mathbb{F}_2)$. The first is due to the author, and the second is a joint work of J. De Cannière and the author.

A few years ago, Choi and Effros [3] and Kirschberg [10] proved independently that nuclearity for a C*-algebra is equivalent to *"the complete positive approximation property"* (cf. § 2 of these notes). In [9] we proved that the non-nuclear C*-algebra has the weaker "metric approximation property". This result has been sharpened in [1] in two directions. The first result (§ 3, Theorem I) is that $C^*_\lambda(\mathbb{F}_2)$ has the *"positive approximation property"*, i.e. the identity on $C^*_\lambda(\mathbb{F}_2)$ can be approximated strongly by positive, finite rank contractions on $C^*_\lambda(\mathbb{F}_2)$. The second result (§3, Theorem II) is that $C^*_\lambda(\mathbb{F}_2)$ has the "*complete bounded approximation property*", i.e. the identity on $C^*_\lambda(\mathbb{F}_2)$ can be approximated strongly, by finite rank contractions $(T_\alpha)$ on $C^*_\lambda(\mathbb{F}_2)$, such that $\|T_\alpha\|_{CB} \leqq 1$ for all $\alpha$ (see § 2 for the definition of $\| \ \|_{CB}$).

The proof of Theorem I is combinatorical and follows the ideas of [9] (cf. § 4). The proof of Theorem II is entirely different, and is based on representation theory for the Lie group $SL(2,\mathbb{R})$ (cf. § 5).

It has been pointed out to us by R. Archbold, that as a corollary of Theorem II, one gets that $C^*_\lambda(\mathbb{F}_2)$ has the *Slice map property* (cf. § 6).

## 1. Group C*-algebras.

Let $G$ be a locally compact group, with (left) Haarmeasure $dx$. The Banach space $L^1(G)$ is an involutive Banach algebra with convolution as product:

$$(f*g)(x) = \int_G f(y)g(y^{-1}x)dy ,$$

and involution

$$f^*(x) = \Delta(x)^{-1} \overline{f(x^{-1})}$$

($\Delta$ is the module function on $G$). To any strongly continuous unitary representation $\pi$ of $G$ on a Hilbert space $H_\pi$, is associated a *-representation $\pi^1$ of $L^1(G)$ given by

$$\pi^1(f) = \int_G f(x)\pi(x)dx .$$

The norm closure of $\pi^1(L^1(G))$ in the set of all bounded operators on $H_\pi$ form a C*-algebra denoted $C^*_\pi(G)$. Note that $\pi(G)$ is not in general contained in $C^*_\pi(G)$. However one has always $\pi(G) \subseteqq \overline{C^*_\pi(G)}^s$ (closure in the strong operator topology). If $G$ is a discrete group, then $\pi(G) \subseteqq C^*_\pi(G)$, and $C^*_\pi(G)$ is simply the norm closed linear span of $\pi(G)$.

Among all unitary representations of the group $G$, there are two particularly important, namely the *universal* representation, and the *regular* representation. The universal representation $\pi_u$ is the sum of all (equivalence classes) of cyclic representations of $G$, and the C*-algebra, associated with $\pi_u$ is denoted $C^*(G)$. For every unitary representation $\pi$ of $G$, $\pi^1$ has a lifting $\tilde{\pi}^1$ to $C^*(G)$:

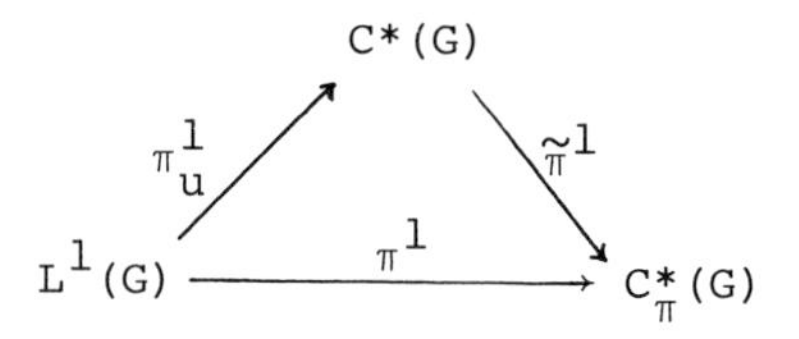

and $\tilde{\pi}^1$ is surjective. The regular representation $\lambda$ of $G$ acts on the Hilbert space $L^2(G)$ by left translations:

$$(\lambda(x)f)(y) = f(x^{-1}y), \quad f \in L^2(G), \qquad x,y \in G.$$

The associated C*-algebra, $C^*_\lambda(G)$ is called the reduced C*-algebra of $G$. It is known that the canonical homomorphism

$$\tilde{\lambda}^1 : C^*(G) \to C^*_\lambda(G)$$

is an isomorphism if and only if $G$ is *amenable* (a group $G$ is called amenable if there exists a positive functional $m$ on $L^\infty(G)$, such that $m(1) = 1$, and such that $m$ is invariant under left translations). Compact groups and solvable groups are amenable, while non compact semisimple Lie groups and free groups on at least two generators are not amenable.

General references to section 1: [5, § 13],[8].

## 2. Nuclear C*-algebras.

Let $A$ be a C*-algebra. We can assume that $A$ is a concrete C*-algebra, i.e. a norm closed *-subalgebra of the set of bounded operators $B(H)$ on some Hilbert space $H$. We let $A_+$ denote the set of positive operators in $A$. The set $M_n(A)$ of $n\times n$-matrices with elements in $A$ is a C*-algebra acting on the Hilbert space

$$H_n = \underbrace{H \oplus \cdots \oplus H}_{n}$$

Let $A$ and $B$ be two C*-algebras, let $T$ be a bounded linear operator from $A$ to $B$, and let $T^{(n)}$ be the linear operator from $M_n(A)$ to $M_n(B)$ obtained by letting $T$ act on each matrix element separately. $T$ is said to be *positive* if $T(A_+) \subseteq B_+$, and *completely positive* if $T^{(n)}(M_n(A)_+) \subseteq M_n(B)_+ \quad \forall n\in\mathbb{N}$. Moreover is said to be *completely bounded* if $\sup_{n\in\mathbb{N}} \|T^{(n)}\| < \infty$. In this case we write $\|T\|_{CB} = \sup_{n\in\mathbb{N}} \|T^{(n)}\|$.

An important class of C*-algebras is the class of *nuclear* C*-algebras.The original definition of a nuclear C*-algebra is related tensor products of C*-algebras. We will not give the full definitic here. Instead we can consider the following characterization of nuclear C*-algebras as a definition([3],[10]):

*A* C*-*algebra* $A$ *is nuclear if and only if there exists a net* $(T_\alpha)$ *of operators on* $A$ *, such that*

1) *Each* $T_\alpha$ *is a completely positive, finite rank operator on* A , *and* $\|T_\alpha\| \leqq 1$ .

2) $\|T_\alpha x - x\| \to 0$ *for all* $x \in A$ .

It is known that the strong closure $\bar{A}^s$ of a nuclear C*-algebra is a *hyperfinite* von Neumann algebra, i.e. there exists an increasing set $(M_\lambda)$ of finite dimensional subalgebras of $\bar{A}^s$ , such that $\bar{A}^s = \overline{\bigcup_\lambda M_\lambda}^s$ .

For any separable connected locally compact group G , and any representation $\pi$ of G it is known that $C^*_\pi(G)$ is nuclear ([4, § 6] and [2]). On the other hand, if G is a discrete group, then

$$C^*(G) \text{ nuclear} \Leftrightarrow C^*_\lambda(G) \text{ nuclear} \Leftrightarrow G \text{ amenable}$$

(cf. [12]). It follows that the reduced C*-algebra of a free group on at least two generators is not nuclear. The algebra of compact operators on a Hilbert space H is always nuclear, but the algebra B(H) of all bounded operators on H is not nuclear, when $\dim H = +\infty$ ([15]) .

## 3. Approximation properties for $C^*_\lambda(\mathbb{F}_2)$.

A Banach space E is said to have the *metric approximation property* (M.A.P.) if there exists a net $(T_\alpha)$ of finite rank operators on E , such that $\|T_\alpha\| \leqq 1$ for all $\alpha$ , and

$$\lim_\alpha \|T_\alpha x - x\| = 0 \qquad \forall x \in E .$$

It follows immediately from the characterization of nuclear C*-algebras stated above, that all nuclear C*-algebras have the M.A.P. The converse implication is false: The reduced C*-algebra of the free group on two generators $\mathbb{F}_2$ is not nuclear, but in [9] we proved that $C^*_\lambda(\mathbb{F}_2)$ has the M.A.P. This result has been sharpened in various directions in a recent joint work [ 1 ] with J. De Cannière. We state the main results in two Theorems:

Theorem I (cf. [1, § 4])

*Let* $A = C^*_\lambda(\mathbb{F}_2)$ . *There exists a sequence* $(T_k)_{k\in\mathbb{N}}$ *of positive finite rank operators on* A , *such that* $T_k(1) = 1$ *and* $\lim_{k\to\infty} \|T_k a - a\| = 0$ *for all* $a \in A$ .

Theorem II (cf. [1, § 3])

*Let* $A = C^*_\lambda(\mathbb{F}_2)$ . *There exists a sequence* $T_k$ *of finite rank operators on* A , *such that* $\|T_k\|_{CB} \leqq 1$ *for all* $k \in \mathbb{N}$, *and*

$$\lim_{k\to\infty} \|T_k a - a\| = 0 , \quad a \in A$$

( $\|T_k\|_{CB}$ *is the "complete bounded norm" of* $T_k$ *defined in section 2).*

Remarks.

a) It is known that if T is a positive operator on a unital C*-algebra, such that $T(1) = 1$ , then $\|T\| = 1$ (cf. [6]). Therefore both Theorem I and Theorem II implies that $C^*_\lambda(\mathbb{F}_2)$ has the metric approximation property.

b) There does not exist a sequence $(T_k)$ of operators on $C^*_\lambda(\mathbb{F}_2)$ satisfying the conditions in Theorem I and Theorem II at the same time. Indeed if $T_k(1) = 1$ and $\|T_k\|_{CB} \leqq 1$ , it is easy to see that $T_k$ is completely positive for all k , which contradicts the fact that $C^*_\lambda(\mathbb{F}_2)$ is not nuclear.

In the following two sections we will sketch parts of the proof of Theorem I and Theorem II.

4. Outline of the proof of Theorem I.

Let $G$ be a discrete group and let $\lambda$ be the regular representation of $G$. From section 1 we know that $\operatorname{span}\{\lambda(x) \mid x \in G\}$ is norm dense in $C^*_\lambda(G)$. We say that a function $\varphi$ on $G$ *multiplies* $C^*_\lambda(G)$ *into itself*, if there exists a bounded operator $M_\varphi$ on $C^*_\lambda(G)$, such that

$$M_\varphi \lambda(x) = \varphi(x)\lambda(x) \ , \quad x \in G \ .$$

Note that $M_\varphi$ is uniquely determined by this property. It is not hard to prove that when $\varphi$ is a *positive definite* function on $G$, then $\varphi$ multiplies $C^*_\lambda(G)$ into itself, $M_\varphi$ is completely positive and $\|M_\varphi\| = \varphi(e)$ (cf. [9, lemma 1.1]). The proof of Theorem I is based on the following two lemmaes:

Lemma A [9, lemma 1.2]

*For* $x \in \mathbb{F}_2$ *, let* $|x|$ *denote the length of the reduced word for* $x$ *. Then for any* $\lambda > 0$ *, the function* $\varphi_\lambda(x) = e^{-\lambda|x|}$ *is positive definite.*

Lemma B [1, § 4]

*Let* $\varphi$ *be a function on* $\mathbb{F}_2$ *, such that* $\varphi(x) = \overline{\varphi(x^{-1})}$ *,* $x \in \mathbb{F}_2$*, and*

$$(1+|x|)^4 |\varphi(x)| \leqq \varphi(e) \ , \quad x \in \mathbb{F}_2 \ ,$$

*then* $\varphi$ *multiplies* $C^*_\lambda(\mathbb{F}_2)$ *into itself, and* $M_\varphi$ *is positive.*

The proofs of these two lemmas are based on rather tricky combinatorical arguments which are not suited for presentation in short form. Therefore we will only show how these two lemmas enter in the proof of Theorem I:

Proof of Theorem I.

Put $\varphi_\lambda(x) = e^{-\lambda|x|}$ as in Lemma A , and define functions $\varphi_{\lambda,m}$ , $\lambda > 0$ , $m \in \mathbb{N}$ on $\mathbb{F}_2$ by

$$\varphi_{\lambda,m}(e) = 1 + \sup_{k>m}(1+k)^4 e^{-\lambda k}$$

$$\varphi_{\lambda,m}(x) = e^{-\lambda|x|} \quad , \quad 1 \leqq |x| \leqq m$$

$$\varphi_{\lambda,m}(x) = 0 \quad , \quad |x| > m \; .$$

For fixed $\lambda$ and $m$ , $\varphi_{\lambda,m}$ is a function on $G$ with finite support. By writing $\varphi_{\lambda,m}$ in the form

$$\varphi_{\lambda,m} = \varphi_\lambda + (\varphi_{\lambda,m} - \varphi_\lambda)$$

one gets from Lemma A and Lemma B, that $\varphi_{\lambda,m}$ multiplies $C^*_\lambda(\mathbb{F}_2)$ into itself, and that $M_{\varphi_{\lambda,m}}$ is positive. Moreover, since $M_{\varphi_{\lambda,m}-\varphi_\lambda}$ is also positive, it follows that for fixed $\lambda > 0$ :

$$||M_{\varphi_{\lambda,m}} - M_{\varphi_\lambda}|| = ||M_{\varphi_{\lambda,m}-\varphi_\lambda}|| = \varphi_{\lambda,m}(e) - \varphi_\lambda(e) \to 0 \quad \text{for} \quad m \to \infty \; .$$

Now put $\psi_{\lambda,m}(x) = \dfrac{\varphi_{\lambda,m}(x)}{\varphi_{\lambda,m}(e)}$ , then $M_{\psi_{\lambda,m}}$ is positive $||M_{\psi_{\lambda,m}}|| = 1$ and $\lim_{m\to\infty} ||M_{\psi_{\lambda,m}} - M_{\varphi_\lambda}|| = 0$ . Since $\varphi_\lambda$ is positive definite we also have $||M_{\varphi_\lambda}|| = \varphi_\lambda(e) = 1$ . Using that $\lim_{\lambda\to 0} \varphi_\lambda(x) = 1$ for all $x$ one gets that $\lim_{\lambda\to 0} ||M_{\varphi_\lambda} a - a|| = 0$ for all $a \in C^*_\lambda(\mathbb{F}_2)$ . Thus the identity on $C^*_\lambda(\mathbb{F}_2)$ belongs to the strong closure of $\{M_{\psi_{\lambda,m}} \mid \lambda>0 \; , \; m\in\mathbb{N}\}$ . Hence we can choose a sequence $T_k$ in this set, such that $\lim_k ||T_k a - a|| = 0$ , $a \in C^*_\lambda(\mathbb{F}_2)$ . Clearly each $T_k$ is positive and $T_k(1) = 1$ . Moreover, since the functions $\psi_{\lambda,m}$ have finite supports the operators $T_k$ are of finite rank. This completes the proof.

Remark.

With a little more effort one can prove that for each fixed $n \in \mathbb{N}$ , there exists a sequence $(T_k)_{k\in\mathbb{N}}$ of *n-positive* operators of finite rank on $A = C^*_\lambda(\mathbb{F}_2)$, such that $T_k(1) = 1$ and $\lim_{k\to\infty} \|T_k a - a\| = 0$ for all $a \in A$ (cf. [1, § 4]).

## 5. Outline of the proof of Theorem II.

The first lemma relates "multipliers" on $C^*_\lambda(G)$ to uniformly bounded (but not necessarily unitary) representations of a discrete group $G$ on a Hilbert space $H$.

Lemma C [1, § 2]

*Let $G$ be a discrete group and let $\pi$ be a uniformly bounded representation of $G$ on a Hilbert space $H$. Let $\xi, \eta \in H$. Then the function*

$$\varphi(x) = (\pi(x)\xi | \eta)$$

*multiplies $C^*_\lambda(G)$ into itself, $M_\varphi$ is completely bounded and* $\|M_\varphi\|_{CB} \leqq \|\pi\|^2 \|\xi\| \|\eta\|$.

Proof (Sketch)

Choose an orthonormal basis $(e_i)_{i\in I}$ in $H$, and put $a_i(x) = (\pi(x^{-1})\eta | e_i)$, $b_i(x) = (\pi(x^{-1})\xi | e_i)$, $x \in G$. Then $a_i, b_i \in \ell^\infty(G)$ and

$$\sum_i |a_i(x)|^2 \leqq \|\pi\|^2 \|\eta\|^2, \quad \sum_i |b_i(x)|^2 \leqq \|\pi\|^2 \|\xi\|^2, \quad x \in G.$$

Now consider the operator $T$ on $B(\ell^2(G))$ given by

$$T(x) = \sum_i a_i^* x b_i, \quad x \in B(\ell^2(G))$$

where $a_i, b_i \in \ell^\infty(G)$ are regarded as multiplication operators on $\ell^2(G)$. Then $T$ is completely bounded, and

$$\|T\|_{CB} \leqq \|\sum_i a_i^* a_i\|^{\frac{1}{2}} \|\sum_i b_i^* b_i\|^{\frac{1}{2}} \leqq \|\pi\|^2 \|\xi\| \|\eta\|.$$

An easy computation shows that $T\lambda(x) = \varphi(x)\lambda(x)$ (compare with the proof of [9, Lemma 1.1]). In particular $T(C^*_\lambda(G)) \subseteq C^*_\lambda(G)$. This proves Lemma C.

In [8,pp.10-11] it is shown that the matrices $\begin{pmatrix} 1 & 2 \\ 0 & 1 \end{pmatrix}$ and $\begin{pmatrix} 1 & 0 \\ 2 & 1 \end{pmatrix}$ generate a closed subgroup of $SL(2,\mathbb{R})$ isomorphic to the free group on two generators $\mathbb{F}_2$. We shall use some facts about unitary and non-unitary representations of $SL(2,\mathbb{R})$. Put $G = SL(2,\mathbb{R})$ and $K = SO(2) \subseteq G$. A continuous function $\varphi$ on $G$ is called a *spherical function* (with respect to $K$) if $\varphi \neq 0$, and

$$\int_K \varphi(xky)\,dk = \varphi(x)\varphi(y) \,, \quad x,y \in G \,.$$

($dk$ = normalized Haarmeasure on $K$). The spherical functions on $SL(2,\mathbb{R})$ are known explicitly. They are indexed by a complex parameter $s \in \mathbb{C}$, such that

$$\varphi_s\begin{pmatrix} a & 0 \\ 0 & a^{-1} \end{pmatrix} = \frac{1}{2\pi}\int_{-\pi}^{\pi} \frac{d\theta}{(a^2\sin^2\theta + a^{-2}\cos^2\theta)^{(s+1)/2}}$$

(cf. [13, p. 84]). This formula determines $\varphi_s$, because $\varphi_s(k_1xk_2) = \varphi_s(x)$, $k_1,k_2 \in K$, $x \in G$. For fixed $x \in G$, the function $s \to \varphi_s(x)$ is analytic. Moreover

1) $\varphi_s = \varphi_{-s}$, $s \in \mathbb{C}$

2) $\varphi_s$ is bounded, when $|\mathrm{Re}\, s| \leqq 1$

3) $\varphi_s$ is positive definite for $s \in i\mathbb{R}$ and for $s \in [-1,1]$

4) $\varphi_1(x) = \varphi_{-1}(x) = 1$, $x \in G$.

The set of unitary representations $(\pi_{it})_{t\in\mathbb{R}}$ induced by $\varphi_{it}$, $t \in \mathbb{R}$, is called the *class 1 principle series* of representations, and the set of unitary representations $\pi_\sigma$ induced by $\varphi_\sigma$, $0 < \sigma < 1$ is called the *complementary series* of representations. In [11] Kunze and Stein showed that there exists a family of continuous uniformly bounded representations $\rho_s$ of $SL(2,\mathbb{R})$, indexed by $s$ in the open strip $-1 < \text{Re}\, s < 1$ , such that all $\rho_s$ act on the same Hilbert space $H$ and such that

5) $\rho_s$ is unitary equivalent to $\pi_s$ for $s \in i\mathbb{R}$ , and $s \in \,]-1,1[$ .

6) For fixed $x \in G$ and $\xi,\eta \in H$ , the function $s \to (\rho_s(x)\xi|\eta)$ is analytic .

7) There exist constants $A_\sigma$ , $\sigma \in \,]-1,1[$ , such that for $s = \sigma+it$ , $|\sigma| < 1$ :

$$\|\rho_s\| \leqq A_\sigma(1+|t|)^{\frac{1}{2}} .$$

Using these facts and Lemma C, we get

Lemma D [1, § 3]

*Let us identify* $\mathbb{F}_2$ *with a closed subgroup of* $SL(2,\mathbb{R})$ *as above, and let* $\varphi_s'$ *be the restriction of the spherical function* $\varphi_s$ *to* $\mathbb{F}_2$, $s \in \mathbb{C}$ . *Then for each* $s$ *in the strip* $-1 < \text{Re}\, s < 1$ , $\varphi_s'$ *multiplies* $C^*_\lambda(\mathbb{F}_2)$ *into itself, and*

$$\|M_{\varphi_s'}\|_{CB} = A_\sigma^3(1+|t|)^{3/2} .$$

Proof (Sketch)

Since $\rho_{it}$ is equivalent to $\pi_{it}$ , $t \in \mathbb{R}$ , there exists $\xi_{it} \in H$ , such that $\varphi_{it}(x) = (\rho_{it}(x)\xi_{it}|\xi_{it})$ , $t \in \mathbb{R}$ , $x \in G$ . Using property 6) above one can prove that there exist $\xi_s,\eta_s \in H$ , $-1 < \text{Re}\, s < 1$ , such that

$$\varphi_s(x) = (\rho_s(x)\xi_s|\eta_s)$$

and $\|\xi_s\|\,\|\eta_s\| \leqq \|\rho_s\|$ . Taking restrictions to $\mathbb{F}_2$ , we get by Lemma C that $\varphi_s'$ multiplies $C^*_\lambda(\mathbb{F}_2)$ into itself and

$$\|M_{\varphi_s'}\|_{CB} \leqq \|\rho_s\|^3 = A_\sigma^3(1+|t|)^{3/2} .$$

Proof of Theorem II (Sketch)

For $0 < \sigma < 1$ , put $\varphi'_{\sigma,n}(x) = \int_{-\infty}^{\infty} f_n(t)\varphi'_{\sigma+it}(x)dt,\ x \in \mathbb{F}_2$ , whe

$f_n(t) = \frac{n}{\sqrt{\pi}} e^{-n^2t^2}$ , $t \in \mathbb{R}$ , $n \in \mathbb{N}$ . Then, using Lemma D and the analycity of the map $s \to \varphi'_s$ , one can prove that

$$\lim_{n\to\infty} \|M_{\varphi'_{\sigma,n}} - M_{\varphi'_\sigma}\|_{CB} = 0 .$$

It turns out that $|\varphi'_{\sigma,n}(x)|$ decreases rapidly when $|x| \to \infty$ , so by modifying the functions $\varphi'_{\sigma,n}$ slightly one can find functions $\psi'_{\sigma,n}$ on $\mathbb{F}_2$ with finite support, such that also

$$\lim_{n\to\infty} \|M_{\psi'_{\sigma,n}} - M_{\varphi'_\sigma}\|_{CB} = 0 .$$

Since $\varphi'_\sigma$ is positive definite, one has $\|M_{\varphi'_\sigma}\|_{CB} = \varphi'_\sigma(e) = 1$ . Therefore it is no loss of generality to assume that $\|M_{\psi'_{\sigma,n}}\|_{CB} \leqq 1$ for all $\sigma \in ]0,1[$ and $n \in \mathbb{N}$ . Finally, since $\lim_{\sigma\to 1} \varphi'_\sigma(x) = 1$ , $x \in G$ , one gets

$$\lim_{\sigma\to 1} \|M_{\varphi'_\sigma}a-a\| = 0 .$$

for all $a \in C^*_\lambda(\mathbb{F}_2)$ . The proof of Theorem II can now be complete by repeating the arguments in the end of the proof of Theorem I.

Remarks.

a) Theorem II is valid for any closed, discrete subgroup of SL(2, by the same proof. By use of [17] it can be extended to any close discrete subgroup of $SO_o(n,1)$ , $n \geqq 2$ .

b) The proofs of Theorem I and II are closely related to the stud of multipliers of the Fourier algebras of $\mathbb{F}_2$ and $SL(2,\mathbb{R})$ : Th Fourier algebra of a locally compact group G is the function al $A(G)$ on G , consisting of functions of the form $\varphi = f*\tilde{g}$ , $f,g \in L^2(G)$ (cf. [7]). $A(G)$ is a Banach algebra with norm

$$\|\varphi\|_{A(G)} = \inf\{ \|f\|_2 \|g\|_2 | \varphi = f*\tilde{g}\} .$$

It is well known that $G$ is amenable if and only if $A(G)$ has a bounded approximative unit. In [9, § 2] we proved that $A(\mathbb{F}_2)$ has an unbounded approximative unit, which is bounded in the multiplier norm

$$\|\psi\|_{MA(G)} = \sup\{ \|\psi\varphi\|_{A(G)} | \ \|\varphi\|_{A(G)} \leqq 1\} .$$

Using the same methods as in the proof of Theorem II, one can show that the Fourier algebra of $SL(2,\mathbb{R})$ (and more generally $SO_o(n,1)$) also admits an unbounded approximative unit, which is bounded in the multiplier norm (cf. [1, § 3]).

## 6. The slice map property.

Let $A \subseteq B(H)$ and $B \subseteq B(K)$ be C*-algebras. The spatial tensor product $A\otimes B$ of $A$ and $B$ is the normclosure of the subalgebra of $B(H\otimes K)$ generated by $\{a\otimes b | a\in A, b\in B\}$. When $\varphi \in A^*$, there exists a unique bounded linear map $R_\varphi : A\otimes B \to B$, such that

$$R_\varphi(a\otimes b) = \varphi(a)b \qquad a \in A , b \in B .$$

$R_\varphi$ is called the *slice map* associated to $\varphi$. A C*-algebra $A$ is said to have the *slice map property* if for all pairs of C*-algebras $(B,D)$ with $D \subseteq B$, one has

$$x \in A\otimes D \Leftrightarrow x \in A\otimes B \quad \text{and} \quad R_\varphi(x) \in D , \ \forall \varphi \in A^* .$$

It is known that all nuclear C*-algebras has the slice map property (cf. [14, Theorem 4.3] or [16, Prop. 10]). On the other hand it easily follows from the results of Wassermann in [15] and [16] that the full C*-algebra $C^*(\mathbb{F}_2)$ of the free group on two generators and the algebra $B(H)$ of all bounded operators on an infinite dimensional Hilbert space do not have the slice map property. The following application of Theorem II was brought to our attention by R. Archbold:

Corollary of Theorem II.

$C^*_\lambda(\mathbb{F}_2)$ *has the slice map property.*

Proof.

The proof of $A$ nuclear $\Rightarrow A$ has the slice map property given in [16, Prop. 10] is based on the existence of a net $(S_\alpha)$ of completely positive, finite rank contractions on $A$, such that $\|S_\alpha x-x\| \to 0$, $x \in A$. However the proof works equally well if one only assumes the existence of a net $(T_\alpha)$ of finite rank operators on $A$, such that $\sup_\alpha \|T_\alpha\|_{CB} < \infty$ and $\lim_\alpha \|T_\alpha a-a\| = 0$, $a \in A$.

References.

[1] J. De Cannière *and* U. Haagerup, *Multipliers of the Fourier algebras on semisimple Liegroups and their discrete subgroups preprint, december 1980 (?).*

[2] M.D. Choi *and* E.G. Effros, *Nuclear C*-algebras and injectivit the general case, Indiana Univ. J. Math. 26 (1977) 443-446.*

[3] M.D. Choi *and* E.G. Effros, *Nuclear C*-algebras and the appro: mation property, Amer. J. Math. 100 (1978) 61-79.*

[4] A. Connes, *Classification of injective factors, Ann. Math. 1 (1976) 73-116.*

[5] J. Dixmier, *Les C*-algèbres et leurs representations, Gauthi Villard, Paris 1969.*

[6] H.A. Dye *and* B. Russo, *A note on unitary operators in C*-algebras, Duke Math. J. 33 (1966) 413-416.*

[7] P. Eymard, *L'algèbre de Fourier d'un groupe localment compact, Bull. Soc. Math. France 92 (1964) 181-236.*

[8] F.P. Greenleaf, *Invariant means on topological groups, Van Nostrand, New York 1969.*

[9] U. Haagerup, *An example of a non nuclear C*-algebra which has the metric approximation property, Inventiones Math. 50 (1979) 279-293.*

[10] E. Kirschberg, *C*-nuclearity implies CPAP, Math. Nachr. 76 (1977) 203-212.*

[11] R.A. Kunze *and* E.M. Stein, *Uniformly bounded representations and Harmonic analysis of the 2×2 real unimodular group, Amer. J. Math. 82 (1960) 1-62.*

[12] E.C. Lance, *On nuclear C*-algebras, J. Funct. Analysis 12 (1973) 157-176.*

[13] S. Lang, $SL_2(\mathbb{R})$, *Addison-Wesley 1975.*

[14] J. Tomiyama, *Some aspects of the commutation Theorem for Tensor products of Operator algebras, Algèbres d'operateurs et leurs applications en physique Théorique, 417-436, Édition C.N.R.S. 1979.*

[15] S. Wassermann, *On tensor products of certain group C*-algebras, J. Funct. Analysis 23 (1976) 239-254.*

[16] S. Wassermann, *A pathology in the ideal space of L(H)⊗L(H), Indiana J. Math. Sci. 1 (1978) 209-215.*

[17] E.N. Wilson, *Uniformly bounded representations of the Lorentz groups, Trans. Amer. Math. Soc. 166 (1972) 431-438.*

J. Haagerup,
Department of Mathematics,
Odense University,
DK-5230 Odense M,
Denmark.

# CONNECTEDNESS THEOREMS IN ALGEBRAIC GEOMETRY

Johan P. Hansen*

## §1. Introduction

The main objectives of this exposition are some recent applications and improvements of the following generalized Bertini theorem, obtained jointly with W. Fulton [5].

Unless mentioned otherwise, varieties will be complete, but possibly singular, over an algebraically closed field of arbitrary characteristic.

Theorem 1. Let $P = \mathbb{P}^m \times \ldots \times \mathbb{P}^m$ be the product of $r$ copies of projective m-space, and let $\Delta$ be the image of the diagonal imbedding of $\mathbb{P}^m$ in $P$. If $X$ is an irreducible variety, $f: X \to P$ a morphism, with $\dim f(X) > \operatorname{codim}(\Delta, P) = (r-1)m$, then $f^{-1}(\Delta)$ is connected. □

An immediate consequence in the following disguised existence assertion.

Proposition ([5, Prop. 2]). Let $X$ be an irreducible variety of dimension $n$, $f: X \to \mathbb{P}^m$ an unramified morphism, with $2n > m$. Then $f$ is a closed imbedding. □

K. Johnson [12] proved this when $f$ is the restriction of a linear projection from a projective space of dimension at most $2n$.

*) Supported by the Danish Natural Science Research Council

We suggested the theorem as an approach to generalize his result.

The theorem and the proposition have several interesting applications to intersections and singularities of mappings for which the reader is referred to the original paper [5].

T. Gaffney and R. Lazarsfeld [6] used the theorem to obtain striking results for branched coverings of projective space. For $x \in X$, they define the local degree $e_f(x)$ of $f$ at $x$, intuitively the number of sheets of the branched covering $f:X \to \mathbb{P}^m$ that come together at $x$. There main assertion is that there exist points $x \in X$, with $e_f(x) \geq \min(d, n+1)$.

Perhaps the strongest application of the theorem is W. Fulton's proof of the algebraic part of O. Zariski's theorem about the fundamental group of the complement of a plane node curve [3]. Subsequentially P. Deligne [1] obtained a topological version of the connectedness theorem implying the topological part of Zariski's theorem. For an account of this, the reader is referred to W. Fulton's article in these proceedings.

In [8] a refinement of the theorem was obtained (see 2.5) and used to study higher order singularities of a finite morphism $f:X^n \to \mathbb{P}^m$.

In a letter to W. Fulton [15], F.L. Zak gives a detailed account of his applications of the connectedness theorem to the question of tangency of a linear space with respect to a subvariety of projective space. He proceeds to outline, among other things, a proof of R. Hartshorne's conjecture on linear normality [10]. Looking forward to F.L. Zak's publication of his work, we refer the reader to W. Fulton and R. Lazarsfeld [4], J. Roberts [13], as well as the last section of this paper for an exposition of these interesting results.

In section 2 we outline further developments of the theorem, including the higher homotopy version due to P. Deligne, M. Goresky and R. Mac Pherson, a connectedness theorem for flagmanifolds and Grassmannians and G. Faltings' recent connectedness theorem for homogeneous spaces.

## §2. Improvements and generalizations.

### 2.1. Flagmanifolds and Grassmannians.

Any sequence of integers $0 \leq a_0 < .. < a_k < m$ determines a flagmanifold $F = F(a_0, \ldots, a_k; m)$ on $\mathbb{P}^m$, whose points are nested sequences $L_0 \subset \ldots \subset L_k$ of linear subspaces of $\mathbb{P}^m$ with $\dim(L_i) = a_i$. In particular $F(n;m)$ is the Grassmannian of projective n-planes in $\mathbb{P}^m$. In [8] the following connectedness theorem was obtained.

Theorem 2. Let $F$ denote any flagmanifold of flags in $\mathbb{P}^m$, and $\Delta_F$ the image of the diagonal embedding of $F$ in $F \times F$. For any morphis[m] $f: X \to F \times F$ from an irreducible variety $X$:

(i) $f^{-1}(\Delta_F) \neq \emptyset$ if $\operatorname{codim}(f(X), FxF) \leq m$

(ii) $f^{-1}(\Delta_F)$ is connected if $\operatorname{codim}(f(X), FxF) < m$. □

In case we take flags to be just points, the theorem is precisely the connectedness theorem for projective space.

Examples show that for any flagmanifold, the theorem is the best possible [8].

The proof is quite geometric and reduces the assertions to a refined version of theorem 1 (see 2.5).

### 2.2. Ampleness and connectivity.

All examples we know of a non-singular subvariety Y in a non-sing[ular] variety Z, such that $f^{-1}(Y)$ is connected for all morphisms $f: X \to Z$ with $\dim f(X) > \operatorname{codim}(Y,Z)$, enjoy the property that the normal bundle of Y in Z is ample in the sense of R. Hartshorne [9]. E.g. (1) $Z = \mathbb{P}^m$, Y arbitrary, (2) $Z = \mathbb{P}^m x \ldots x \mathbb{P}^m$, $Y = \Delta_{\mathbb{P}^m}$, (3) Y a diviso[r] on $Z$ with ample normal bundle.

In [5] it was conjectured that in the situation above $f^{-1}(Y)$ is always connected, provided the normal bundle of $Y$ in $Z$ is ample. However examples of R. Hartshorne [9] and H. Hironaka [11] are also counterexamples to this conjecture.

Example. For any prime $p$, consider the group action of the cyclic group $G$ of order $2p$ on $\mathbb{P}^{2p-1}$ (over $\mathbb{C}$)

$$(x_0:\ldots:x_i:\ldots:x_{2p-1}) \to (x_0:\ldots:\xi^i x_i:\ldots:\xi^{2p-1}x_{2p-1}),$$

where $\xi$ is a $2p$'th primitive rooth of unity.

$X = \mathbb{P}^{2p-1}$, $f:X \to X/G = Z$ the quotient map and if $L = V(x_0-x_1,\ldots,x_{2p-2}-x_{2p-1})$, let $Y = f(L)$. Then $f$ restricts to an isomorphism $L \simeq Y$ and therefore induces an isomorphism on normal bundles $N_{Z/Y} \simeq N_{\mathbb{P}^{2p-1}/L}$, hence $N_{Z/Y}$ is ample. Also $\operatorname{codim}(Y,Z) = p$. However, the inverse image of $Y$

$$f^{-1}(Y) = L \cup \xi(L) \cup \ldots \cup \xi^{2p-1}(L)$$

is the union of $2p$ mutually disjoint $(p-1)$-planes.

Finally we note that $Y$ does not meet the singular part of $Z$, consisting of images under $f$ of points where $G$ does not act freely, i.e., points on one of the $(p-1)$-planes $V(x_0=x_2=\ldots=x_{2p-2})$ or $V(x_1=x_3=\ldots=x_{2p-1})$. By resolution of singularities, the above example modifies to one where all the spaces are non-singular. □

In spite of this example, the possitivity of the normal bundle of $Y$ in $Z$ appears to be intimately related to the connectivity of $f^{-1}(Y)$ for morphisms $f:X \to Z$.

A. Sommese [14] has the notion of k-ample bundles. One expects that $f^{-1}(Y)$ would be connected if $\dim f(X) > \operatorname{codim}(Y,Z) + k$ where the normal bundle of Y in Z is k-ample.

Although false, even for $k = 0$, this does give the precise estimate for $Y = \Delta_G$ in $Z = G \times G$, G a Grassmannian [8].

Below we discuss G. Faltings' recent connectedness theorem for homogenous spaces [2] - also in that case, the degree of ampleness of the normalbundle provides the right estimate.

## 2.3. Homogenous spaces. (char. $k = 0$)

Let Z be an irreducible variety, defined over a field k of characteristic 0, equipped with a linear algebraic group acting transitively. For an irreducible subvariety $Y \subset Z$, $\hat{Z}$ denotes the formal completion of Z along Y.

Assume that $H^q(\hat{Z}, \mathcal{O}_{\hat{Z}})$ is finite dimensional for $0 \leq q < s$ - in case Y and Z are smooth, this amounts to the normal bundle of Y in Z being $(\dim(Y) - s)$ - ample in the sense of A. Sommese [14] (see [9]).

G. Faltings [2, p. 32] then proves the following theorem.

Theorem 3. In the notation above, let $f: X \to Z$ be a morphism from an irreducible variety X.

(i) If $\Gamma(\hat{Z}, \mathcal{O}_{\hat{Z}}) \neq 0$ and $\dim f(x) \geq \dim(Z) - s$, then $f^{-1}(Y) \neq \emptyset$.

(ii) If $\Gamma(\hat{Z}, \mathcal{O}_{\hat{Z}}) = k$ and $\dim f(x) > \dim(Z) - s$, then $f^{-1}(Y)$ is connected. □

This result recovers theorem 1 and 2, at least in characteristic 0. Most interesting, it is further evidence of the importance of ampleness in connectivity questions as discussed in 2.2.

## 2.4. Higher homotopy.

In [1], P. Deligne suggested and proved a topological version of theorem 1 over $\mathbb{C}$; he also conjectured the following higher homotopy statement for complex varieties.

Theorem 4. Let $X$ be a local complete intersection of pure dimension $n$. Let $f: X \to \mathbb{P}^m \times \mathbb{P}^m$ be a finite morphism. Denote by $\Delta$ the diagonal in $\mathbb{P}^m \times \mathbb{P}^m$.

a) If $n - m \geq 1$, then $\pi_1(X, f^{-1}(\Delta))$ is trivial

b) If $n - m \geq 2$, one has an exact sequence

$$\pi_2(f^{-1}(\Delta)) \to \pi_2(X) \to \mathbb{Z} \to \pi_1(f^{-1}(\Delta)) \to \pi_1(X) \to 1$$

c) If $2 < i \leq n - m$, then $\pi_i(x, f^{-1}(\Delta)) = 0$. □

The theorem is implied by a result of M. Goresky and R. Mac Pherson [7], conjectured by P. Deligne in the smooth case. For a detailed and excellent account of this, the reader is referred to [4].

## 2.5. An $\ell$-connectedness theorem

In [8] a refinement of theorem 1 is obtained in terms of $\ell$-connectedness. This is essential to the study of higher order singularities of finite morphisms to projective space [8], as well as to the proof of theorem 2.

Definition. Fix $\ell \geq 0$. A variety $X$ is said to be $\ell$-connected, if all its irreducible components have dimension greater or equal to $\ell + 1$, and for all its closed subvarieties $V$, with $\dim(V) < \ell$, $X \setminus V$ is connected. □

With this definition, 0-connected means that all irreducible components of $X$ are at least curves and $X$ is connected. It is important to note that an irreducible variety $X$ of dimension $\geq 1$ is $(\dim(X)-1)$-connected, but not $\dim(X)$-connected.

In this notation theorem 1 improves as follows.

Theorem 5. Let $P = \mathbb{P}^m \times \ldots \times \mathbb{P}^m$ be the product of $r$ copies of projective m-space, and let $\Delta$ be the image of the diagonal embedding of $\mathbb{P}^m$ in $P$. Let $d = \operatorname{codim}(\Delta, \mathbb{P}^m)$. Assume $X$ is $\ell$-connected, $\ell \geq d$, and consider a morphism $f: X \to P$.

(i) If $f$ is finite, then $f^{-1}(\Delta)$ is $(\ell-d)$-connected.

(ii) If for all $p \in P \setminus \Delta$: $\dim(f^{-1}(p)) \leq \ell-d$, then $F^{-1}(\Delta)$ is 0-connected. □

## §3. On F.L. Zak's work

In this section we will assume that $X \subseteq \mathbb{P}^m$ is an irreducible, non-singular subvariety of dimension $n$ in projective m-space.

For a linear subspace $L \subseteq \mathbb{P}^m$ of dimension $k$ $(n \leq k < m)$, F.L. Zak [15] studies the loci of tangency with respect to X, i.e.,

$$\{x \in X | T_{X,x} \subseteq L\},$$

where $T_{X,x}$ is the projective tangent space to $X$ at $x$. We will be particularly interested in the following theorem. For more detailed expositions of F.L. Zak's work see [4] and [13].

Theorem 6. Assume $X$ is not contained in a k-plane, then in the notion above, $\{x \in X | T_{X,x} \subseteq L\}$ has dimension at most k-n. □

The proof is an elegant application of the connectedness theorem. Namely let $Y$ be an irreducible component of $\{x \in X | T_{X,x} \subseteq L\}$ and let $y \in Y$. Pick a point $x \in X \setminus L$, which can be done by assumption. The secant line $\overline{xy}$ is not contained in $X$ as $T_{X,y}$ is contained in $L$ whereas $\overline{xy}$ is not. Finally choose a linear space $M \subseteq \mathbb{P}^m$ of codimension $k+1$ subject to the three conditions: $M \cap L = \emptyset$, $M \cap X = \emptyset$, $M \cap \overline{xy} \neq \emptyset$. Projection from $M$ then induces a finite morphism $P_M: X \to \mathbb{P}^k$ such that $P_M$ does not ramify along $Y$ and

$P_M(x) = P_M(y)$.

Consider the morphism

$$f = P_M \times P_{M|Y} : X \times Y \to \mathbb{P}^k \times \mathbb{P}^k.$$

Clearly $\Delta_Y \subset f^{-1}(\Delta_{\mathbb{P}^k})$ and we have by construction $(x,y) \in f^{-1}(\Delta_{\mathbb{P}^k}) \smallsetminus \Delta_Y$. Assuming $\dim Y > k-n$, the connectedness theorem applies to give that $f^{-1}(\Delta_{\mathbb{P}^k})$ is connected. In particular, a component of $f^{-1}(\Delta_{\mathbb{P}^k})$ other than $\Delta_Y$ must meet $\Delta_Y$. I.e., double points of $P_M$ come together at Y, giving rise to a ramification point of $P_M$ on Y. This contradicts the fact that $T_{X,y} \subseteq L$ for all $y \in Y$.

Corollary. Let $X^n$ be an irreducible, non-singular subvariety of $\mathbb{P}^m$. Assume X is not linear, then the Gauss-map

$$X \to G(n,m),$$

defined by $X \to T_{X,x}$, is finite. □

Corollary. Let X be an irreducible, non-singular subvariety that generates $\mathbb{P}^m$. Let $X^\vee \subset \mathbb{P}^{m^\vee}$ be the dual variety of X. Then $\dim X^\vee \geqq \dim X$. □

This is certainly not true for singular varieties. It would be interesting to understand the interplay between the singularieties of X, $X^\vee$ and the fibres of the Gauss-map.

Remark. In case $X^n$ generates $\mathbb{P}^m$, then an arbitrary hyperplane $H \subset \mathbb{P}^m$ is by the theorem tangent to $X$ in dimension at most $m-n-1$, i.e., $\dim \operatorname{sing}(X \cap H) \leq m-n-1$.

In particular, if $2n > m$ then $X \cap H$ is reduced, and if $2n > m+1$ then $X \cap H$ is normal, hence irreducible.

In case $X$ is a non-degenerate complete intersection, then R. Lazarsfeld obtains a much stronger result [4]. Namely a hyperplane can be tangent to $X$ at only finitely many points.

F. L. Zak proceeds to prove the following assertion, which is R. Hartshorne's conjecture on linear normality [10].

Theorem 7. Let $X \subset \mathbb{P}^m$ be an irreducible, non-singular subvariety that generates $\mathbb{P}^m$. If $3 \dim X > 2(m-2)$, then $X$ can not be projected to $\mathbb{P}^{m-1}$ without acquiring singularities. □

The main line of proof is that to a variety, which can be projected isomorphically onto its image, one can exhibit hyperplanes having large loci of tangency with respect to $X$. On the other hand theorem 6 bounds the dimension of such loci.

R. Lazarsfeld (see [4]) has given an alternate proof of the theorem. His proof is an intriguing - but direct - application of the connectedness theorem.

## R E F E R E N C E S


1. P. Delique, Le groupe fondamental du complément d'une courbe plane n'ayant que des points doubles ordinaires est abélien, Seminaire Bourbaki, 1979/80, n° 543.

2. G. Faltings, Formale Geometrie und homogene Räume, preprint Münster (Westf.) 1980.

3. W. Fulton, On the fundamental group of the complement of a node curve, Ann. of Math. 111 (1980), 407-409.

4. W. Fulton and R. Lazarsfeld, Connectivity and its applications in algebraic geometry, to appear in the Proc. of the Midwest Alg. Geo. Conf. 1980.

5. W. Fulton and J. Hansen, A connectedness theorem for projective varieties, with applications to intersections and singularities of mappings, Ann. of Math. 110 (1979), 159-166.

6. T. Gaffney and R. Lazarsfeld, On the ramification of branched coverings of $\mathbb{P}^m$, Inventiones Math. 59 (1980), 53-58.

7. M. Goresky and R. MacPherson, Stratified Morse Theory, preprint.

8. J. Hansen, A connectedness theorem for flagmanifolds and Grassmannians and singularieties of morphisms to $\mathbb{P}^m$, Thesis, Brown University (1980).

9. R. Hartshorne, Ample Subvarieties of Algebraic Varieties, Springer Lecture Notes 156, Springer Verlag, Berlin -Heidelberg-New York, 1970.

10. R. Hartshorne, Varieties of small codimension in projective space, BAMS. 80 (1974), 1017-1032.

11. H. Hironaka, On some formal imbeddings, Ill. J. Math. 12 (1968), 587-602.

12. K. Johnson, Immersion and embedding of projective varieties, Thesis, Brown University (1976), revised in Acta Math. 140 (1978), 49-74.

13. J. Roberts, On Zak's proof of Hartshorne's conjecture on linear normality, preprint.

14. A. J. Sommese, Submanifolds of Abelian Varieties, Math. Ann. 233 (1978), 229-256.

15. F. L. Zak, Letter to W. Fulton, dated Dec. 26, 1979.


/EM/AP


Johan P. Hansen
Matematisk Institut
Aarhus Universitet
8000 Aarhus C
Denmark

# CLASSIFICATION IN PROJECTIVE ALGEBRAIC GEOMETRY

By Audun Holme (Bergen).

## §1. The basic problem.

Our starting point is the following general

CLASSIFICATION PROBLEM. *For all positive integers* $n$ *and* $N > n$, *classify all non singular, closed subvarieties of* $\mathbb{P}_k^N$.

Here $k$ denotes an algebraically closed field, for instance the complex numbers $\mathbb{C}$. No essential aspects of this talk will be lost by replacing $k$ by $\mathbb{C}$ everywhere.

There exist many approaches to this problem, and of course the word "classify" may be understood in different ways.

Let $V(N,n)$ denote the subvarieties referred to, and consider the following version of the problem, consisting of three parts:

1. GOOD INVARIANTS. *Find a good set of numerical invariants for the set* $V(N,n)$,

$$i: \; V(N,n) \longrightarrow \mathbb{Z}^{m(N,n)} = \mathbb{Z}^m$$

We put $i(X) = (i_o(X), \ldots, i_{m-1}(X))$.

2. ALLOWABLE RANGE. *Describe explicitly the subset of* $\mathbb{Z}^m$:

$$I(N,n) = \{i(X) \mid X \varepsilon V(N,n)\}$$

We refer to points in $I(N,n)$ as *allowable points*.

3. ENUMERATION. <u>For all allowable points</u> $i$ , <u>give a good description of the set</u>

$$V(N,n)_i = \{X \varepsilon V(N,n) | i(X) = i\}$$

If the invariants are taken to be the coefficients of the Hilbert polynomial, then $V(N,n)_i$ is parametrized by an open subset of the Hilbert scheme.

A minimal requirement for 3 would then be to know the number of irreducible components of this scheme and their dimensions. Even for $n = 1$ , no general answer to this appears to be known, although some information is available [1).]

In the case of curves, these invariants are equivalent to

$$d = \deg(X) , \quad g = \mathrm{genus}(X) .$$

Thus part 2 above is now equivalent to the following problem [2)]:

4. GENERA OF SMOOTH CURVES IN $P^3$ . <u>Determine all points</u> $(d,g) \varepsilon \mathbb{Z}^2$ <u>such that there exists a non singular curve</u> $X \subset \mathbb{P}^3$ <u>with</u>

$$d = \deg(X) , \quad g = \mathrm{genus}(X) .$$

Here there is a conjecture, actually stated as a theorem by G. Halphen in 1882, but appearently without an adequate proof. The situation may be summarized in the following diagram:

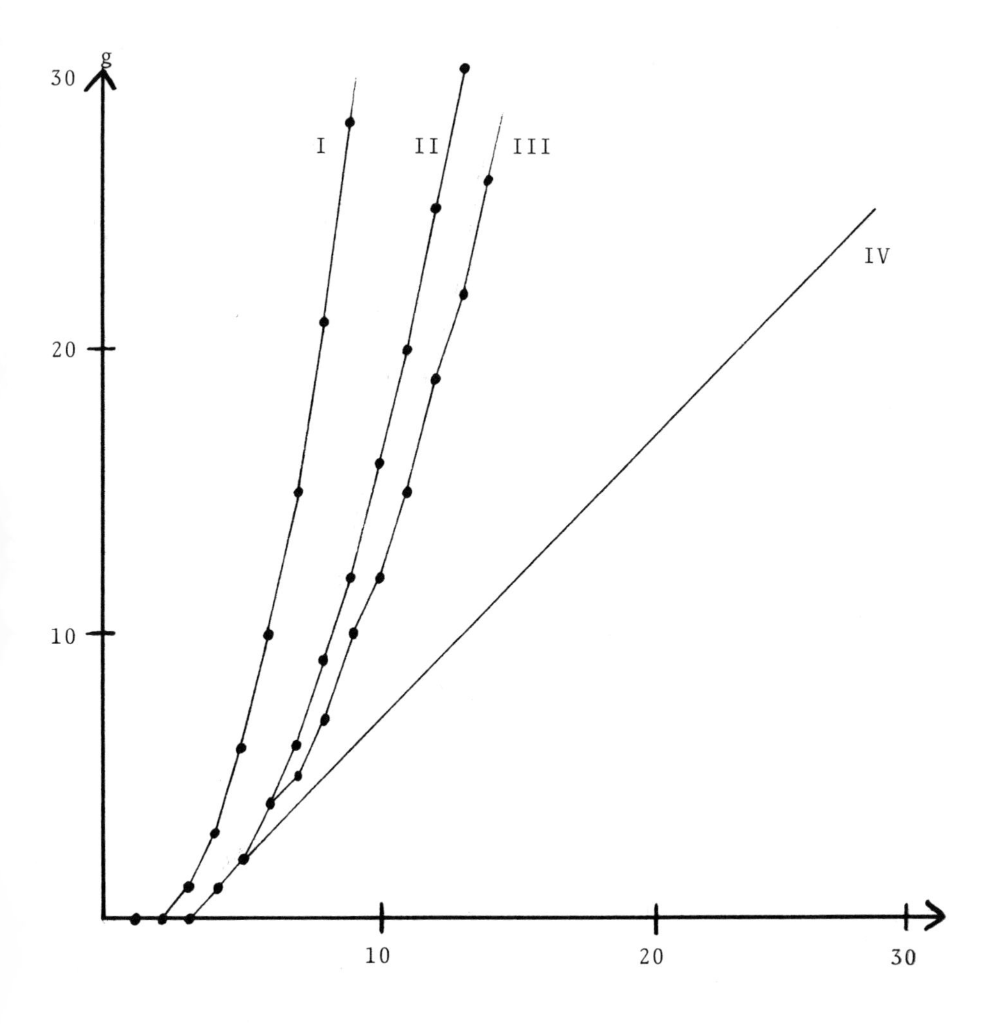


I is the curve $g = \frac{1}{2}(d-1)(d-2)$ , II is given by $g = \left[\frac{1}{4}d^2-d+1\right]$ and III is defined by $g = \left[\frac{1}{6}d(d-3)+1\right]$ . IV is the line $g = d - 3$ .

Clearly there are no allowable points to the left of I . On I we have the plane curves, and between I and II there are again no allowable points. On II all points are allowable, and between II and III there are some allowable points, they can be described explicitly. In fact, these curves all lie on the quadratic surface $Q \simeq \mathbb{P}^1 \times \mathbb{P}^1$ , and up to linear equivalence such curves are parametrized by $\mathbb{Z}^+ \times \mathbb{Z}^+$ . If $(a,b) \in \mathbb{Z}^+ \times \mathbb{Z}^+$ , then a corresponding curve has

$$d = a + b , \quad g = (a-1)(b-1) .$$

However, there are non singular curves on $Q$ outside the given area.

For $d \leq 30$ , we have the following allowable points between II and III :

| d | Values of g |
|---|---|
| 8 | 8 |
| 9 | None |
| 10 | 15 |
| 11 | 18 |
| 12 | 21, 24 |
| 13 | 24, 28 |
| 14 | 27, 32, 35 |
| 15 | 36, 40 |
| 16 | 40, 45, 48 |
| 17 | 44, 50, 54 |
| 18 | 48, 55, 60, 63 |
| 19 | 52, 60, 66, 70 |
| 20 | 65, 72, 77, 80 |
| 21 | 70, 78, 84, 88 |
| 22 | 75, 84, 91, 96, 99 |
| 23 | 80, 90, 98, 104, 108 |
| 24 | 96, 105, 112, 117, 120 |
| 25 | 102, 112, 120, 126, 130 |
| 26 | 108, 119, 128, 135, 140, 143 |
| 27 | 114, 126, 136, 144, 150, 154 |
| 28 | 120, 133, 144, 153, 160, 165, 168 |
| 29 | 140, 152, 162, 170, 176, 180 |
| 30 | 147, 160, 171, 180, 187, 192, 195 |

On and below the line IV all points are allowable. We are left with the

5. OPEN QUESTION. _Are all points between_ III _and_ IV _allowable_?

We shall return to this question in section 3.

Ideally the set of invariants asked for in question 1 should be so large that $V(N,n)_i$ is reasonably small, say parametrized by a variety. Since it is sensible to require the coefficients of the Hilbert polynomial or an equivalent set of invariants as part of our system, we are faced with the problem of adding new invariants which separate the irreducible components of the open subset of the Hilbert scheme referred to above. This appears to be difficult, even for curves [3].

From a different point of view, one could hope for such an adequate set of invariants that $V(N,n)_i$ could be classified up to projective equivalence by a (possibly infinite) list of canonical forms [4].

## §2. Numerical invariants.

My own interest in numerical projective invariants comes from computing the minimal dimension of the projective space into which a given projective variety can be embedded. For simplicity we keep the assumption of non-singularity for $X$ .

There are two rather different problems:

1. PROJECTIVE EMBEDDING OBSTRUCTION. *For* $X \in V(N,n)$ , *and* $N > m \geq n$ , *give a numerical criterion for when* $X$ *can be embedded into* $\mathbb{P}^m$ *by a generic projection from* $P^N$ .

2. ABSOLUTE EMBEDDING OBSTRUCTION. *With notation as in* 1, *give a numerical criterion for when* $X$ *can be embedded into* $\mathbb{P}^m$ .

Problem 1 has a simple answer. In fact, let $c(X) = c(T_X)$ be the total Chern class of $X$ and $s(X) = 1/c(X)$ the total Segre class. Put

$$d_i = \deg(s_i(X)) = s_i(X).D^{n-i}$$

where the divisor $D$ is a hyperplane section. Then

$$d_0 = \deg(X) , \ d_1 = -\deg c_1(X) , \ d_2,\ldots,d_n$$

are projective invariants of $X$ , easily expressed in terms of the degrees of the monomials in the Chern classes, namely the following set of projective invariants:

$$\left\{ c_1(X)^{i_1} \cdots c_n(X)^{i_n} D^{n-(i_1+\cdots+i_n)} \;\middle|\; \begin{array}{l} i_1,\ldots,i_n \geq 0 \\ i_1+2i_2+\cdots+ni_n \leq n \end{array} \right\}$$

We shall refer to this as the <u>projective Chern numbers</u>. These include in particular the usual Chern numbers $c_1(X)^n, \ldots, c_n(X)$, and also satisfy the requirement that the Hilbert polynomial can be recovered from them.

Now write

$$\beta_m = d_o^2 - \sum_{\ell=0}^{m-n} \binom{m+1}{m-n-\ell} d_\ell$$

Then we have the following [5]:

THEOREM. X *may be embedded into* $\mathbb{P}^m$ *by a generic projection from* $\mathbb{P}^N$ *if and only if* $\beta_m = 0$.

Today this is understood as a corollary of an assertion on the degree of the double point locus of the morphism induced by the generic projection. I.e., as part of the theory of singularities of morphisms [6].

As for the second problem, the projective Chern numbers are probably not sufficient in general. New, additional invariants seem to be required. In fact, it appears to be necessary to study singularities of morphisms which are non-generic in a certain sense [7].

## §3. Computing d,g for curves.

We now concentrate on the problem of finding the allowable range for the projective Chern numbers. In order to get numerical information on this range, we proceed as follows:

1. Fix the integers $n \geq 1$ and $N \in [n+1, 2n+1]$ .

2. Fix a general family $F \subset V(N,n)$ , parametrized up to an available equivalence by some set of parameters, say positive integers.

3. Express the projective Chern numbers in terms of the parameters.

4. Fix some bounded domain $B \subset \mathbb{Z}^m$ , and compute all projecti Chern numbers from $F$ in $B$ .

This was carried out in section 1, to give all allowable (d,g) between II and III .

In order to obtain information on the allowable points between III and IV , we study the families of curves in V(3,1) listed below. Of course there are many other similar families which one may think of, but the numerical tables which follow seem to indicate that the present list contains the most interesting ones.

At least from the point of view of finding the allowable range for $(d,g)$. Actually A and B can probably be disregarded, perhaps even C.

A. COMPLETE INTERSECTIONS. Here $d = ab$, $g = \frac{1}{2}d(a+b-4)+1$. For $d \leq 50$ we have the following list of such curves in the area between III and IV:

| d \ g | | d \ g | | d \ g | |
|---|---|---|---|---|---|
| | | 32 | 129 | 48 | 241, 289 |
| 16 | 33 | 35 | 141 | 49 | 246 |
| 20 | 51 | 36 | 145, 163 | 50 | 276 |
| 24 | 73 | 40 | 181, 201 | | |
| 25 | 76 | 42 | 190 | | |
| 28 | 99 | 44 | 243 | | |
| 30 | 106 | 45 | 226 | | |

As we see, there are remarkably few complete intersection curves in this area, and those which do exist are in a rather uninteresting part of it, as we shall see below.

B. CURVES ON THE QUADRIC SURFACE. These curves were described in section 1. In the area between III and IV we find the following points corresponding to such curves for $d \leq 50$.

| d \ g | | | | | |
|---|---|---|---|---|---|
| 11 | 14 | | | | |
| 12 | 16 | | | | |
| 13 | 18 | | | | |
| 14 | 20 | 27 | | | |
| 15 | 22 | 30 | | | |
| 16 | 24 | 33 | | | |
| 17 | 26 | 26 | | | |
| 18 | 28 | 39 | | | |
| 19 | 30 | 42 | | | |
| 20 | 32 | 45 | | | |
| 21 | 34 | 48 | 60 | | |
| 22 | 36 | 51 | 64 | | |
| 23 | 38 | 54 | 68 | | |
| 24 | 40 | 57 | 72 | | |
| 25 | 42 | 60 | 76 | | |
| 26 | 44 | 63 | 80 | 95 | |
| 27 | 46 | 66 | 84 | 100 | |
| 28 | 48 | 69 | 88 | 105 | |
| 29 | 50 | 72 | 92 | 110 | 126 |
| 30 | 52 | 75 | 96 | 115 | 132 |

| d \ g | | | | | | | | | |
|---|---|---|---|---|---|---|---|---|---|
| 31 | 54 | 78 | 100 | 120 | 138 | | | | |
| 32 | 56 | 81 | 104 | 125 | 144 | | | | |
| 33 | 58 | 84 | 108 | 130 | 150 | | | | |
| 34 | 60 | 87 | 112 | 135 | 156 | 175 | | | |
| 35 | 62 | 90 | 116 | 140 | 162 | 182 | | | |
| 36 | 64 | 93 | 120 | 145 | 168 | 189 | | | |
| 37 | 66 | 96 | 124 | 150 | 174 | 196 | | | |
| 38 | 68 | 99 | 128 | 155 | 180 | 203 | | | |
| 39 | 70 | 102 | 132 | 160 | 186 | 210 | 232 | | |
| 40 | 72 | 105 | 136 | 165 | 192 | 217 | 240 | | |
| 41 | 74 | 108 | 140 | 170 | 198 | 224 | 248 | | |
| 42 | 76 | 111 | 144 | 175 | 204 | 231 | 256 | | |
| 43 | 78 | 114 | 148 | 180 | 210 | 238 | 264 | | |
| 44 | 80 | 117 | 152 | 185 | 216 | 245 | 272 | 297 | |
| 45 | 82 | 120 | 156 | 190 | 222 | 252 | 280 | 306 | |
| 46 | 84 | 123 | 160 | 195 | 228 | 259 | 288 | 315 | |
| 47 | 86 | 126 | 164 | 200 | 234 | 266 | 296 | 324 | |
| 48 | 88 | 129 | 168 | 205 | 240 | 273 | 304 | 333 | |
| 49 | 90 | 132 | 172 | 210 | 246 | 280 | 312 | 342 | 370 |
| 50 | 92 | 135 | 176 | 215 | 252 | 287 | 320 | 351 | 380 |

This is still a relatively small number of curves in the area under examination. Also, as we shall see below, most of them are not located in the interesting part of that area.

C. CURVES ON THE NON SINGULAR CUBIC. [8] The non singular curves of degree $d \geq 3$ on the non singular cubic surface in $\mathbb{P}^3$ have genus $g$ given by:

$$d = 3a - \sum_{i=1}^{6} b_i$$

$$g = \binom{a-1}{2} - \sum_{i=1}^{6} \binom{b_i}{2}$$

where

$$a > 0$$

$$b_1 \geq b_2 \geq \dots \geq b_6 \geq 0$$

$$a \geq b_1 + b_2$$

$$2a \geq b_1 + \dots + b_5$$

$$a^2 > b_1^2 + \dots + b_6^2 \ .$$

It is easily checked that all these points lie on or below III and that all points on III are attained. We now get a large number of curves between III and IV . Below we list the gaps, i.e. those points which are not attained by curves of this type.

We write $m - n$ for all integers from and including $m$ to and including $n$ .

| d | Values of g not attained between III and IV. |
|---|---|
| 21 | 19 |
| 24 | 22 |
| 25 | 23, 25 |
| 26 | 25 |
| 27 | 25, 26 |
| 28 | 26, 28 |
| 29 | 27-29, 31 |
| 30 | 28, 29, 31 |
| 31 | 29-32, 34 |
| 32 | 30-32, 34 |
| 33 | 31-35, 37 |
| 34 | 32-35, 37 |
| 35 | 33-38, 40 |
| 36 | 34-38, 40 |
| 37 | 35-41, 43, 51 |
| 38 | 36-41, 43, 53 |
| 39 | 37-44, 46, 55 |
| 40 | 38-44, 46, 57 |
| 41 | 39-47, 49, 57, 59 |
| 42 | 40-47, 49, 59, 61, 76 |
| 43 | 41-50, 52, 60, 61, 63, 78 |
| 44 | 42-50, 52, 62, 63, 65, 81 |
| 45 | 43-53, 55, 63-65, 67, 83 |
| 46 | 44-53, 55, 65-67, 69, 86 |
| 47 | 45-56, 58, 66-69, 70, 86, 88 |
| 48 | 46-56, 58, 68-71, 73, 88, 91 |
| 49 | 47-59, 61, 69-73, 75, 91, 93 |
| 50 | 48-59, 61, 71-75, 77, 92, 93, 96 |

One immediately observes remarkable regularities in these gap-values, and it is reasonable to conjecture that some of this holds in general. The largest gap is particularly interesting. It seems to increase by small, regular jumps by 2 or 3, interrupte occasionally by larger ones. Hartshorne has conjectured an approx mate upper bound of $\frac{1}{3}d^{3/2}$ for this gap. We have:

| d | 21 | 22 | 23 | 24 | 25 | 26 | 27 | 28 | 29 | 30 | 31 | 32 | 33 | 34 | 35 | 36 | 37 | 38 | 39 | 40 |
|---|---|---|---|---|---|---|---|---|---|---|---|---|---|---|---|---|---|---|---|---|
| $\left[\frac{1}{3} d^{3/2}\right]$ | 32 | 34 | 36 | 39 | 41 | 44 | 46 | 49 | 52 | 54 | 57 | 60 | 63 | 66 | 69 | 71 | 75 | 78 | 81 | 84 |

| d | 41 | 42 | 43 | 44 | 45 | 46 | 47 | 48 | 49 | 50 |
|---|---|---|---|---|---|---|---|---|---|---|
| $\left[\frac{1}{3} d^{3/2}\right]$ | 87 | 90 | 93 | 97 | 100 | 103 | 107 | 110 | 114 | 117 |

Thus we see all the computed values are well below Hartshorne's estimate.

Moreover, we now see that A yields rather uninteresting points in the area under examination. B actually starts to become interesting for $d = 42$, when the upper gap is filled. From then on, B yields occasional new points in addition to those from C.

For the purpose of filling the gaps left by C, B can be improved considerably by reembedding $Q = \mathbb{P}^1 \times \mathbb{P}^1$ in some projective space as follows:

$$Q \overset{i}{\hookrightarrow} \mathbb{P}^{m_1} \times \mathbb{P}^{m_2} \overset{j}{\hookrightarrow} \mathbb{P}^m$$

where $j$ is the Segre embedding and $i$ is the product of the $\alpha$-uple and the $\beta$-uple embedding. Then the curves of B are reembedded with a new degree. We get:

D. CURVES ON RE-EMBEDDINGS OF Q. Curves in this category have degree $d$ and genus $g$ given by

$$d = a\beta + b\alpha$$

$$g = (a-1)(b-1)$$

where a, b, $\alpha$, $\beta$ are positive integers. B is of course a special case, but in addition to the allowable points of B in our area we now obtain additional values. Below we list the gaps from C which are filled by B and D , those from B being marked with an *:

GAPS FILLED BY B AND D .

| d | g | d | g | d | g |
|---|---|---|---|---|---|
| 24 | 22 | 35 | 36,38 | 42 | 40,42,44,45,46,76* |
| 28 | 28 | 36 | 34,35,40 | 43 | 42,48,52,60,78* |
| 29 | 28 | 37 | 36,40 | 44 | 42,44,45,48,50,52,62,63,65,8 |
| 30 | 28 | 38 | 36,38,39,40 | 45 | 46,48,50,52,64 |
| 31 | 30 | 39 | 40,42,44,46 | 46 | 44,45,48,50,51,65,66 |
| 32 | 30,32 | 40 | 39,42,57 | 47 | 50,52,54,56,66,68,86* |
| 33 | 32,34 | 41 | 40,44 | 48 | 46,48,49,51,52,54,55,58,68,69,70,8 |
| 34 | 32,35 | | | 49 | 48,52,54,56,70,72 |
| | | | | 50 | 48,50,51,54,56,74,75,92*,96 |

We are thus left with the table:

REMAINING GAPS AFTER B, C AND D .

| d | g | d | g |
|---|---|---|---|
| 21 | 19 | 37 | 35,37-39,41,43,51 |
| 25 | 23,25 | 38 | 37,41,43,53 |
| 26 | 25 | 39 | 37-39,41,43,55 |
| 27 | 25,26 | 40 | 38,40,41,43,44,46 |
| 28 | 26 | 41 | 39,41-43,45-47,49,57,59 |
| 29 | 27,29,31 | 42 | 41,43,47,49,59,61 |
| 30 | 29,31 | 43 | 41,43,44-47,49,50,61,63 |
| 31 | 29,31,32,34 | 44 | 43,46,47,49 |
| 32 | 31,34 | 45 | 43-45,47,49,51,53,55,63,65,67,83 |
| 33 | 31,33,35,37 | 46 | 46,47,49,52,53,55,67,69,86 |
| 34 | 33,34,37 | 47 | 45-49,51,53,55,58,67,69,70,88 |
| 35 | 33-35,37,40 | 48 | 47,50,53,56,71,73,91 |
| 36 | 36-38 | 49 | 47,49-51,53,55,57-59,61,69,71,73,75,91,93 |
| | | 50 | 49,52,53,55,57-59,61,71-73,77,93 |

As we see, we are still left with a large number of gaps, starting already at $d = 21$. However, using a refinement of method D, we get additional points as follows:

E. CURVES ON THE PRODUCT OF TWO NON SINGULAR CURVES.[9)] Let $d_i, g_i, i = 1,2$ be integers such that there exists a non singular curve of degree $d_i$ and genus $g_i$ in $\mathbb{P}^3$, and let

$$a_i \geq \begin{cases} g_i + 3 & \text{if } g_i \geq 2 \\ 3 & \text{if } g_i = 1 \\ 1 & \text{if } g_i = 0 \end{cases}$$

Then there exists a non singular curve in $\mathbb{P}^3$ of degree

$$d = d_2 a_1 + d_1 a_2$$

and genus

$$g = g_1 + g_2 - g_1 g_2 + (a_1 + g_1 - 1)(a_2 + g_2 - 1) .$$

When the new points which correspond to this method are generated, we are left with the following gaps:

REMAINING GAPS AFTER B - E .

| d | g | d | g |
|---|---|---|---|
| 25 | 23 | 41 | 39 |
| 29 | 27 | 43 | 47 |
| 31 | 29 | 45 | 47, 63, 83 |
| 33 | 35 | 46 | 86 |
| 36 | 38 | 48 | 50 |
| 37 | 35 | 49 | 59 |
| 39 | 39 | 50 | 72 |
| 40 | 44 | | |

Of these remaining gaps, some may obviously be deleted by taking the 2-uple embedding of curves of lower degrees. In fact, we may thus remove the gaps for

$$d = 36,\ 40,\ 48,\ 50 \quad .$$

Finally we consider the following class of curves:

F. CURVES ON REEMBEDDINGS OF THE CUBIC SURFACE.[10)]

These curves have degree $d$ and genus $g$ given as follows:

$$d = a\alpha - \sum_{i=1}^{6} b_i \beta\sigma(i)$$

$$g = \binom{a-1}{2} - \sum_{i=1}^{6} \binom{b_i}{2}$$

where $\sigma$ is any permutation of $\{1,\dots,6\}$ ,

$$\beta_1 \geq \beta_2 \geq \dots \geq \beta_6 > 0$$

$$\alpha > 0$$

$$\alpha > \beta_1 + \beta_2$$

$$2\alpha > \beta_1 + \dots + \beta_6$$

and $a, b_1, \dots, b_6$ satisfy the requirement in C. Here C corresponds to

$$\alpha,\ \beta_1,\dots,\beta_6 = 3,\ 1,\dots,\ 1 \ .$$

This yields a large number of admissible points. Thus for instance the twisting referred to above is a special case corresponding to

$$\alpha,\ \beta_1,\ldots,\beta_6 = 6,\ 2,\ldots,\ 2\ .$$

In particular all remaining gaps listed above can be removed by this method.[11)]

## §4. On the Horrochs - Mumford conjecture.[12)]

If $X \subset P^N$ is a complete intersection of hypersurfaces of degrees $\delta_1,\ldots,\delta_r$, then the canonical sheaf is given by

$$w_X = 0_X(\Sigma\delta_i - N - 1)$$

as is easily seen.[13)] Hence

$$d_1(X) = \deg(c_1(w_X)) = d(\Sigma\delta_i - N - 1) .$$

Thus in particular we see that if all non singular 4-dimensional subvarieties of $\mathbb{P}^6$ are complete intersections, then their invariant $d_1$ has a very restricted allowable range. Of course this is also true for $d_2$, $d_3$ and $d_4$. In fact, we have the following general computation:

If $X$ is a complete intersection as above, then the total Segre class is given by

$$s(X) = (1+\delta_1 i^*(b))\cdot\ldots\cdot(1+\delta_r i^*(h))(1+i^*(h))^{-N-1}$$

which is easily seen by the usual direct sum decomposition of the normal bundle. Here $i$ denotes the canonical embedding of $X$ in $\mathbb{P}^N$.

For $X \in V(6,4)$ we have

$$s(X) = (1+\delta_1 i^*(h))(1+\delta_2 i^*(h))(1+i^*(h))^{-7}$$

$$= (1+(\delta_1+\delta_2)i^*(h) + \delta_1\delta_2 i^*(h^2))(1-7i^*(h) + 28i^*(h^2) - 84i^*(h^3) + 210i^*(h^4))$$

$$= 1 + (\delta_1+\delta_2-7)i^*(h) + (\delta_1\delta_2-7(\delta_1+\delta_2) + 28)i^*(h^2) + (-7\delta_1\delta_2 + 28(\delta_1+\delta_2) - 84)i^*(h^3) + (28\delta_1\delta_2 - 84(\delta_1+\delta_2) + 210)i^*(h^4)$$

Writing $d = d_o = \deg(X)$ , we now get:

$$d_1 = d(\delta_1+\delta_2-7)$$

$$d_2 = d(d-7(\delta_1+\delta_2) + 28)$$

$$d_3 = d(-7d+28(\delta_1+\delta_2) - 84)$$

$$d_4 = d(28d-84(\delta_1+\delta_2) + 210)$$

For $d \leq 15$ the allowable range for the complete intersections in $V(6,4)$ thus becomes

| d | $d_1$ | $d_2$ | $d_3$ | $d_4$ |
|---|---|---|---|---|
| 1 | -5 | 15 | -35 | 70 |
| 2 | -8 | 18 | -28 | 28 |
| 3 | -9 | 9 | 21 | -126 |
| 4 | -8 | -12 | 112 | -392 |
| 4 | -12 | 16 | 0 | -56 |
| 5 | -5 | -45 | 245 | -770 |
| 6 | 0 | -90 | 420 | -1260 |
| 6 | -12 | -6 | 84 | -252 |
| 7 | 7 | -147 | 637 | -1862 |
| 8 | 16 | -216 | 896 | -2576 |
| 8 | -8 | -48 | 224 | -560 |
| 9 | 27 | -297 | 1197 | -3402 |
| 9 | -9 | -45 | 189 | -378 |
| 10 | 40 | -390 | 1540 | -4340 |
| 10 | 0 | -110 | 420 | -980 |
| 11 | 55 | -495 | 1925 | -5390 |
| 12 | 72 | -612 | 2352 | -6552 |
| 12 | 0 | -108 | 336 | -504 |
| 12 | 12 | -192 | 672 | -1512 |
| 13 | 91 | -741 | 2821 | -7826 |
| 14 | 112 | -882 | 3332 | -9212 |
| 14 | 28 | -294 | 980 | -2156 |
| 15 | 135 | -1035 | 3885 | -10710 |
| 15 | 15 | -195 | 525 | -630 |

In view of all the convincing evidence which exist for the conjecture, it is reasonable to expect that it should be possible to prove directly that this is in fact the allowable range for the d-invariants on $V(6,4)$ . But as far as I know, no such theorem has been proven.

## §5. Computations for Grassmanians.[14)]

Let $G = G(r,n)$ be the Grassmanian which parametrizes linear r-subspaces of $\mathbb{P}^n$. Let $Q$ be the universal quotient, $V = k^{n+1}$ and define $M$ by

$$0 \to M \to V_G \to Q \to 0 \quad .$$

Then we have the relation [15)]

$$\Omega^1_{G/k} \simeq Q^\vee \otimes M \quad .$$

Putting

$$c(Q) = 1 + c_1 + \ldots + c_{r+1}$$

we now get

$$s(G) = \frac{c(Q \otimes Q^\vee)}{(1+c_1+\ldots+c_{r+1})^{n+1}}$$

Once we have a formula for the numerator in terms of $c_1,\ldots,c_{r+1}$, we thus may compute $s(G)$. And then $d_1,\ldots,d_m$ $(m = (n-r)(r+1))$ are computed for instance by a repeated application of Pieri's formula.[16)]

Thus the hard part is to find $c(Q \otimes Q^\vee)$ in terms of $c_1, \ldots, c_{r+1}$ . The advantage with this method is that once we know $c(Q \otimes Q^\vee)$ for some $r$ , then we may treat all grassmanians $G(r,n)$, $n = r+2, \ldots$ .

Using the MACSYMA [17)] system for Symbolic Manipulation at MIT in 1978, I got the following initial expressions for $c(Q \otimes Q^\vee)$ :

| r | $c(Q \otimes Q^\vee)$ |
|---|---|
| 1 | $1 + 4c_2 - c_1^2$ |
| 2 | $1 + 6c_2 - 2c_1^2 + 9c_2^2 - 6c_1^2c_2 + c_1^4$<br>$+ 4c_1^3c_3 + 4c_2 - c_1^2c_2^2 + 27c_3^2 - 18c_1c_2c_3$ |
| 3 | $1 + 8c_2 - 3c_1^2$<br>$+ 8c_4 - 2c_1c_3 + 22c_2^2 - 16c_1^2c_2 + 3c_1^4$<br>$+ 16c_2c_4 - 6c_1^2c_4 + 26c_3^2 - 30c_1c_2c_3 + 8c_1^3c_3$<br>$+ 28c_2^3 - 24c_1^2c_2^2 + 8c_1^4c_2 - c_1^6$<br>$- 112c_4^2 + 56c_1c_3c_4 + 24c_2^2c_4 - 32c_1^2c_2c_4 + 6c_1^4c_4$<br>$+ 48c_2c_3^2 - 25c_1^2c_3^2 - 54c_1c_2^2c_3 + 38c_1^3c_2c_3$<br>$- 6c_1^5c + 17c_2^4 - 12c_1^4c_2^2 + 2c_1^4c_2^2$<br>$-192c_2c_4^2 + 72c_1^2c_4 + 216c_3^2c_4 - 120c_1c_2c_3c_4$<br>$+ 18c_1^3c_3c_4 + 32c_2^3c_4 - 6c_1^2c_2^2c_4 - 54c_1c_3^3$<br>$+ 18c_2^2c_3^2 + 42c_1^2c_2c_3^2 - 9c_1^4c_3^2 - 26c_1c_2^3c_3$<br>$+ 6c_1^3c_2^2c_3 + 4c_2^5 - c_1^2c_2^4$<br>$+ 256c_4^3 - 192c_1c_3c_4^2 - 128c_2^2c_4^2 + 144c_1^2c_2c_4^2$<br>$- 27c_1^4c_4^2 + 144c_2c_3^2c_4 - 6c_1^2c_3^2c_4 - 80c_1c_2^2c_3c_4$<br>$+ 18c_1^3c_2c_3c_4 + 16c_2^4c_4 - 4c_1^2c_2^3c_4 - 27c_3^4$<br>$+ 18c_1c_2c_3^3 - 4c_1^3c_3^3 - 4c_2^3c_3^2 + c_1^2c_2^2c_3^2$ |

$r = 4$ was more than the system would handle in normal, interactive use.

## Notes

1. See for instance [El] and [Kle] for some recent results on this.

2. The main reference is the exellent survey [Ha 3]. See also [G-P], as well as [Ha 1].

3. According to [Ha 3] this was attempted unsuccessfully already by Halphen.

4. Perhaps something along the lines of [S].

5. In this form the theorem is an improvement of a result from [Ho 1], due to D.Laksov. See [Ho 2], [Ho 3], [H-R], [Lk 1]-[Lk 3] and [Kln] for more information, including the singular case.

6. See [Kln] for a survey of this and related developments.

7. Loosely speaking, the "double point loci" and ramification loci are larger than they should be. We hope to address these questions in a forthcoming paper.

8. The general theory in what follows may be found in [Ha 1] and [Ha 3].

9. This and the remaining part of section 3 will appear in [Ho 8] in greater detail and with more extensive computations.

10. See [Ha 1], where the very ample divisors on the non singular cubic are determined as indicated below.

11. In fact, as of this lecture I can say that there are no remaining gaps after A - F for $d \leq 56$ .

12. See [Ha 2] for this and related conjectures.

13. See for instance [Ha 1].

14. For details and related results, see [Ho 4] and [Ho 5]. For some further developments, see [Ho 9].

15. For references and more background, see [Ho 6] and [Ho 7].

16. See [Ho 5] and [Ho 7].

17. I would like to express my appreciation to the Math. lab group at the Laboratory for Computer Science at MIT for letting me use their facilities during my visit.

# R E F E R E N C E S

[El] Ellingsrud, G.: "Sur le schéma de Hilbert des variétés de codimension 2 dans $\mathbb{P}^e$ á cône de Cohen-Macauley." Ann.Sc.Ec.Norm.Sup. 1975, pp. 423-431.

[G-P] Gruson, L. and Peskine, C.: "Genre de courbes de l'espace projectif." In Algebraic Geometry. Proceedings, Tromsø, Norway 1977. Olson, L., Editor, pp. 31-59.

[Ha 1] Hartshorne, R.: Algebraic Geometry. Graduate Texts in Mathematics. Springer-Verlag 1977.

[Ha 2] Hartshorne, R.: "Algebraic vector bundles on projective spaces: A problem list." Topology Vol. 18, pp. 117-128, 1979.

[Ha 3] Hartshorne, R.: "On the classification of algebraic space curves." Preprint, 1979.

[Ho 1] Holme, A.: "Embedding-Obstruction for Smooth, Projective Varieties, I." Advances in Mathematics Supplementary Studies, Volume 5, pp. 39-67. Studies in Algebraic Topology, Academic Press 1979. (Preprint from 1973).

[Ho 2] Holme, A.: "Embedding-Obstruction for Singular Algebraic Varieties in $\mathbb{P}^N$" Acta mathematicae, Vol. 135, 1975, pp. 155-185.

[Ho 3] Holme, A.: "Deformation and stratification of secant structure." Algebraic Geometry. Proceeding Tromsø, Norway 1977. L. Olson, Editor. Springer Lecture Notes in Mathematics 688, 19

[Ho 4] Holme, A.: "On the dual of a smooth variety." Algebraic Geometry. Proceedings, Copenhagen 1978. K. Lönsted, Editor. Springer Lecture Notes in Mathematics 732, 1979.

[Ho 5] Holme, A.: "Some Computing Aspects of Projective Geometry." Preprint Series, Department of Mathematics, University of Bergen, 1980.

[Ho 6] Holme, A.: Introduccion a la Teoria de las Clases Caracteristicas en la Geometria Algebraica (in Spanish). Monografias del Instituto de Matemáticas 6, UNAM, Mexico, 1978.

[Ho 7] Holme, A.: Embeddings, Projective Invariants and Classifications. Monografias del Instituto de Matemáticas 7, UNAM, Mexico, 1979.

[Ho 8] Holme, A.: "A numerical study of Algebraic Space Curves." To appear.

[Ho 9] Holme, A.: "Computer Algebra and Algebraic Geometry." To appear.

[H-R] Holme, A. and Roberts, J.: "Pinch-Points and Multiple Locus of Generic Projections of Singular Varieties." Advances in Mathematics, Vol. 33, 1979, pp. 212-256.

[Kln] Kleiman, S.L.: "The enumerative theory of singularities." In Real and complex singularities, Oslo 1976, pp. 297-396. P. Holm, Editor. Sijthoff & Noordhof International Publishers. Alphen aan der Rijn, The Netherlands 1977.

[Kle] Kleppe, J.O.: "The Hilbert flag functor and classification." Preprint, Oslo 1980.

[Lk 1] Laksov, D.: "Some enumeritive properties of secants to non-singular projective schemes." Mathematica Scandinavicae 39 (1976) pp. 171-190.

[Lk 2] Laksov, D.: "Secant bundles and Todd's formula for the double points of maps into $\mathbb{P}^N$" Proc.Lond. Math.Soc. 1978 pp. 120-142.

[Lk 3] Laksov, D.: "Residual Intersections and Todd's formula for the double locus of a morphism." Acta mathematicae Col. 140, 1978, pp. 75-92.

[S] Swinneston - Dyer, H.P.F.: "An enumeration of all varieties of degree 4." Amer. J. of Math., 95 (1973), pp. 403-418.

Distribution of primes in intervals and arithmetic progressions

Matti Jutila

1. Introduction. In analytic number theory it is customary to distinguish elementary and non-elementary (or analytic) methods. The meaning of the term "elementary" is more or less intuitive, anyway it is usually understood that the non-elementary methods are those applying the complex function theory.

In prime number theory, which is the topic of this talk, the classical analytic methods are based on the theory of Riemann's zeta-function and Dirichlet's L-functions, while the most important elementary methods, the sieves, go back to the fundamental work of Viggo Brun. Other sieves of great importance were presented by Atle Selberg and J. Berkeley Rosser (we leave here aside the "large sieve" of Yu. V. Linnik since it is not a sieve in the same sense as the others). For an excellent account of sieve methods and their development, we refer to the book of Halberstam and Richert [4]. Rosser's work remained unfortunately unpublished and its influence was therefore delayed; however, H. Iwaniec has recently found that Rosser's sieve opens some new interesting possibilities which will be explained later in this talk.

The elementary and analytic methods have their merits and defects. The sieves are of great generality, but just because of this they often fail to give the optimal result in a particular problem. As for limitations of sieve methods, we refer to the lecture of Atle Selberg [25] at the Trondheim congress in 1949. Typically a sieve result can be viewed as a compromise; for instance, we get an upper or lower bound instead of an asymptotic equality, or a statement concerning almost primes

(i. e. numbers having at most a finite prescribed number of prime divisors) instead of primes. On the other hand, an analytic approach is usually well-focused, taking into account the specific features of the problem, and it leads to a result of the expected form - or to nothing. In the end of his Trondheim-talk, Selberg points out that sieves could be valuable in combination with analytic methods. Some recent advances in prime number theory have indeed been obtained in this way. We shall consider three such instances:

(i) the difference between consecutive primes,

(ii) the Brun-Titchmarsh theorem,

(iii) Linnik's theorem on the least prime in an arithmetic progression.

In the problem (i) the appearance of the sieve (in the form of the Rosser-Iwaniec sieve) is a new feature, while in (ii) the classical treatment is by the sieve alone and the application of analytic arguments is a novelty. In Linnik's proof of his famous theorem (iii) the sieve method (in the form of the Brun-Titchmarsh theorem) played already an essential rôle; in a new approach, suggested by Atle Selberg [26], the argument is related to Selberg's sieve.

<u>2. The difference between consecutive primes</u>. The prime number theorem $\pi(x) \sim x/\log x$ implies that the mean difference between consecutive primes near x is about log x. It has been conjectured by H. Cramér that there always exists a prime between x and $x + A\log^2 x$ if $x \geqq 2$ and A is a suitable constant. While this conjecture can be supported by a statistical reasoning, its proof or disproof seems hopelessly difficult (maybe it is not decideable at all), for even on the Riemann hypothesis we

can only prove the existence of a prime between x and $x + Ax^{\frac{1}{2}}\log x$.

The classical approach to this problem, due to Hoheisel and Ingham, is via the relation

$$(2.1) \qquad \pi(x) - \pi(x-y) \sim y/\log x$$

with $y = x^{\theta}$, $\theta < 1$. The proof of (2.1) requires information about the zeros of the zeta-function, namely a zero-free region near the line Re s = 1 and a zero-density theorem. After the work of I. M. Vinogradov on exponential sums the zero-free region was no problem any more. Denote by $N(\alpha,T)$ the number of zeros $\varrho = \beta + i\gamma$ of $\zeta(s)$ such that $\beta \geqq \alpha$ and $|\gamma| \leqq T$. The estimate

$$N(\alpha,T) \ll T^{b(1-\alpha)}\log^{c}T$$

for $T \geqq 2$, $\frac{1}{2} \leqq \alpha \leqq 1$ implies (2.1) for $y = x^{\theta}$ with $\theta > \theta_0 = 1 - b^{-1}$. In particular, the "density hypothesis" b = 2 gives practically the same result as Riemann's hypothesis. In this manner the results $\theta_0 = 5/8, 3/5, 7/12$ were obtained resp. by A. E. Ingham [9], H. L. Montgomery [18] and M. N. Huxley [8].

However, so far as the difference between consecutive primes is concerned, the relation (2.1) is unnecessarily strong, since it is enough that the left hand side of (2.1) be positive. This is certainly true if there exists a positive constant c such that

$$(2.2) \qquad \pi(x) - \pi(x-y) \geq cy/\log x.$$

But the classical method gives either (2.1) or nothing. By a combination of analytic and sieve methods, a result of the type (2.2) was proved by Iwaniec and myself [14] for $y = x^{13/23}$. Subsequently the exponent was improved to $11/20+\varepsilon$ by

Heath-Brown and Iwaniec [6]. For comparison, $7/12 = 0.5833..$, $13/23 = 0.5652..$, $11/20 = 0.5500...$ Thus, denoting by $p_n$ the n'th prime, we have

$$p_{n+1} - p_n \ll p_n^{11/20+\varepsilon}$$

for any $\varepsilon > 0$.

We now give a sketch of the proof of (2.2). Taking first a general point of view, let $\mathcal{A}$ be any finite set of integers and denote by $S(\mathcal{A},z)$ the number of elements of $\mathcal{A}$ which are not divisible by any prime less than z. The central problem of the sieve theory is to deduce upper and lower estimates for $S(\mathcal{A},z)$ from the available information about the set $\mathcal{A}$. In our problem, $\mathcal{A}$ is the set of integers n with $x-y < n \leq x$. Let $x^{1/3} < z_1 < z_2 < x^{1/2}$, and denote by $\mathcal{A}_d$ the set of the elements of $\mathcal{A}$ which are divisible by d. It is easy to verify that if x is not the square of a prime, then

$$(2.3)\quad \pi(x) - \pi(x-y) = S(\mathcal{A},x^{1/2})$$

$$= S(\mathcal{A},z_1) - \sum_{z_1 \leq p < z_2} S(\mathcal{A}_p,p) - \sum_{z_2 \leq p < x^{1/2}} S(\mathcal{A}_p,p)$$

$$= S_1 - S_2 - S_3,$$

say, by classifying the numbers counted in $S(\mathcal{A},z_1)$ but not in $S(\mathcal{A},x^{1/2})$ according to the least prime factor $p \in [z_1,x^{1/2})$. The principle contained in the identity (2.3) is due to A. A. Buchstab. In our case the numbers n counted in $S(\mathcal{A},z_1)$ are either primes or products of two primes, $n = pq$ with $z_1 \leq p \leq q$, $p < x^{1/2}$. Now, in order to get a lower bound in (2.3), the sum $S_1$ is estimated from below and the sum $S_3$ from above by the sieve, while for $S_2$ it is possible to find an asymptotic formula if $z_1$ and $z_2$ are suitably chosen. The sum $S_2$ is dealt with by an analytic argument similar to that used in the proof of (2.1). The estimation of $S_2$ being sharp, the success of the whole reasoning depends on how much (or rather how little) we lose in the estimations of $S_1$ and $S_3$. Here it is essential that the linear sieve of Iwaniec enables us to make use of the special structure of our set $\mathcal{A}$.

The general scheme of proving lower or upper estimates for $S(\mathcal{A},z)$ by a sieve is as follows. Let the numbers $\Lambda_d^+$ and $\Lambda_d^-$ be defined for integers $d \geqq 1$, and suppose that

$$\Lambda_1^+ = \Lambda_1^- = 1, \quad \sum_{d|n} \Lambda_d^- \leqq 0 \leqq \sum_{d|n} \Lambda_d^+ \text{ for } n \geqq 2.$$

Then trivially, putting $P(z) = \prod_{p<z} p$, we have

$$\sum_{n\in\mathcal{A}} \sum_{d|(n,P(z))} \Lambda_d^- \leqq S(\mathcal{A},z) \leqq \sum_{n\in\mathcal{A}} \sum_{d|(n,P(z))} \Lambda_d^+,$$

or

$$\sum_{d|P(z)} \Lambda_d^- |\mathcal{A}_d| \leqq S(\mathcal{A},z) \leqq \sum_{d|P(z)} \Lambda_d^+ |\mathcal{A}_d|, \tag{2.4}$$

where $|\mathcal{A}_d|$ means the cardinality of the set $\mathcal{A}_d$. In the so-called combinatorial sieve methods like those of Brun and Rosser, the numbers $\Lambda_d^{\pm}$ are chosen to be $\mu(d)$ (the Möbius function) or 0 according to whether or not d belongs to a certain set $\mathcal{D}^-$ or $\mathcal{D}^+$, satisfying

$$\sum_{\substack{d|n \\ d\in\mathcal{D}^-}} \mu(d) \leqq \sum_{d|n} \mu(d) \leqq \sum_{\substack{d|n \\ d\in\mathcal{D}^+}} \mu(d), \quad n \geqq 1.$$

In Brun's sieve, $\mathcal{D}^-$ (resp. $\mathcal{D}^+$) consists of integers having at most a given odd (resp. even) number of prime factors. In order to define these sets in the case of the "linear" Rosser sieve, write $d = p_1 \cdots p_r$ with $p_1 \geqq p_2 \geqq \ldots \geqq p_r$, and let D be a parameter at our disposal. Then $\mathcal{D}^-$ is defined by the conditions

$$p_1p_2\cdots p_{2k-1}p_{2k}^3 \leq D \text{ for } 0 \leq k \leq \tfrac{1}{2}r,$$

and $\mathcal{D}^+$ by the conditions

$$p_1p_2\cdots p_{2k}p_{2k+1}^3 \leq D \text{ for } 0 \leq k \leq \tfrac{1}{2}(r-1).$$

In applying the inequalities (2.4), we need information about the quantities $|\mathcal{A}_d|$. If the set $\mathcal{A}$ is well-behaved, one may approximate to $|\mathcal{A}_d|$ by a simple expression, with an error $r(\mathcal{A},d)$. For instance, for our set $\mathcal{A}$, the natural approximation to $|\mathcal{A}_d|$ is $y/d$, the error being

$$r(\mathcal{A},d) = \left[\frac{x}{d}\right] - \left[\frac{x-y}{d}\right] - y/d.$$

According to the decomposition

$$|\mathcal{A}_d| = y/d + r(\mathcal{A},d),$$

the inequalities (2.4) lead to the main term and the error term of the lower or upper sieve estimate. An analysis of the main terms has been given by Iwaniec [10,12]. As to the error terms, a recent fundamental discovery of Iwaniec [11] is that in the linear lower and upper bound sieve the error has the shape

$$\sum_{m<M}\ \sum_{n<N} a_m b_n r(\mathcal{A},mn) \tag{2.5}$$

with bounded coefficients $a_m$, $b_n$, where M and N are parameters which are chosen suitably. It is favourable to choose M and N so that MN is as large as possible under the requirement that the sum (2.5) still remains small in comparison with the main term. By analytic means the error term (2.5) can be estimated satisfactorily with MN somewhat larger than y; note that the trivial estimate $|r(\mathcal{A},mn)| \leq 1$ does not allow this. Increasing the quantity MN improves the main term and leads thus to a sharper sieve estimate.

The derivation of the mentioned sieve result of Iwaniec has been simplified by Y. Motohashi [23]; a detailed account of these questions with some generalizations will be given in his forthcoming book [24].

Though there may be big differences between consecutive primes, these are anyway rare, for it has been proved by Heath-Brown [5] that with $d_n = p_{n+1} - p_n$

$$\sum_{p_n \leq x} d_n^2 \ll x^{23/18+\varepsilon},$$

$$\sum_{p_n \leq x,\ d_n \geq p_n^{\frac{1}{2}}} d_n \ll x^{3/4+\varepsilon}.$$

3. The Brun-Titchmarsh theorem. Let a and q be relatively prime integers and denote by $\pi(x;q,a)$ the number of primes $p \leq x$ with $p \equiv a \pmod q$. In 1930 E. C. Titchmarsh [27] proved by Brun's sieve that if $q < x^{1-\varepsilon}$, then

$$\pi(x;\ q,a) \ll \frac{x}{\varphi(q)\log x}, \tag{3.1}$$

where $\varphi$ is Euler's function. Up to a constant factor, this estimate is of the expected order of magnitude. Its most significant aspect is the very weak restriction for q. For comparison, in the function-theoretic approach it is presently necessary to suppose that $q \ll (\log x)^A$, in which case an asymptotic formula for $\pi(x;\ q,a)$ can be proved. Even on the Riemann hypothesis for L-functions we can manage only the case $q \ll x^{\frac{1}{2}}(\log x)^{-2}$. This example shows clearly the characteristic differences between analytic and sieve methods.

By Selberg's sieve (van Lint and Richert [17]) or by Rosser's sieve (Iwaniec [10]) it can be proved a more precise form of (3.1), namely that

$$(3.2) \qquad \pi(x;q,a) < \frac{2x}{\varphi(q)\log(x/q)}\left(1 + O\left(\frac{1}{\log(x/q)}\right)\right) \text{ for } q < x.$$

The same result can be obtained also by using large sieve inequalities, in fact without the term $O(\cdots)$ (Montgomery and Vaughan [19]). Nevertheless, this result becomes weak when q approaches x. The reason is that in this case we can sieve with small primes only, at least if the error term in the sieve is estimated trivially. However, as in the problem on differences between primes, it is possible to sharpen the estimate of the error. The first non-trivial results in this respect were given by C. Hooley [7]. His results were of statistical nature, i. e., he considered $\pi(x;q,a)$ either for fixed a and varying q, or vice versa, and obtained improved estimates of "almost all" type. His method depended on Weil's estimates of Kloosterman sums. Motohashi [20] used the generating Dirichlet series method and estimates of L-functions in combination with Selberg's sieve to prove, beside statistical results, also definite improvements on (3.2) for individual q and a when $q < x^{\frac{1}{2}-\varepsilon}$. Further improvements were given by Iwaniec by using Rosser's sieve instead of Selberg's. Here one example of his results:

$$(3.3) \qquad \pi(x;q,a) < \frac{(2+\varepsilon)x}{\varphi(q)\log(xq^{-\frac{3}{8}})} \qquad \text{for } q \leqq x^{\frac{2}{5}}$$

if x is sufficiently large. In a recent long paper, Iwaniec [13] gives a thorough treatment of the Brun-Titchmarsh theorem via the character sum approach and the Kloosterman sum approach, discussing also consequences of various hypotheses such as the Lindelöf hypothesis

$$L(\tfrac{1}{2} + it, \chi) \ll (q(|t|+1))^{\varepsilon}$$

for all characters $\chi$ mod q and real t, or Hooley's hypothesis on incomplete Kloosterman sums:

$$\sum_{\nu_1 < \nu \leq \nu_2} e^{2\pi i b \bar{\nu}/q} \ll (b,q)^{\frac{1}{2}}(\nu_2 - \nu_1)^{\frac{1}{2}} q^{\varepsilon},$$

where $1 \leq \nu_2 - \nu_1 < q$ and $\nu\bar{\nu} \equiv 1 \pmod q$.

Looking at the results (3.2) and (3.3), one finds that the estimates are at least twice as large as one would expect. But replacing the coefficient 2 by any smaller number is a very deep problem since such a result would have drastic consequences on the so-called Siegel-zero in the theory of L-functions.

4. Linnik's theorem. Let q be a positive integer, $(a,q) = 1$ and $p(q,a)$ the least prime $p \equiv a \pmod q$. Linnik [16] proved in 1944 the existence of a constant C such that $p(q,a) \ll q^C$; nowadays C is called Linnik's constant. Since Linnik's proof was extremely complicated, it was difficult to work out any numerical estimate for C. The Chinese mathematicians Pan and Chen proved respectively $C \leq 5448, 770$. Ten years ago I proved by Turan's method that $C \leq 550$. A new approach to Linnik's theorem, based on ideas related to Selberg's sieve, was indicated by Atle Selberg [26] in 1972. By this method the following substantially improved estimates have been obtained: $C \leq 80$, 20, 17 by Jutila [15], Graham [3] and Chen [1], respectively.

All these estimates of Linnik's constant are based on an analysis of the zeros of L-functions

$$L(s,\chi) = \sum_{n=1}^{\infty} \chi(n) n^{-s}$$

near the point $s = 1$; here $\chi$ is a Dirichlet character (mod q). No elementary proof of Linnik's theorem is known. Define $N(\alpha,T,\chi)$ for $L(s,\chi)$ similarly as $N(\alpha,T)$ was defined for $\zeta(s)$. The so-called Linnik's density theorem states, in a variant due to E. Fogels, that for $\frac{1}{2} \leq \alpha \leq 1$ and $T \geq 1$

$$\sum_{\chi \bmod q} N(\alpha, T, \chi) \ll (qT)^{c(1-\alpha)},$$

where c is a constant. This is enough for a proof of Linnik's theorem unless there exists a real "Siegel-zero" near s = 1 of a function $L(s, \chi_1)$ with a real "exceptional" character $\chi_1$ (mod q); in this case we also need Linnik's theorem on the Deuring-Heilbronn phenomenon, to the effect that if a Siegel-zero exists, then there is a certain region near s = 1 free of other zeros of L-functions (mod q).

We briefly consider the proof of Linnik's density theorem; the Deuring-Heilbronn phenomenon was dealt with independently by Motohashi [21] and myself [15] by somewhat similar arguments. We follow here S. Graham [2-3]. Selberg's sieve method has its origin in Selberg's investigations of the zeros of the zeta-function near (or on) the critical line, and the present application is in the same spirit.

In studying zeros of L-functions, a standard device is to go over to the function

$$F(s, \chi) = M(s, \chi) L(s, \chi),$$

where

$$M(s, \chi) = \sum_{n \leq z} \alpha_n \chi(n) n^{-s}$$

is a "mollifier". The zeros of $L(s, \chi)$ are included among those of $F(s, \chi)$. Supposing that $\alpha_1 = 1$, we have for Re $s > 1$

$$F(s, \chi) = \sum_{n=1}^{\infty} a_n \chi(n) n^{-s},$$

where $a_1 = 1$. The mollifier is chosen so as to make the coefficients $a_n$ ($n \geq 2$) in a certain mean value sense small. If

$$M(s, \chi) = \sum_{d,e} \theta_d \lambda_e \chi([d,e])[d,e]^{-s},$$

where $\theta_1 = \lambda_1 = 1$ and $\theta_d = \lambda_e = 0$ for sufficiently large d, e, then for Re $s > 1$

$$F(s,\chi) = \sum_{n=1}^{\infty} (\sum_{d|n} \theta_d)(\sum_{e|n} \lambda_e)\chi(n)n^{-s}.$$

Choosing

$$\lambda_d = \begin{cases} \mu(d), & d \leq z_1, \\ \mu(d)\log(z_2/d)/\log(z_2/z_1), & z_1 < d \leq z_2, \\ 0, & d > z_2, \end{cases}$$

where $1 \leq z_1 < z_2$, we have $a_n = 0$ for $2 \leq n \leq z_1$. The coefficients $\theta_d$ are chosen as in Selberg's sieve in order to minimize the quadratic form

$$\sum_{\substack{d,e \leq R \\ (de,q)=1}} \theta_d \, \theta_e / [d,e].$$

If now $L(\varrho,\chi) = 0$ for a zero $\varrho$ counted in $N(\alpha,T,\chi)$, then for suitable values of the parameters $R$, $z_1$, $z_2$, $y$ (some powers of $qT$) we have

$$\left| \sum_{z_1 \leq n \leq y} (\sum_{d|n} \theta_d)(\sum_{e|n} \lambda_e)\chi(n)n^{-\varrho} \right| \gg 1.$$

By a suitable averaging procedure and using some auxiliary results it is found that this cannot happen too often, and Linnik's density theorem follows.

It should be also noted that Motohashi [22] has proved Linnik's theorem (or actually a more general result due to P. X. Gallagher) without the zero-density argument and the Deuring-Heilbronn phenomenon. His arguments are, however. analogous to those used in proving these results about the zeros.

The Riemann hypothesis for L-functions would give Linnik's constant $C = 2 + \varepsilon$, but in all probability even $C = 1 + \varepsilon$ is true. Thus the gap between these and Chen's estimate $C \leq 17$ is still considerable. Though the possible existence of a Siegel-zero makes some trouble, its non-existence would not imply any new estimate for C. Of crucial importance in this respect would be widening the zero-free region near s=1, even allowing the possible existence of a Siegel-zero.

References

[1] Chen Jingrun, On the least prime in an arithmetical progression and theorems concerning the zeros of Dirichlet's L-functions II. Sci. Sinica 22 (1979), 859-889.

[2] S. Graham, Applications of Sieve Methods. Dissertation, University of Michigan, Ann Arbor 1977.

[3] S. Graham, On Linnik's Constant. To appear.

[4] H. Halberstam and H.-E. Richert, Sieve Methods. Academic Press, London 1974.

[5] D. R. Heath-Brown, The differences between consecutive primes. J. London Math. Soc. (2) 20 (1979), 177-178.

[6] D. R. Heath-Brown and H. Iwaniec, On the Difference Between Consecutive Primes. Invent. Math. 55 (1979), 49-69.

[7] C. Hooley, On the Brun-Titchmarsh theorem. J. reine angew. Math. 255(1972), 60-79.

[8] M. N. Huxley, On the difference between consecutive primes. Invent. Math. 15 (1972), 164-170.

[9] A. E. Ingham, On the difference between consecutive primes. Quart. J. Math. Oxford 8 (1937), 255-266.

[10] H. Iwaniec, On the error term in the linear sieve. Acta Arith. 19 (1971), 1-30.

[11] H. Iwaniec, A new form of the error term in the linear sieve. Acta Arith. (to appear).

[12] H. Iwaniec, Rosser's sieve. Acta Arith. (to appear).

[13] H. Iwaniec, On the Brun-Titchmarsh theorem. To appear.

[14] H. Iwaniec and M. Jutila, Primes in short intervals. Arkiv för Matematik 17 (1979), 167-176.

[15] M. Jutila, On Linnik's constant. Math. Scand. 41 (1977), 45-62.

[16] Yu. V. Linnik, On the least prime in an arithmetic progression, I-II. Mat. Sb. 15 (1944), 139-172 and 347-368 (in Russian).

[17] J. H. van Lint and H.-E. Richert, On primes in arithmetic progressions. Acta Arith. 11 (1965), 209-216.

[18] H. L. Montgomery, Zeros of L-functions. Invent. Math. 81 (1969), 346-354.

[19] H. L. Montgomery and R. C. Vaughan, On the large sieve. Mathematika 20 (1973), 119-134.

[20] Y. Motohashi, On some improvements of the Brun-Titchmarsh theorem. J. Math. Soc. Japan 26 (1974), 306-323.

[21] Y. Motohashi, On the Deuring-Heilbronn phenomenon, I-II. Proc. Japan Acad. Ser. A 53 (1977), 1-2 and 25-27.

[22] Y. Motohashi, Primes in Arithmetic Progressions. Invent. Math. 44 (1978), 163-178.

[23] Y. Motohashi, On the linear sieve, I-II. To appear.

[24] Y. Motohashi, Lectures on sieve methods and prime number theorems. To appear.

[25] A. Selberg, On elementary methods in primenumber theory and their limitations. 11th Skand. Mat. Kongr. Trondheim 1949, 13-22.

[26] A. Selberg, Remarks on sieves. Proc. of the 1972 Number Theory Conference, 205-216. Boulder, Colorado University 1972.

[27] E. C. Titchmarsh, A divisor problem. Rend. Circ. Mat. Palermo 54 (1930), 414-429.

Department of Mathematics
University of Turku
SF-20500 Turku 50
Finland

# CONCERNING THE DUAL VARIETY

Steven L. Kleiman*

Around 1800 at the Ecole Polytechnique in Paris, geometry was revolutionized, as descriptive and projective geometry were founded, by Gaspard Monge and his students, particularly, Jean-Victor Poncelet. Among other things, imaginary points were used as if they were real, and the principle of duality was discovered. As to the latter, it was found that interchanging the words "point" and "line" in a theorem about rectilinear plane figures led to a new theorem, which was equally valid. This duality was extended to curvilinear figures by the classical process of rectification, in which a curve $X$ is approximated by a sequence of points connected by lines; dualizing the points and passing to the limit leads to the dual curve $X^{\perp}$, whose points correspond to the tangent lines of $X$ and whose tangent lines correspond to the points of $X$.

At first, duality was understood in terms of polar reciprocation with respect to a conic. However, in 1830 Julius Plücker (then in Bonn after studying in Paris and elsewhere) introduced homogeneous coordinates for points and for lines.

---

* It is a pleasure to acknowledge the financial support offered during the course of this work in part by the J. S. Guggenheim Foundation, the (US)NSF under MCS-7906895, this congress, and Aarhus University, and to thank the University's Mathematics Institute for its hospitality.

By using the coefficients $y_i$ of the equation,

$$\Sigma x_i y_i = 0,$$

of a line as coordinates for the line and by viewing the equation itself as simply expressing incidence of the point $(x_i)$ and the line $(y_i)$, Plücker put the notions of point and line formally on an equal footing. Moreover, he put a curve $X$ and its dual $X^{\perp}$ on an equal footing, noting that the tangents determine the shape as much as the points do and that the analysis of $X$ and that of $X^{\perp}$ is formally the same.

Philosophically it is clear that "biduality" must hold, that the dual $(X^{\perp})^{\perp}$ of the "line" curve $X^{\perp}$ coincides with the original "point" curve $X$. It is also clear that the two basic projective characters of $X$, its degree $\mu_0$ and its class $\mu_1$ (the number of tangent lines through a general point), and the analogous characters $\mu_0^{\perp}$ and $\mu_1^{\perp}$ of $X^{\perp}$ are equal in reverse order,

$$(1) \qquad \mu_0 = \mu_1^{\perp}, \qquad \mu_1 = \mu_0^{\perp},$$

provided that $X$ is not a line. If $X$ is a line, then biduality holds, although $X^{\perp}$ is now a point, and (1) must be replaced by

$$(2) \qquad \mu_0 = \mu_0^{\perp}, \qquad \mu_1 = \mu_1^{\perp};$$

however, the second of these relations reduces to $0 = 0$. The rigorous development of these intuitive observations and the generalization of them to varieties of arbitrary dimension over fields of arbitrary characteristic, a topic of current interest, is the subject of this report.

In what follows, $X$ will denote a (fixed) closed, $n$-dimensional subvariety (reduced and irreducible) of a projective $m$-space $P = \mathbb{P}(V)$ where $V$ is a vector space over an algebraically closed ground field. The dual variety $X^{\perp}$ of $X$ is defined as the closure of the locus of hyperplanes that are tangent at smooth points; thus $X^{\perp}$ is a closed subvariety of the dual projective space $P^{\perp} = \mathbb{P}(V^{\perp})$ where $V^{\perp}$ is the dual vector space. Denote the dimension of $X^{\perp}$ by $n^{\perp}$.

The "biduality" theorem, a fundamental theorem, asserts that the dual variety $(X^{\perp})^{\perp}$ of $X^{\perp}$ is equal to $X$ provided that the characteristic is zero or, more generally, a certain separability condition holds (namely, the natural map $q: X^{\sim} \rightarrow X^{\perp}$, described below, should be separable or, what amounts to the same, smooth on a dense open subset $U^{\sim}$ of $X^{\sim}$). The theorem was first proved for arbitrary $n$ and $m$ over the complex numbers by C. Segre [1910], no. 16; the general theorem with the separability condition was proved by Wallace [1956]. Wallace's proof and the proofs that followed (Moisezon [1967], §1; Kay [1969], Exposé XVII, pp. 212-253; Urabe [1980]) involved the equations and are fairly complicated. A simple proof, whose essence is easy to recall, will be given below; it involves 1-parameter families of points and is basically Segre's original proof*.

In 1902 Severi introduced and studied certain projective characters, the classes $\mu_k$ of $X$. While the $\mu_k$ are natural

---

* On returning to MIT, the author found waiting for him a Berkeley preprint by T. Fujita and J. Roberts, "Varieties with small secant varieties: the extremal case", in which these authors had independently given in Prop. 3.1 a similar proof.

generalizations of the degree and class of a curve, the lovely relation between the $\mu_k$ and the analogous characters $\mu_l^\perp$ of $X^\perp$, which generalizes the relations (1) and (2) for curves, went undiscovered until remarkably recently. Piene [1978], Prop.(3.6.) obtained the case in which both $X$ and $X^\perp$ are hypersurfaces; Urabe [1980] obtained the general case. Urabe's proof is fairly involved. However, when the theory of polar loci is setup as below, the proof reduces to a few lines of computation and it is much like Piene's original proof. The theorem may be stated as follows.

Theorem. (Piene - Urabe): _If the characteristic is zero or Wallace's separability condition holds_, then

$$\mu_k = \mu_l^\perp \quad \text{for} \quad (k+1) = (n+n^\perp)-(m-1)$$

The $k$th class $\mu_k$ of $X$ is defined as the degree of the $k$th polar locus $M_k$,

$$\mu_k = \deg(M_k).$$

The locus $M_k$ is defined as follows. Let $A$ be a general linear subspace of $P$ with dimension $a$,

$$a = m-n+k-2.$$

Let $M_k^0$ denote the locus of smooth points $x$ of $X$ such that the embedded tangent space $T_x$ intersects $A$ in a linear space of dimension at least $(k-1)$. (The numbers are chosen so that $M_k^0$ has codimension $k$, see below). Finally, $M_k$ is defined as the closure of $M_k^0$.

If $k > n$, then $\mu_k = 0$, because $M_k$ is obviously empty. Obviously, $\mu_0 = \deg(X)$, because $M_0 = X$. It is convenient to set $\mu_k = 0$ for $k < 0$. It is intuitively clear and not hard to prove formally (see Piene [1978], Thm. (4.2), p. 270) that $\mu_k$ for $k < n$ is equal to the kth class of a general hyperplane section of $X$.

Notice that $M_k$ may be expressed as the closure of the locus of smooth points $x$ of $X$ such that the dual linear space $T_x^{\perp}$ in $P^{\perp}$ simply intersects the dual linear space $A^{\perp}$, which has codimension $(a+1)$. Thus, in particular, the top class $\mu_n$ is equal to the number of hyperplanes in the general pencil $A^{\perp}$ that are tangent to $X$ (at smooth points).

Let $X^{\sim}$ denote the closure in $P \times P^{\perp}$ of the locus of pairs $(x,y)$ such that $x$ is a smooth point of $X$ and $y$ is a point of $P^{\perp}$ representing a hyperplane containing $T_x$. Consider the fundamental diagram,

$$\begin{array}{ccc} & X^{\sim} & \\ {}^{p}\swarrow & & \searrow^{q} \\ X & & X^{\perp}, \end{array}$$

in which $p$ and $q$ are the projections. Then $p$ induces a map,

$$b\colon q^{-1}(X^{\perp} \cap A^{\perp}) \to M_k,$$

which is birational, because its fiber over a point $x$ of $M_k^0$ is the linear space $T_x^{\perp} \cap A^{\perp}$ and because $T_x^{\perp} \cap A^{\perp}$ is

0-dimensional for all $x$ in a dense open subset since $A^{\perp}$ is in general position. It follows in particular that $M_k$ has pure codimension $k$ or is empty.

To prove the theorem, set

$$h = c_1(O_X(1)), \qquad h^{\perp} = c_1(O_{X^{\perp}}(1)).$$

Then, since $q$ is birational, we have

$$\mu_k = \int_X h^{n-k} p_*[q^{-1}(X^{\perp} \cap A^{\perp})].$$

Since $q$ is generically flat and $A^{\perp}$ is in general position, we have

$$[q^{-1}(X^{\perp} \cap A^{\perp})] = q^*[X^{\perp} \cap A^{\perp}] = q^*(h^{\perp})^{a+1}.$$

Hence the projection formula yields

$$\mu_k = \int_{X^{\sim}} (p^*h^{n-k})(q^*(h^{\perp})^{a+1}).$$

By symmetry, therefore, $\mu_k = \mu_1^{\perp}$ as asserted, provided that $(X^{\perp})^{\sim}$ and $X^{\sim}$ are equal. However, they are equal if Wallace's separability condition holds (namely, $q$ is separable); this stronger form of the biduality theorem was proved by Wallace and will be reproved below.

Note that, if $X^{\sim}$ is equal to $(X^{\perp})^{\sim}$, then $q$ is a bundle-map over the smooth locus of $X^{\perp}$ and so the scheme $q^{-1}(X^{\perp} \cap A^{\perp})$ is reduced on a dense open subset; in particular, $\mu_k$ is just the naive (unweighted) geometric number. In general, however, it is necessary to deal with the natural scheme structure on $M_k$ and to take more care in establishing the birationality of $b$, which is still true. Although the details will not be given here, these matters can be handled conveniently, in what by now is a straightforward way, using the sheaf of first order principal parts of $0_X(1)$. (This sheaf can be defined abstractly using the first infinitesmal neighborhood of the diagonal and then related via a local computation to the family of embedded tangent spaces; this approach is followed in Piene [1978],§1, in Piene [1977],§2,§6, and in Kleiman [1977], IV, A, B. Alternatively, the consideration of the variety of chords with endpoints leads to a definition of the same sheaf as a certain pushout; this more geometric approach is followed in Laksov [1976], pp. 175-6 and in Laksov [1979], pp. 8-9.)

There are birational modifications $X'$ of $X$ along the singular locus over which the Gauss map $x \to T_x$ extends. The consideration of these $X'$ leads in certain cases to generalized Plücker formulas, which relate the $M_k$ and characters of the singular locus; see Piene [1978], §2,3, and Kleiman [1977], IV,C, and Urabe [1979]. Plücker in 1834 and more explicitly in 1839 gave such formulas for plane curves with simple singularities; this work was the greatest advance made in the theory of higher plane curves in nearly 200 years.) On the other hand, by drawing attention away from $X^{\sim}$, these considerations delayed the discovery of the equality in reverse order of the $\mu$'s and $\mu^{\perp}$'s for arbitrary $n$ and $n^{\perp}$ and they account for the complexity of Urabe's proof. At this congress, Piene told the author that Teissier had also made this same point to her, and Holme told the author that he had independently shortened

Urabe's proof and he pointed out the connections with Holme [1978].

The key to the simple proof of biduality is the following observation. Let

$$(x(t), y(t)) \in P \times P^{\vee}$$

be a 1-parameter family of points, given in homogeneous coordinates,

$$x(t) = (x_0(t), \cdots, x_m(t), \quad y(t) = (y_0(t), \cdots, y_m(t)),$$

by formal power series (which may be truncated at $t^2$ if desired),

$$x_i(t) = x_i(0) + \dot{x}_i(0)t + \cdots, \; y_i(t) = \dot{y}_i(0) + y_i(0)t + \cdots .$$

Then the vanishing of the two dot products,

$$x(t) \cdot y(t) \;\; (= \Sigma\, x_i(t) y_i(t)) = 0 \tag{3}$$

$$\dot{x}(0) \cdot y(0) = 0, \tag{4}$$

implies the vanishing of the third,

$$x(0) \cdot \dot{y}(0) = 0. \tag{5}$$

Indeed, the formula for the derivative of a product,

$$(x(t) \cdot y(t))^{\bullet}(0) = \dot{x}(0) \cdot y(0) + x(0) \cdot \dot{y}(0),$$

is obviously valid and immediately yields the assertion.

For any point $y$ of $Y$, the $y$-<u>contact</u> <u>locus</u> is defined as the closed subvariety $p(q^{-1}(y))$ of $X$. It is the locus of points of tangency of the hyperplane represented by $y$.

<u>Theorem</u>. (C. Segre-Wallace): If $q$ is separable, then

$$(X^{\perp})^{\sim} = X^{\sim}.$$

Before beginning the proof, let us note that, conversely, if this equation holds, then $q$ is separable; in fact, $q$ is a bundle-map over the smooth locus of $X^{\perp}$. Moreover, if the equation holds, then for each smooth point $y$ of $Y$ the y-contact locus is a linear space; in fact, it is the dual space $T_y^{\perp}$.

Assume now that $q$ is separable. It will be shown that for all y in a certain dense, open subset $W$ of $X$ the y-contact locus is equal to $T_y^{\perp}$. Then $X^{\sim}$ and $(X^{\perp})^{\sim}$ must coincide, for they are both closed subsets of $P \times P^{\perp}$ and they will have the same fibers over $W$.

The set $W$ is obtained as follows. Since $q$ is separable, $q$ is smooth on a dense, open subset $V^{\sim}$ of $X^{\sim}$. Let $V$ and $V^{\perp}$ denote the smooth loci of $X$ and $X^{\perp}$. Set

$$F = X^{\sim} - (V^{\sim} \cap p^{-1}V)$$

and consider the open set $W_1$ of $X^{\perp}$ over which the fibers of $F$ have dimension at most $(m - n^{\perp} - 2)$. Clearly $W_1$ is nonempty because $V^{\sim}$ and $p^{-1}V$ are dense, open subsets of $X^{\sim}$ and the fibers of $q|V^{\sim}$ have dimension $(m - n^{\perp} - 1)$. Let

$$W_2 = q(V^{\sim} \cap p^{-1}V)$$

and note that $W_2$ is open because $q|V^{\sim}$ is smooth. Finally, set

$$W = W_1 \cap W_2 \cap V^{\perp}$$

and note that $W$ is open and dense in $X^{\perp}$.

Fix $y \in W$. Then clearly both the y-contact locus and $T_y^{\perp}$ have dimension $(m - n^{\perp} - 1)$. Moreover, $T_y^{\perp}$ is irreducible, being a linear space. Hence if it is proved that the y-contact locus is contained in $T_y^{\perp}$, then the two will coincide.

It is clear from the construction of $W$ that the intersection $S$ of the y-contact locus and $p(V^{\sim} \cap p^{-1}V)$ is dense in the y-contact locus. Hence it suffices to show that $S$ is contained in $T_y^{\perp}$ or, equivalently, that for each $x \in S$ and $\tau \in T_y$ the dot product vanishes, $x \cdot \tau = 0$.

Since $q|V^{\sim}$ is smooth, tangent vectors can be lifted. So there is a 1-parameter family of points,

$$(x(t), y(t)) \in V^{\sim},$$

such that

$$x(0) = x, \quad y(0) = y, \quad \dot{y}(0) = \tau.$$

Finally, apply the observation made before the statement of the theorem. Equation (3) holds by the very definition of $X^{\sim}$. Equation (4) holds because $X$ is in the y-contact locus. Hence (5) holds; in other words, $x \cdot \tau = 0$, q.e.d.

Wallace himself emphasized that the proper condition of biduality is the coincidence of $(X^{\perp})^{\sim}$ and $X^{\sim}$. This coincidence obviously implies the coincidence of $(X^{\perp})^{\perp}$ and $X$. However, the converse is false. Wallace [1956], 7.2, illustrates this failure with the curve,

$$X: u - u^{p} - v^{p+1} = 0$$

for which the natural map from $X$ to $X^{\perp\perp}$ is equal to the square of the Frobenius endomorphism.

## R E F E R E N C E S


1. A. Holme, [1978], Deformation and Stratification of Secant Structure, Algebraic Geometry, Proceedings, Tromsø, Norway 1977, L. Olson, ed., pp. 60-91, Lecture Notes in Math., 687, Springer (1978).

2. N. Katz, [1969], Groupes de monodromie en géométrie algébrique, P. Deligne and N. Katz editors. Springer Lecture Notes in Math. 340 (1973).

3. S. Kleiman, [1977], The enumerative theory of singularities, Real and complex singularities, Oslo 1976, P. HOlm, ed., pp. 297-396, Sijthoff & Noordhooff (1977).

4. D. Laksov, [1976], Some enumerative properties of secants to non-singular projective schemes, Math. Scand. 39 (1976), 171-190.

5. D. Laksov, [1979], Indecomposability of restricted tangent bundles, Institut Mittag-Leffler (Auravägen 17, S - 182 62 Djursholm, Sweden). Preprint 1979 nr. 10.

6. B. Moisezon, [1967], Algebraic homology classes on algebraic varieties, Igv. Akad. Nauk. SSSR Ser. Mat. 31 (1967) = Math. Ussr-Igvestija 1 (1967), 209-251.

7. R. Piene, [1977], Numerical characters of a curve in projective n-space, Real and complex singularities, Oslo 1976, P. Holm, ed., pp. 475-496, Sijthoff & Noordhooff (1977).

8. R. Piene, [1978], Polar classes of singular varieties, Ann. S. Ec. N. Sup. 11 (1978), 247-276.

9. C. Segre, [1910], Preliminari de una teoria delle varietà luoghi di spazi, Rendiconti Circolo Mat. Palermo XXX (1910), 87-121 = Opere vol. II, Cremonese, Roma (1958), 71-114.

10. T. Urabe, [1979], Preprint (1979).

11. T. Urabe, [1980], Duality of numerical characters of polar loci, Preprint (1980).

12. A. Wallace, [1956], Tangency and duality over arbitrary fields, Proc. Lond. Math. Soc. (3)6 (1956), 321-342.


/AP

# Burnside ring and Segal's conjecture

Erkki Laitinen

## 1. The Burnside ring.

Let $G$ be a finite group. A finite $G$-set is a finite set with G-action. Two G-sets $S$ and $T$ are isomorphic if there is an equivariant bijection $f: S \to T$. The equivalence classes of finite G-sets form a commutative semiring $A^+(G)$ with disjoint union as addition and cartesian product as multiplication. The associated Grothendieck ring is denoted by $A(G)$ and called the Burnside ring of $G$.

The Burnside ring $A(G)$ is easily computable from the group theoretic structure of $G$. Its elements are formal differences of finite G-sets. Each G-set is a disjoint union of its orbits, which are isomorphic to $G/H$, where $H$ is a subgroup of $G$, and $G/H_1 \cong G/H_2$ if and only if $H_1$ and $H_2$ are conjugate in $G$. Additively, $A(G)$ is a free abelian group with basis $\{G/H\}$, one for each conjugacy class of subgroups. Fix a set $C(G)$ of representatives of $H$.

For the multiplicative structure we introduce characters. Let $H$ be a subgroup of $G$. If we define for any finite G-set $S$ $\chi_H(S) = |S^H|$, the number of points in $S$ left fixed by $H$, then $\chi_H$ extends to a ring homomorphism $\chi_H: A(G) \to \mathbb{Z}$. Together the $\chi_H$, $H \in C(G)$, give a ring homomorphism

$$\chi: A(G) \to \prod_{C(G)} \mathbb{Z}$$

Proposition 1. $\chi$ is an injective homomorphism with finite cokernel. The image is characterized by certain congruences.

Hence, to describe $A(G)$ as a ring one has to compute the "character table" $\{\chi_{H_1}(G/H_2)\}$, $H_1, H_2 \in C(G)$.

The Burnside ring is a convenient device to keep track on the lattice of subgroups of a given group. In algebra it plays a role in axiomatic representation theory and in integral representation theory. In topology it has applications to transformation groups.

For our problem we have to complete the Burnside ring. Let $I(G) = \chi_e^{-1}(0)$ be the ideal of virtual G-sets with cardinality zero and define the I(G)-adic completion of $A(G)$ as the inverse limit

$$\hat{A}(G) = \lim_{\leftarrow} A(G)/I(G)^n.$$

For p-groups $G_p$ the completion is the p-adic one on $I(G_p)$: $\hat{A}(G_p) = \mathbb{Z} \oplus (\hat{\mathbb{Z}}_p \otimes I(G_p))$. It follows that they have characters $\chi_H: \hat{A}(G_p) \to \hat{\mathbb{Z}}_p$ which embed $\hat{A}(G_p)$ into $\prod_{C(G_p)} \hat{\mathbb{Z}}_p$.

For any injection $i: H \to G$ there are restriction and induction homomorphisms

$$\mathrm{Res}_H^G: A(G) \to A(H), \ \mathrm{Ind}_H^G: A(H) \to A(G)$$

defined regarding G-sets as H-sets via $i$ and sending $H/U$ to $G/i(U)$, respectively. $\mathrm{Res}_H^G$ is a ring homomorphism with $\mathrm{Res}_H^G(I(G)) \subset I(H)$ so it induces $\hat{\mathrm{Res}}_H^G: \hat{A}(G) \to \hat{A}(H)$.

Proposition 2. Let $\{G_p\}$ be the Sylow subgroups of a finite group $G$. The restrictions give an embedding

$$0 \to \hat{A}(G) \to \bigoplus_p \hat{A}(G_p)$$

whose image consists of families $(x_p)$ which satisfy

(i) $\chi_e(x_p) = \chi_e(x_q)$

(ii) $\chi_{H_1}(x_p) = \chi_{H_2}(x_p)$ if $H_1$ and $H_2$ are conjugate in $G$

for all primes $p,q$ and subgroups $H_1$, $H_2$ of $G_p$.

The recipe for computing $\hat{A}(G)$ is thus: for each Sylow subgroup $G_p$ of $G$, find out $I(G_p)$ and form $\hat{I}(G_p) = \mathbb{Z}_p \otimes I(G_p)$. Let $\hat{I}(G)_p$ be the subset satisfying (ii). Then $\hat{A}(G) = \mathbb{Z} \oplus \bigoplus_p \hat{I}(G)_p$.

As the proof of Prop.2 has not yet appeared in printed form, we give it here:

<u>Proof</u>. The fact that $\hat{A}(G)$ embeds in $\bigoplus_p \hat{A}(G_p)$ follows by commutative algebra, see [2, 1.15]. For the rest we have to describe the prime ideals in $\hat{A}(G)$. From proposition 1 and Cohen-Seidenbergs theorem the prime ideals of $A(G)$ are

$$p_{H,p} = \{x \in A(G) \mid \chi_H(x) \equiv 0(p)\},$$

$$p_{H,0} = \{x \in A(G) \mid \chi_H(x) = 0\}$$

for all primes $p$ and $H \in C(G)$. As the kernel of $A(G) \to \hat{A}(G)$ is $I(G)^\infty = \bigcap_{H\ p\text{-group}} p_{H,0}$ [2,1.10], the prime ideals of $\hat{p}$ of $\hat{A}(G)$ must satisfy $p \supset I(G)^\infty$, so ther are included in

$$\hat{p}_{H,p},\ \hat{p}_{H,0},\ H\ p\text{-group for any } p.$$

We claim that the induction maps from Sylow subgroup induce $\widehat{\mathrm{Ind}}_{G_p}^G : \hat{A}(G_p) \to \hat{A}(G)$. Indeed, $\hat{A}(G_p) = \mathbb{Z} \oplus \hat{I}(G_p)$ and it is enough to consider $\hat{I}(G_p)$. As $\hat{A}(G) \subset \oplus \hat{A}(G_p)$, it is detected by characters $\chi_H: \hat{A}(G_p) \to \hat{\mathbb{Z}}_p$. Clearly $\mathrm{Ind}_{G_p}^G(I(G_p))$ lies in $\hat{I}(G)_p = \mathrm{Ker}(\hat{I}(G) \to \bigoplus_{q \neq p} \hat{I}(G_q))$ [2,1.5] and on $\hat{I}(G)_p$ the characters $\chi_H$, $H$ p-group are divisible by $p$ [2, p.45]. Hence the

$\hat{I}(G)$-adic topology on $\hat{I}(G)_p$ is p-adic by the same proof as [2,1.12]. As the $I(G_p)$-adic topology on $\hat{I}(G_p)$ is p-adic, too [2,1.2] and Ind: $I(G_p) \to \hat{I}(G)_p$ is linear, it is continuous and induces $\hat{\mathrm{Ind}}_{G_p}^G : \hat{I}(G_p) \to \hat{I}(G)_p \subset \hat{A}(G)$.

Next we claim that $\oplus \hat{\mathrm{Ind}}_{G_p}^G : \oplus_p \hat{A}(G_p) \to A(G)$ is surjective. The image is an ideal by the reciprocity [2,1.6]. If it is proper, it is contained in a maximal ideal $\hat{p}_{H,p}$, $H$ a p-group. But $\hat{p}_{H,p} = \hat{p}_{e,p}$ by [2,1.9] and $S = \mathrm{Ind}_{G_p}^G(1) = G/G_p$ satisfies $\chi_e(S) = |G|/|G_p| \not\equiv 0 \pmod p$, so $S \notin \hat{p}_{e,p}$.

The abstract induction theory now gives the claim. Indeed, let $(x_p)$ be a family satisfying (i) and (ii) and let $a_p \in \hat{A}(G_p)$ satisfy $\Sigma \mathrm{Ind}_{G_p}^G(a_p) = 1$. Then $x = \Sigma \hat{\mathrm{Ind}}_{G_p}^G(x_p a_p) \in \hat{A}(G)$ has restrictions $x_p$, as can be verified using the reciprocity and double coset formulas [2,1.6,1.7]. □

## 2. Segal's conjecture.

Let $G$ be a finite group. The n+1-fold join $E_nG = G * \ldots * G$ consists of elements $(g,t) = (g_0,t_0,\ldots,g_n,t_n)$ where $g_i \in G$, $t_i \in [0,1]$ and $\Sigma t_i = 1$ with the relation $(g,t) \sim (g',t')$ if $t_i = t_i'$ for all $i$ and $g_i = g_i'$ for all $i$ such that $t_i = t_i' > 0$, topologized as a quotient space of $G^{n+1} \times \Delta^n$.

Clearly $E_nG \subset E_{n+1}G$ and the union

$$EG = \bigcup_{n \geq 0} E_nG = G*G*\ldots*G*\ldots ,$$

is called the universal G.space. H is a contractible infinite-dimensional CW-complex whose cellular chain complex forms the standard resolution of G. G acts freely on EG by $(h,t)g = (hg,t)$ and $BG = EG/G$ is called the <u>classifying space of</u> G. It is characterized up to homotopy by the property that its homotopy groups vanish except $\pi_1(BG) = G$. For $\mathbb{Z}/2$, $E\mathbb{Z}/2 = S^\infty$ with the antipodal action and $B\mathbb{Z}/2 = \mathbb{R}P^\infty$.

The philosophy is now that topological invariants of BG should be obtained purely algebraically from G. For example, the cohomology of BG is the usual group cohomology. Another example is K-theory. The stable equivalence classes of complex vector bundles on a finite CW-complex X form a ring $K^0(X)$, and we define $K^0(BG) = \varprojlim K^0(B_nG)$ where $B_nG = E_nG/G$. M. Atiyah studied in [1] the K-theory of BG and representations of G by the following construction: if $\rho: G \to \mathbb{C}^n$ is a representation then $EG\times_G(\mathbb{C}^n,\rho)$ is a complex vector bundle on BG and defines an element of $K^0(BG)$. One gets a ring homomorphism $R(G) \to K^0(BG)$, and Atiyah proves that the completion $\hat{R}(G) \to K^0(BG)$ is an isomorphism.

Analogously, each G-set S gives rise to a finite covering $EG\times_G S$ of BG with fibre S and we get a semiring homomorphism $\alpha: A^+(G) \to C(BG)$ where $C(X) = [X, \coprod_{n\geq 0} B\Sigma_n]$ is the set of finite coverings of X, a semiring under disjoint union and fibrewise cartesian product of total spaces. There is a map $\coprod_{n\geq 0} B\Sigma_n \to QS^0$ to the space $QS^0 = \varinjlim_n \Omega^n S^n$ of stable self-maps of spheres, giving rise to a transformation $C \to \pi_S^0$ to the stable cohomotopy of degree 0. $\pi_S^0(X) = [X, QS^0]$ is a ring under loop addition and smash product and $C \to \pi_S^0$ is universal among transformations of semiring valued functors from C to a ring valued functor. $\alpha$ therefore gives rise to a ring homomorphism

$$\alpha_G: A(G) \to \pi_S^0(BG)$$

and around 1970 G. Segal made the conjecture

$$\text{(SC)} \quad \hat{\alpha}_G: \hat{A}(G) \to \pi_S^0(BG) \quad \text{is an isomorphism.}$$

This conjecture can be reduced to p-groups

Theorem 3. Let $G$ be a finite group and $G_p$ its Sylow subgroups. If $\hat{\alpha}_{G_p}$ is injective (bijective) for all $p$, then $\hat{\alpha}_G$ is injective (bijective).

Proof. As $\hat{A}(G) = \mathbb{Z} \oplus \hat{I}(G)$ and $\pi_S^0(BG) = \mathbb{Z} \oplus \pi_S^0(BG)$, we can consider $\hat{\alpha}_G: \hat{I}(G) \to \tilde{\pi}_S^0(BG)$. In the diagram

$$\begin{array}{ccccccc}
0 \to & \hat{I}(G) & \to & \bigoplus_p \hat{I}(G_p) & \to & \bigoplus_{g,p} \hat{I}(G_p \cap gG_pg^{-1}) \\
& \downarrow \hat{\alpha}_G & & \downarrow \oplus\hat{\alpha}_{G_p} & & \downarrow \\
0 \to & \tilde{\pi}_S^0(BG) & \to & \bigoplus_p \tilde{\pi}_S^0(BG_p) & \to & \bigoplus_{g,p} \tilde{\pi}_S^0(B(G_p \cap gG_pg^{-1})
\end{array}$$

where the last horizontal maps are induced from the two embeddings $i,j: G_p \cap gG_pg^{-1} \to G_p$, $i(x) = x$, $j(x) = g^{-1}xg$, the upper row is exact by proposition 2 and the lower row by general properties of cohomology theories. The 5-lemma now gives the claim. □

There are results on two classes of p-groups.

Theorem 4. If $G$ is elementary abelian group $(\mathbb{Z}/p)^n$ then $\hat{\alpha}_G$ is injective.

This is proven by induction on the order of the group. In the induction step, maps are detected by the singular homology with $\mathbb{Z}/p$-coefficients. See [2].

Quite different techniques are required in the stronger result

Theorem 5. (Lin, Davis, Mahowald, Adams, Gunawardene, Ravene

If $G = \mathbb{Z}/p^n$ then $\hat{\alpha}_G$ is bijective.

W.-H. Lin proved the result for $\mathbb{Z}/2$ by computing the Adams spectral sequence; Davis, Mahowald and Adams simplified the proof

See [3],[4]. A student of Adams, Gunawardena showed the result for $\mathbb{Z}/p^n$, p odd and finally Ravenel generalized it to $\mathbb{Z}/p^n$.

Corollary. Segal's conjecture holds for cyclic groups and the injectiveness part holds for groups with cyclic or elementary abelian Sylow subgroups.

References.


1. M.F. Atiyah, Characters and cohomology of finite groups, Publ.Math.IHES 9(1961), 247-288.
2. E. Laitinen, On the Burnside ring and stable cohomotopy of a finite group, Math.Scand.44(1979), 37-72.
3. W.-H. Lin, On conjectures of Mahowald, Segal and Sullivan, Math.Proc.Camb.Phil.Soc. 87(1980), 449-458.
4. W.-H. Lin, D.M. Davis, M.E. Mahowald and J.F. Adams, Calculation of Lin's Ext groups, Math.Proc.Camb.Phil. Soc. 87(1980), 459-469.



University of Helsinki
Department of Mathematics
Hallituskatu 15
SF-00100 Helsinki 10
Finland

# Geometry of Banach spaces and intersection properties of balls.

Åsvald Lima

Let X, Y and Z be Banach spaces with $Y \subseteq Z$ and let $T: Y \to X$ be a bounded linear operator. In 1950, Nachbin [15] showed that the existence of a normpreserving extension of T to Z is related to the non-void intersection of some families of balls in X.

Let us demonstrate this relationship. We assume all spaces are real Banach spaces.

First assume $z \in Z \setminus Y$. Then if $y_1, y_2 \in Y$ we have

$$(*) \quad ||Ty_1 - Ty_2|| \leq ||T|| \; ||y_1 - y_2|| \leq ||T|| \; ||y_1 - z|| + ||T|| \; ||y_2 - z||.$$

This means that the family of balls $\{B(Ty, ||T|| \; ||y-z||) : y \in Y\}$ intersect in pairs.

If $\tilde{T}$ is a norm-preserving extension of T to $Y \oplus R\{z\}$, then

$$\tilde{T}z \in \bigcap_{y \in Y} B(Ty, ||T|| \; ||y-z||).$$

In fact, for any $y \in Y$ we have

$$||Ty - \tilde{T}z|| = ||\tilde{T}(y-z)|| \leq ||\tilde{T}|| \cdot ||y-z|| = ||T|| \cdot ||y-z||.$$

On the other side, if there exists some $u \in \bigcap_{y \in Y} B(Ty, ||T|| \; ||y-z||)$ then T admits a norm-preserving extension to $Y \oplus R\{z\}$. We only define for $y \in Y$ and $\lambda \in R$

$$\tilde{T}(y + \lambda z) = Ty + \lambda u$$

and observe that (for $\lambda \neq 0$)

$$||\tilde{T}(y+\lambda z)||=|\lambda|\ ||\tilde{T}(\frac{y}{\lambda}+z)||$$

$$=|\lambda|\ ||T(\frac{y}{\lambda})+u||=|\lambda|\ ||u-T(-\frac{y}{\lambda})||$$

$$\leq|\lambda|\ ||T||\ ||z-(-\frac{y}{\lambda})||=||T||\ ||y+\lambda z||.$$

By a Zorn's lemma argument we can proceed from the case $\dim Z/Y=1$ to the general case.

In $C(K)$ spaces order completeness can be expressed by intersection properties of balls. Nachbin proved the following theorem for real spaces in [15].

THEOREM 1. The following statements are equivalent for a Banach space $X$.

1) For all pairs of Banach spaces $Y\subseteq Z$ every bounded linear operator $T:Y\to X$, $T$ admits a norm-preserving extension $\tilde{T}:Z\to X$.

2) If $X$ is isometrically imbedded into a space $Y$, then there exists a norm-one projection from $Y$ onto $X$.

3) If $\{B(x_\alpha,r_\alpha)\}_{\alpha\in A}$ is a family of closed balls in $X$ such that

$$(*)\qquad ||x_\alpha-x_\beta||\leq r_\alpha+r_\beta \qquad \forall\alpha,\beta\in A$$

then there exists an $x\in\bigcap_{\alpha\in A}B(x_\alpha,r_\alpha)$,

4) $X$ is isometric to $C(K)$ for some Hausdorff and extremally disconnected compact set $K$.

If $\dim X<\infty$, we can add:

5) $X$ is isometric to $l_\infty^n$ for some $n$.

In the proof of 1)$\Rightarrow$4) Nachbin assumed that the unit ball of $X$ contains an extreme point. Later on, Kelley [7] showed that this extra assumption was superfluous. In the same volume of the same journal that Nachbin published his results, similar results was proved by Goodner [1].

In 1958 Hasumi [4] extended part of the theorem to the complex case, but it was not before 1973 that Hustad [5] found the correct formulation of 3) in the complex case.

Observe that we have (in the real case)

(*) in 3) if and only if $\sup_{\substack{f\in X^* \\ ||f||\leq 1}} |f(x_\alpha)-f(x_\alpha)| \leq r_\alpha + r_\beta$

for all $\alpha,\beta\in A$ which in turn is equivalent to

(#) $\bigcap_{\alpha\in A} B(f(x_\alpha), r_\alpha) \neq \emptyset$ for all $f\in X^*, ||f||\leq 1$.

Hustad showed that the correct formulation of 3) in the complex case is obtained by replacing (*) by (#).

We shall follow Hustad and say that a family of balls has the <u>weak intersection property</u> if (#) is satisfied by the family.

In 1964 Lindenstrauss [13] studied the problem of extending compact operators. Some of his results can be summarized in the following theorem.

THEOREM 2. The following statements are equivalent for a Banach space X.

1) For all pairs of Banach spaces $Y\subseteq Z$, all compact operators $T:Y\to X$ and all $\varepsilon>0$, there exists a compact extension $\tilde{T}:Z\to X$ of T with $||\tilde{T}||\leq ||T||+\varepsilon$.

2) If $\{B(x_\alpha, r_\alpha)\}_{\alpha\in A}$ is a family of closed balls in X with the weak intersection property and $\{x_\alpha\}_{\alpha\in A}$ is relatively norm-compact, then $\bigcap_{\alpha\in A} B(x_\alpha, r_\alpha)$ is non-empty.

3) Same as 2) but with cardinality of A equal 4.

4) $X^*$ is isometric to an $L_1(\mu)$-space.

5) $X^{**}$ satifies 1) to 4) in Theorem 1.

Hustad [6] extended this theorem to the complex case, but he needed 7 balls in 3). Later on, Lima [9] showed that it suffices with 4 balls. In another paper, Lima [10] showed that in the complex case it even suffices with 3 balls. This is false in the real case as we shall show below.

Before we proceed to consider intersection of 3 balls, let us look at 1) of Theorem 2. It would be nice if we could take $\varepsilon=0$. It was realized by Lindenstrauss that this is not the case. Then Lazar [8] in the real case and Olsen and Nielsen [14] in the complex case, showed that we can take $\varepsilon=0$ in 1) precisely when the sequence space c does not imbed isometrically into X.

Let us say that a real space X has the 3.2. intersection property (3.2.I.P.) if it satisfied 2) in Theorem 2 for families of 3 balls. Already in 1956 Hanner [2] characterized finite dimensional spaces with the 3.2.I.P. His results were extended to the general case by Lima [11], [12].

THEOREM 3. For a real Banach space the following statements are equivalent:

1) X has the 3.2.I.P.

2) $X^*$ has the 3.2.I.P.

3) If G, H are disjoint faces of the unit ball $B(0,1)$, there exist a proper face F of $B(0,1)$ such that $G\subseteq F$ and $H\subseteq -F$.

The theorem shows that real $L_1(\mu)$-spaces has the 3.2.I.P. Hanner showed that if $\dim X\leq 5$, then X has the 3.2.I.P. if and only if X can be obtained by forming $l_1$- and $l_\infty$-sums of the real line. 15 years later Hansen and Lima [3] proved the following result.

THEOREM 4. If $\dim X<\infty$, the following statements are equivalent:

1) X has the 3.2.I.P.

2) X can be obtained by forming $l_1$- and $l_\infty$-sums of the real line.

The main step in the proof is to show that if X has the 3.2.I.P. and $2<\dim X<\infty$, then X has a proper decomposition $X=(Y\oplus Z)_{l_1}$ or $X=(Y\oplus Z)_{l_\infty}$. To distinguish between the case with $l_\infty$-norm on the direct sum from the case with $l_1$-norm on the direct sum, is used the notion of an <u>M-face.</u> A proper face F of $B(0,1)$ is called an M-face if there exist disjoint subfaces

G and H of F such that dimG=dimH=dimF-1. Then let

$$\underline{m(X)}=\max\{\dim\operatorname{span}F : F \text{ is an M-face of } B(0,1)\}$$

We always have $2\leq m(X)\leq \dim X$ when $3\leq \dim X<\infty$ and X has the 3.2.I.P. Furthermore $m(X)=\dim X$ if and only if X has a proper decomposition $X=(Y\oplus Z)_{1_\infty}$ and $m(X)<\dim X$ if and only if X has a proper decomposition $X=(Y\oplus Z)_{1_1}$.

# REFERENCES

[1] D.B. Goodner, Projections in normed linear spaces, Trans. Amer. Math. Soc. 69 (1950), 89-108.

[2] O. Hanner, Intersections of translates of convex bodies, Math. Scand. 4 (1956), 65-87.

[3] A. Hansen and Å. Lima, The structure of finite-dimensional Banach spaces with the 3.2. intersection property. (To appear in Acta. Math.)

[4] M. Hasumi, The extension property of complex Banach spaces, Tôhoku Math. J. (2) 10 (1958), 135-142.

[5] O. Hustad. A note on complex $P_1$-spaces, Israel J. Math. 16 (1973), 117-119.

[6] O. Hustad, Intersection properties of balls in complex Banach spaces whose duals are $L_1$-spaces, Acta. Math. 132 (1974), 283-313.

[7] J.L. Kelley, Banach spaces with the extension property, Trans. Amer. Math. Soc. 72 (1952), 323-326.

[8] A. Lazar, Polyhedral Banach spaces and extensions of compact operators, Israel J. Math. 7 (1969), 357-364.

[9] Å. Lima, Complex Banach spaces whose duals are $L_1$-spaces, Israel J. Math. 24 (1976), 59-72.

[10] Å. Lima, An application of a theorem of Hirsberg and Lazar, Math. Scand. 38 (1976), 325-340.

[11] Å. Lima, Intersection properties of balls and subspaces in Banach spaces, Trans. Amer. Math. Soc. 227 (1977), 1-62.

[12] Å. Lima, Intersection properties of balls in spaces of compact operators, Ann. Inst. Fourier, 28 (1978), 35-65.

[13] J. Lindenstrauss, Extension of compact operators, Mem. Amer. Math. Soc. No. 48 (1964).

[14] G. Olsen and N.J. Nielsen, Complex preduals of $L_1$ and subspaces of $l_\infty^n(C)$, Math. Scand. 40 (1977), 271-287.

[15] L. Nachbin, A theorem of the Hahn-Banach type for linear transformations, Trans. Amer. Math. Soc. 68 (1950), 28-46.

# New methods in injectivity theorems connected with Schwarzian derivative

O. Martio

## 1. Schwarzian derivative

In 1869, H. A. Schwarz eliminated constants in a Möbius transformation of the plane and obtained the following definition.

1.1. Definition. Let $f : D \to C$ be analytic in a domain $D \subset C$ and $f'(z_0) \neq 0$. The Schwarzian derivative of $f$ at $z_0$ is

$$S_f(z_0) = (f''/f')' - (f''/f')^2/2.$$

The basic property of the Schwarzian derivative is that $S_f = 0$ iff $f$ is a restriction of a Möbius transformation to $D$.

There is another way to look at $S_f$ which is easy to prove by computation.

1.2. Proposition. *Let $f$ be as above. There is a unique Möbius transformation $T$ such that*

$$T \circ f(z) - z = \frac{S_f(z_0)}{6}(z - z_0)^3 + \left[(z - z_0)^4\right]$$

*for all* $z \in D$.

This can also be taken as a definition for the Schwarzian derivative. It has several advantages.

a) It relates $f$ and an approximation for its inverse.

b) It generalizes to much wider class of mappings in the sense that

$$T \circ f(z) - z \longrightarrow \text{small}.$$

## 2. John-domains

Here we consider a concept introduced by F. John [J] in 1961.

2.1. Definition. Let $D \subset R^n$ be a domain, $n \geq 2$. $D$ is called an $(\alpha,\beta)$-John domain, $0 < \alpha \leq \beta < \infty$, if there is a point $x_0 \in D$ such that for all $x \in D$ there exists a rectifiable path $\gamma : [0,d] \longrightarrow D$ (arc length as parameter) such that

a) $\gamma(0) = x$, $\gamma(d) = x_o$,

b) $d \leq \beta$,

c) $\mathrm{dist}(\gamma(t), \partial D) \geq \alpha t/d$, $t \in [0,d]$.

It is easy to see that all bounded convex domains are John-domains. John called $\alpha$ the inner and $\beta$ the outer radius of D.

## 3. Stability theorems

3.1. General stability problem. For which classes of mappings f and domains $D \subset R^n$ is it possible to have an estimate of the form

$$T \circ f(x) - x \longrightarrow \text{small}$$

in all of D? Here T is some Möbius transformation of $\bar{R}^n$.

We consider some answers to this problem.

a) John showed in [J], although he did not state it in this form, that if f is locally L-bi-lipschitz in an $(\alpha,\beta)$-John domain D of $R^n$, $n \geq 2$, then there is a rigid motion T of $R^n$ such that

$$|T \circ f(x) - x| \leq C_n \frac{\beta^2}{\alpha} (L - 1),$$

if L is smaller than some universal constant.

b) Yu. Rešetnjak [R] proved that if $n \geq 3$ and $f : D \rightarrow R^n$ is K-quasiregular (f is $ACL^n$ and $|f'(x)|^n \leq K\, J(x,f)$ a.e.) and non-constant in an $(\alpha,\beta)$-John domain D, then there is $T \in GM(n)$ (= the general Möbius group of $\bar{R}^n$) with

$$|T \circ f(x) - x| \leq C_n \frac{\beta^2}{\alpha} \sigma(K-1) .$$

Here $\sigma : [0, \infty) \rightarrow [0, \infty]$ with $\lim_{t \searrow 0} \sigma(t) = 0$ and $\sigma$ depends only on n. The proof makes essential use of the Liouville theorem: For $n \geq 3$ the only 1-quasiregular mappings are Möbius transformations or constants.

c) For n = 2 the following explicit result can be proved [MS]. If $f : D \rightarrow C$ is analytic and $f'(z) \neq 0$ in an $(\alpha,\beta)$-John domain D, then there is $T \in GM(2)$ such that

$$|T \circ f(z) - z| \leq \frac{k(\beta/\alpha)^2}{1-k(\beta/\alpha)^2} \beta$$

for all $z \in D$ whenever

$$2|S_f(z)|\,\mathrm{dist}(z, \partial D)^2 < k \quad (\leq (\alpha/\beta)^2).$$

The result in (c) suggests the much used definition.

3.2. Definition. The Schwarzian norm of an analytic function $f : D \to C$, $f'(z) \neq 0$, is

$$\|f\| = \sup_{z \in D} |S_f(z)| \operatorname{dist}(z, \partial D)^2.$$

4. Uniform domains

For injectivity theorems the concept of a John domain must be modified.

4.1. Definition. [MS] A domain $G \subset R^n$ is called an $(\alpha,\beta)$-uniform domain, $0 < \alpha \leq \beta < \infty$, if for all $x_1, x_2 \in G$, $x_1 \neq x_2$, there is an $(\alpha|x_1 - x_2|, \beta|x_1 - x_2|)$-John domain $D$ such that $x_1, x_2 \in D \subset G$.

For $n = 2$ it turns out that a Jordan domain is uniform iff it is a quasiconformal disc, i.e. the image of the unit disc under some quasiconformal mapping of $\bar{C}$, see [MS]. Hence it seems obvious that uniform domains should be conformally invariant in $R^n$, $n \geq 2$. The following proposition gives the answer.

4.2. Theorem. [M1] A domain $G \subset R^n$ is uniform iff for all $x_1, x_2 \in G$, $x_1 \neq x_2$, there is a continuum $K$ joining $x_1$ to $x_2$ in $G$ such that

$$\frac{|x - y|}{|x - x_i|} \frac{|x_1 - x_2|}{|y - x_j|} \geq t, \quad i \neq j, \ i,j = 1,2, \tag{4.3}$$

$x \in K$, $y \in \bar{R}^n \setminus G$, for some $t > 0$.

Clearly the cross-ratio in (4.3) satisfies the invariance property. However, although not conformally invariant the definition 4.1 is usually easier to apply.

5. Injectivity theorems

5.1. Proposition. Let $G \subset R^n$ and let $f : G \to R^n$ be a map. Suppose that for all $x_1, x_2 \in G$, $x_1 \neq x_2$, there is a set $D \subset G$ and a map $T : fD \to R^n$ such that $x_1, x_2 \in D$ and

$$|T \circ f(x) - x| < |x_1 - x_2|/2$$

for all $x \in D$. Then $f$ is injective.

Proof. Let $x_1, x_2 \in G$, $x_1 \neq x_2$, and suppose $f(x_1) = f(x_2)$. Then

$$\begin{aligned} |x_1 - x_2| &\leq |T \circ f(x_1) - x_1| + |T \circ f(x_2) - x_2| \\ &< |x_1 - x_2|/2 + |x_1 - x_2|/2 = |x_1 - x_2| \end{aligned}$$

is a contradiction.

5.2. Theorem. [MS] Let $f : G \to C$ be analytic, $f'(z) \neq 0$, and $G$ an $(\alpha,\beta)$-uniform domain. If

$$\|f\| < C(\alpha,\beta) = \frac{1}{2} (\alpha/\beta)^2 (2\beta + 1)^{-1} ,$$

then $f$ is injective in $G$.

Proof. Let $z_1, z_2 \in G$, $z_1 \neq z_2$. Then there is an $(\alpha|z_1 - z_2|, \beta|z_1 - z_2|)$-John domain $D \subset G$ containing $z_1$ and $z_2$. Apply 3 (c) in D. If $\|f\|$ is small and thus $k$ is small, then

$$|T \circ f(z) - z| < |z_1 - z_2|/2$$

and 5.1 gives the result.

This theorem is a generalization of a result of L. Ahlfors [A]: Suppose that D is a quasiconformal disc and $f : D \to C$ is analytic. Then there is $\delta > 0$ such that $\|f\| < \delta$ implies f injective in D.

6. Extensions and applications

6.1. Quasiregular mappings. Using 3 (b) and 5.1 it can be proved as above.

6.2. Theorem. [MS] Suppose that $G \subset R^n$, $n \geq 3$, is a uniform domain. Then there is $\delta > 1$ such that if $f : G \to R^n$ is K-quasiregular, non-constant, and $K < \delta$, then f is injective in G.

The theorem 6.2 is an extension of the theorem 5.2 and the result of Ahlfors to higher dimensional euclidean spaces. Observe that if $f : G \to R^n$, $n \geq 3$, is quasiregular, then the Schwarzian norm can be defined on G as $\|f\| = \log K(f)$, here $K(f)$ is the dilatation of f. Observe that $\|f\| = 0$ iff f is a restriction of a Möbius transformation to G or a constant.

6.3. Locally bi-lipschitz mappings. Using 3 (a) and 5.1 a similar theory for locally bi-lipschitz mappings in uniform domains can be developed, see [MS]. F. John [J] has earlier proved an injectivity theorem for these mappings in bounded convex domains (which are uniform).

6.4. New applications. The concept of a uniform domain is also useful in other classical problems. A well-known theorem of G. Hardy and J. Littlewood states that if $f : B \to C$, B the unit disc, is analytic, then f belongs to the Hölder-class $\Lambda_t(\bar{B})$, i.e.

$$|f(z_1) - f(z_2)| \leq C|z_1 - z_2|^t, \ z_1, z_2 \in \bar{B},$$

$0 < t \leq 1$, iff

$$(6.5) \qquad |f'(z)| \le C' \operatorname{dist}(z,\partial B)^{t-1}$$

in B. Moreover, $\Lambda_t(\bar{B})$ can be replaced by $\Lambda_t(\partial B)$ if f, in addition, is continuous on $\bar{B}$. Now these results have natural extensions as follows [M2].

6.6. Theorem. *Let $G \subset C$, $\partial G \neq \emptyset$, be a uniform domain and $f : G \to C$ analytic. Then $f \in \Lambda_t(\bar{G})$ iff (6.5) holds in G.*

6.7. Theorem. *Suppose that $G \subset C$ is a uniform domain such that $C \setminus G$ has a finite number of components. Let $f : \bar{G} \to C$ be continuous, bounded, and analytic in G. Then $f \in \Lambda_t(\partial G)$ iff (6.5) holds in G.*

## References

[A] Ahlfors, L.: Quasiconformal reflections.-Acta Math. 109, 1963, 291-301.

[J] John, F.: Rotation and strain.-Comm. Pure Appl. Math. 14, 1961, 391-413.

[M1] Martio, O.: Definitions for uniform domains.-Ann. Acad. Sci. Fenn. A I (to appear).

[M2] — " — : Note on a theorem of Hardy and Littlewood. (to appear).

[MS] Martio, O. and J. Sarvas: Injectivity theorems in plane and space.-Ann. Acad. Sci. Fenn. A I 4, 1978/1979, 383-401.

[R] Rešetnjak, Yu.: Stability in Liouville's theorem on conformal mappings of a space for domains with non-smooth boundary.-Sibirsk Mat. Z. 17, 1976, 361-369 (Russian).

University of Jyväskylä
Department of Mathematics
Jyväskylä, Finland

# THE DETERMINATION OF BOUNDARY COLLOCATION POINTS FOR SOLVING SOME PROBLEMS FOR THE LAPLACE OPERATOR

## Introduction

Many questions of applied mathematics give rise to boundary problems for linear partial differential equations, for which complete families of particular solutions are explicitly known. Boundary collocation is a method to obtain an approximate solution to such a problem, where one

a) chooses a finite-dimensional subspace of solutions to the differential equation (or several different subspaces for different regions),

b) requires that the given boundary or continuity conditions be satisfied in a finite number of selected points, collocation points.

This yields a finite set of simultaneous equations.

Boundary collocation has been known and used for a long time, see e.g. Collats [3, pp 28 ff, pp 413 ff]. More recently, applications to questions in microwave technology have been reported, see e.g. Bates [1].

We are, however, not aware of any investigations of the numerical aspects of this method. The method contains some decisions, which are, indeed, important for its performance. These important factors are:

1) Choice of subspace, with respect to good accuracy with few terms and easy handling. The choice will depend on the geometry.

2) Choice of basis for the subspace so that the simultaneous linear equations do not become unnecessarily ill-conditioned. Also here the easy handling aspect is of importance.

3) Choice of collocation points, which is of importance for the accuracy of the approximation as well as for the condition of the linear system of equations.

The situation is somewhat analogous to the more familiar problem of approximating a function by a set of simple functions on the interval $[-1,1]$, where we make decisions like "polynomials or something else", "power basis or Chebyshev polynomial basis", "interpolation in equidistant points or Chebyshev points", etc.

In this paper we shall in some detail consider the last mentioned choice (the collocation points), but we shall of course try to avoid obviously bad alternatives in the first choices as well.

Our investigation will be in the form of illustrative model problems. We hope that the experience will be suggestive also for more complicated situations.

Model problem I

Let $\Omega$ be a simply connected open limited point set in the complex plane $\mathbb{C}$. Let $\partial\Omega$ be its boundary, and let $\Omega_c := \mathbb{C} \setminus (\Omega \cup \partial\Omega \cup \{\infty\})$. Find $u = u(x,y)$ such that

$$\begin{cases} u \text{ is twice continuously differentiable in } \Omega \text{ and } \Omega_c. \\ \Delta u(x,y) = 1 \quad (x,y)\in\Omega \\ \Delta u(x,y) = 0 \quad (x,y)\in\Omega_c \\ u \text{ and } \frac{\partial u}{\partial n} \text{ are continuous across the boundary } \partial\Omega,\ \frac{\partial}{\partial n} \\ \text{being the normal derivative.} \\ \text{There is a constant } \alpha \text{ so that } u(x,y)\ O\ \alpha\log|z| + O(|z|^{-1}), \\ z = x + iy \text{ when } |z|\ O\ \infty. \end{cases}$$

Ex.1 Let $\Omega$ be the open unit disk. Then Problem I has the solution, with $r = \sqrt{x^2 + y^2}$,

$$\begin{cases} u(r) = \frac{r^2}{4} - \frac{1}{4}, & 0\le r\le 1 \\ u(r) = \frac{1}{2}\log(r), & r>1 \end{cases}$$

■

One advantage of the above model problem is that it is related to the following problem, which can be conveniently discussed by the theory of analytic functions.

Model problem II

For a given function $\psi$, continuous on $\partial\Omega$, find functions $f$ and $g$ such that

$$\begin{cases} f \text{ is analytic in } \Omega, \text{ continuous on } \Omega \cup \partial\Omega \\ g \text{ is analytic on } \Omega_c, \text{ continuous on } \Omega_c \cup \partial\Omega, \text{ and vanishes at infinity} \\ f - g = \psi \text{ on } \partial\Omega. \end{cases}$$

The solution of this problem is given by the following formula,

$$\frac{1}{2\pi i}\int_{\partial\Omega} \frac{\psi(\zeta)\,d\zeta}{\zeta - z} = \begin{cases} f(z) & \text{for } z\in\Omega \\ g(z) & \text{for } z\in\Omega_c \cup\{\infty\} \end{cases}$$

Proposition

Model problem I is equivalent to Model problem II when $\psi(z)=\frac{1}{2}\bar{z}$.

Let $u_{int}(x,y)$ and $u_{ext}(x,y)$ be the restrictions of u defined on $\Omega \cup \partial\Omega$ and $\Omega_c \cup \partial\Omega$ respectively. For $z = x + iy$ let

$$\begin{cases} E_{int}(z) := \left(\frac{\partial}{\partial x} - i\frac{\partial}{\partial y}\right) u_{int}(x,y) & (x,y)\in\Omega \\ E_{ext}(z) := \left(\frac{\partial}{\partial x} - i\frac{\partial}{\partial y}\right) u_{ext}(x,y) & (x,y)\in\Omega_c \end{cases}$$

If u is a solution of Model problem I, and if the boundary is smooth, we have

$$E_{int}(z) = E_{ext}(z) \text{ on } \partial\Omega$$

$E_{ext}(z)$ is regular on $\Omega_c$ and we choose

$$g(z) := -E_{ext}(z).$$

$E_{int}(z) - \frac{\bar{z}}{2}$ is analytic in $\Omega$ and we let

$$f(z) := -E_{int}(z) + \frac{\bar{z}}{2}.$$

Then f and g satisfy Model problem II. Conversely, given functions f,g satisfying Model problem II, we can by integration obtain a solution of Model problem I.

Ex.2 Let $\Omega$ be the interior of the ellipse $\{a \cos t + ib \sin t,\ 0\le t<2\pi\}$, $a>b>0$, with foci at $\pm c$, $c = \sqrt{a^2-b^2}$. One can verify that, with $z = x + iy$,

$$u_{int}(x,y) = \frac{1}{4}(x^2+y^2) - \frac{1}{4}\frac{a-b}{a+b}(x^2-y^2) - \frac{1}{4}ab + \frac{1}{2}ab\log\left(\frac{a+b}{2}\right)$$

$$u_{ext}(x,y) = \frac{1}{2}ab\left(\frac{1}{c^2}(x^2-y^2) - \frac{1}{c^2}\operatorname{Re}(z\sqrt{z^2-c^2}) + \log|z+\sqrt{z^2-c^2}| - \frac{1}{2} - \log(2)\right)$$

$$E_{int}(z) = \frac{1}{2}\bar{z} - \frac{1}{2}\frac{a-b}{a+b}z$$

$$E_{ext}(z) = ab\,\frac{1}{z+\sqrt{z^2-c^2}}$$

Notice that $E_{ext}(z)$ and $E_{int}(z) - \frac{1}{2}\bar{z}$ can be continued analytically across $\partial\Omega$.

Interpolation

Instead of trying to solve Model problem I, we compute an approximate solution of Model problem II and integrate it. Thus we are to find approximations of f on $\Omega$ and of g on $\Omega_c$ where $f(z)-g(z) = \frac{1}{2}\bar{z}$ on $\partial\Omega$.

According to a theorem in Walsh [4], p 36, we can for f analytic in $\Omega$ and continuous on $\bar{\Omega} = \Omega \cup \partial\Omega$ find a polynomial P(z) approximating f(z) arbitrarily well on $\bar{\Omega}$ in the uniform sense. Since g analytic in $\Omega_c$, continuous on $\bar{\Omega}_c = \Omega_c \cup \partial\Omega$, and regular at infinity, we can by the same theorem find a polynomial Q(1/z) in 1/z, approximating g arbitrarily well on $\bar{\Omega}_c$ in the uniform sense. We determine P(z) and Q(1/z) by interpolation of $\psi(z) = \frac{1}{2}\bar{z}$ on $\partial\Omega$.

Ex.3. For unit disk-like regions $\Omega$ an expansion in powers of z and 1/z is appropriate.

$$P(z) := \sum_{0}^{M} A_k z^k$$

$$Q(1/z) := \sum_{1}^{N} B_k z^{-k}$$

Let $z_0, z_1, \ldots, z_{N+M}$ be distinct, approximately equidistant, points on $\partial\Omega$ and solve

$$(1) \qquad \psi(z_j) = -\sum_{0}^{M} A_k z_j^k + \sum_{1}^{N} B_k z_j^{-k}, \quad j = 0,1,\ldots,N+M$$

for the coefficients $A_k$, $B_k$. ∎

The above example raises a couple of questions:

a) For more general regions $\Omega$ the basis $\{1, z, z^2, \ldots, 1/z, 1/z^2, \ldots\}$ chosen above will often lead to a very ill-conditioned linear system (1), for all choices of interpolation points $z_j$. In these cases the basis is ill-conditioned with respect to the regions $\Omega$ and $\Omega_c$.

b) To obtain fast convergence to f and g for increasing M and N, it is necessary for the basis to span an appropriate subspace. The functions used to approximate g in $\Omega_c$ should in some sense be able to represent g also in parts of $\Omega$.

c) Turning to the interpolation points, we notice that their placement certainly affects the numerical condition of (1).

These questions will be treated in some detail in a forthcoming paper. For the moment it suffices to say that the basis functions chosen below give reasonable convergence and make the linear system corresponding to (1) quite well-conditioned.

We introduce the new basis $\{p_0(z), p_1(z), \ldots, 1/q_1(z), 1/q_2(z) \ldots\}$ where

$$p_k(z) := \begin{cases} 1 & k = 0 \\ \gamma_k \prod_{j=1}^{k} (z - w_j^{(p)}) & k = 1,2,\ldots \end{cases}$$

$$q_k(z) := \delta_k \prod_{j=1}^{k} (z - w_j^{(q)}) \qquad k = 1,2,\ldots$$

The coefficients $\gamma_k, \delta_k$ are scaling factors for instance chosen to make

$$\max_{z \in \partial\Omega} |p_k(z)| = 1, \quad \max_{z \in \partial\Omega} |q_k(z)| = 1.$$

The $w_j^{(q)}$:s are placed in $\Omega$. Note that for $w_1^{(q)}, \ldots, w_\ell^{(q)}$ distinct

$$\mathrm{span}\{1/q_1(z), 1/q_2(z), \ldots\} = \mathrm{span}\left\{\frac{1}{z-w_1^{(q)}}, \ldots, \frac{1}{z-w_\ell^{(q)}}, \frac{1}{(z-w_1^{(q)})^2}, \ldots\right\}$$

The latter basis is not chosen since the former is better conditioned.

For later reference we define

$$(2)\qquad W(z) := \begin{cases} \prod_{1}^{N}(z-w_j^{(q)}) & N \geq 1 \\ 1 & N = 0 \end{cases}$$

The equations corresponding to (1) are

$$(1')\qquad \psi(z_j) = -\sum_{k=0}^{M} A_k p_k(z_j) + \sum_{k=1}^{N} B_k/q_k(z_j) \qquad j=0,1,\dots,N+M$$

Multiplying (1') by W(z) we obtain a formula for polynomial interpolation

$$(1'')\quad W(z_j)\psi(z_j) = -\sum_{0}^{M} A_k p_k(z_j)W(z_j) + \sum_{k=1}^{N-1} B_k \frac{1}{\delta_k} \prod_{\ell=k+1}^{N} (z_j - w_\ell^{(q)}) + B_N \cdot \frac{1}{\delta_N}$$

If we now assume that f(z)-g(z) is analytic in a region containing $\partial\Omega$ in its interior, then the interpolation error can be expressed with the Hermite rest term formula. Let c be the boundary of a closed point set having $\partial\Omega$ as a subset and such that f and g are analytic in the interior of c and continuous on c. Then for z in the interior of c we can express the interpolation error in (1") as (see Walsh [4], §3.1)

$$(3)\quad W(z)(f(z)-g(z)) - \left(-\sum_{k=0}^{M} A_k p_k(z)W(z) + \sum_{k=1}^{N-1} B_k \frac{1}{\delta_k} \prod_{\ell=k+1}^{N} (z - w_\ell^{(q)}) + B_N \cdot \frac{1}{\delta_N}\right) =$$

$$= \prod_{0}^{N+M} (z - z_k) \cdot \frac{1}{2\pi i} \int_c \frac{W(\zeta)(f(\zeta)-g(\zeta))}{\prod_{0}^{N+M} (\zeta - z_k)(\zeta - z)} d\zeta \,, \quad z \in \partial\Omega$$

Since f and g are one-valued, the boundary c may consist of several contours.

<u>Ex. 4</u> Let $\Omega$ be the open unit disk and let $\psi(z) := \frac{1}{2}\bar{z}$. Then

$$\begin{cases} f(z) = 0 \\ g(z) = -\dfrac{1}{2z} \end{cases}$$

is a solution of Model problem II. This solution corresponds to the solution of Model problem I in Ex.1. $g(z)$ has a singularity at $z = 0$ and we can choose the boundary $c$ as shown

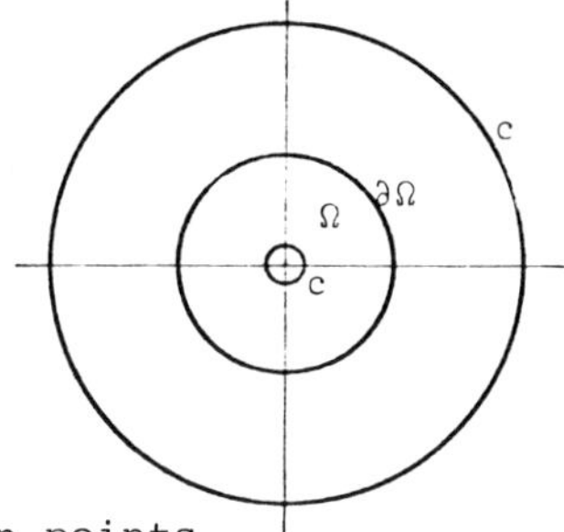


## Placing collocation points

We are now able to obtain an upper bound for the interpolation error. Let

$$s = \min_{\substack{\zeta \in c \\ z, z_k \in \partial\Omega}} \left( \prod_{0}^{N+M} |\zeta - z_k| \cdot |\zeta - z| \right)$$

An upper bound for the integral in (3), independent of $z_0, z_1, \dots, z_{N+M}$ and $z$ is given by

$$\left| \frac{1}{2\pi i} \int_c \frac{W(\zeta)(f(\zeta)-g(\zeta))}{\prod_0^{N+M} (\zeta - z_k)(\zeta - z)} d\zeta \right| \le \frac{1}{2\pi s} \int_c |W(\zeta)| \, |f(\zeta) - g(\zeta)| \, |d\zeta| =: K(f,g)$$

From (3) we obtain for $z \in \partial\Omega$:

$$\left| \psi(z) + \sum_0^M A_k p_k(z) - \sum_1^N B_k / q_k(z) \right| \le \frac{\prod_0^{N+M} |z - z_k|}{|W(z)|} K(f,g).$$

(4) We want to place collocation points $z_0, z_1, \dots, z_{N+M}$ on $\partial\Omega$ so that

$$\max_{z \in \partial\Omega} \frac{\prod_0^{N+M} |z - z_k|}{|W(z)|}$$

is minimized with respect to $z_0, z_1, \dots, z_{N+M}$.

Before proceeding we remark that (4) is independent of $\psi$, i.e. other right hand sides for $\Delta u$ on $\Omega$ than in Model problem I yield the same problem (4). Pure interior/exterior problems can also be fitted in the above frame with N=0/M=-1 respectively.

Problem (4) has the same solution $\hat{z}_0, \hat{z}_1, \ldots, \hat{z}_{N+M}$ as

$$(5) \qquad \min_{\substack{z_k \in \partial\Omega \\ k=0,1,\ldots,N+M}} \; \max_{z \in \partial\Omega} \frac{1}{N+M+1} \left( \sum_0^{N+M} \ln|z-z_k| - \sum_1^N \ln|z-w_k^{(q)}| \right)$$

(The normalization factor will be commented on below.) This minimization problem can be regarded as a problem of electrostatics: Given positive unit point charges at $z = w_k^{(q)}$, $k = 1, 2, \ldots, N$, in the interior of $\Omega$, we are to place negative unit point charges at $z = z_k, k = 0,1,\ldots,N+M$ on $\partial\Omega$ so that the maximum of the potential on $\partial\Omega$ is minimized.

Instead of solving the non-linear problem of determining the optimal discrete charge distribution $\hat{z}_0, \hat{z}_1, \ldots, \hat{z}_{N+M}$ in (5), we ask for a non-negative single layer distribution, which is the solution of a modified problem. This modified problem is linear.

To be more precise, let $\tilde{\sigma}$ be a non-negative measure on $\partial\Omega$ such that

$$\int_{\partial\Omega} d\tilde{\sigma}(\zeta) = 1$$

and let

$$(6) \qquad \tilde{\phi}(z;\tilde{\phi}) := \int_{\partial\Omega} \log|z-\zeta|\, d\tilde{\sigma}(\zeta) - \frac{1}{N+M+1} \sum_{k=1}^{N} \log|z-w_k^{(q)}| \, .$$

Then problem (5) will be put in the following <u>modified</u> form:

$$(7)\quad \begin{cases} \min\limits_{\tilde\sigma}\ \max\limits_{z\in\partial\Omega}\ \tilde\phi(z;\tilde\sigma) \\ \text{subject to the constraints} \\ \tilde\sigma \text{ non-decreasing} \\ \int_{\partial\Omega} d\tilde\sigma(\zeta) = 1 \end{cases}$$

Remark.

1) If we require $\tilde\sigma$ to have a finite number of points of increase $z_0, z_1, \ldots, z_{N+M}$, and that $\tilde\sigma$ has jumps $\frac{1}{N+M+1}$ at each $z_k$, then the modified problem (7) is equivalent to the original problem (5).
2) Having, for a finite N+M, determined a continuous approximate solution of (7), we will replace it by a staircase function with jumps $\frac{1}{N+M+1}$ at $z_k$. Although this staircase function may not be the exact solution of the original problem (5), we can expect it to give a reasonably good choice of collocation points, at least if M+N is not too small.

The computational method to solve the minimization problem (7) is based on that the solution of (7) can also be obtained by solving a system of integral equations. We formulate this as a theorem.

Theorem. The solution $\{\sigma^*, q^*\}$ of the minimization problem

$$(8)\quad \begin{cases} \min\limits_{\tilde\sigma} \tilde q \\ \tilde\sigma \text{ non-decreasing} \\ \int_{\partial\Omega} d\tilde\sigma(\zeta) = 1 \end{cases}$$

is also the unique solution of the system

$$(9)\quad \begin{cases} \tilde\phi(z;\tilde\sigma) = \tilde q \\ \int_{\partial\Omega} d\tilde\sigma(\zeta) = 1. \end{cases}$$

Proof. In Reichel [3], we first show that (9) has a unique solution for all scalings of the contour $\partial\Omega$. Then it is shown that this solution also satisfies (8).

∎

We next describe a method to solve the system of integral equations (9).

Fourier-Galerkin formulation

Let $t \mapsto \zeta(t)$, $0 \le t < 2\pi$ be a parametric representation of the boundary $\partial\Omega$, and let $\{\sigma, q\}$ be a solution of (9). With

$$\varphi'(t) := \sigma(\zeta(t))\,|\zeta'(t)|$$

$$\hat{q} := q$$

$$\hat{\phi}(t) := \frac{1}{N+M+1} \sum_{k+1}^{N} \log |\zeta(t) - w_k^{(q)}|$$

the system of integral equations (9) becomes

$$(10)\quad \begin{cases} -\hat{q} + \int_0^{2\pi} \log|\zeta(s)-\zeta(t)|\,\varphi'(t)\,dt = \hat{\phi}(s), \quad 0 \le s < 2\pi \\ \int_0^{2\pi} \varphi'(t)\,dt = 1 \end{cases}$$

Approximate $\varphi'(t)$ by the trigonometric polynomial

$$\varphi_p'(t) := \frac{1}{2\pi} + \sum_{j=1}^{p-1} (\alpha_j \cos jt + \beta_j \sin jt) + \frac{\alpha_p}{2} \cos pt$$

Then $\varphi_p'(t)$ satisfies the constraint in (10).

Let

$$F(s) := \hat{q} + \int_0^{2\pi} \log \frac{1}{|\zeta(s)-\zeta(t)|}\,\varphi'(t)\,dt + \hat{\phi}(s), \quad 0 \le s < 2\pi$$

and find $\hat{q}, \alpha_1, \alpha_2, \ldots, \alpha_p, \beta_1, \ldots, \beta_{p-1}$ satisfying

$$\begin{cases} \int_0^{2\pi} F(s) \cos js\,ds = 0, & j = 0(1)p \\ \int_0^{2\pi} F(s) \sin js\,ds = 0, & j = 1(1)p-1 \end{cases}$$

From the orthogonality of the trigonometric functions it follows that $\hat{q}$ just appears in one equation. Since we do not need the value of $\hat{q}$, this equation can be discarded. A linear system with a $(2p-1)\times(2p-1)$ matrix is obtained. It can be shown that the matrix is symmetric and positive definite. The elements of the matrix and of the right hand side are computed by the fast Fourier transform method. More details are found in Reichel [3].

Discretization of the charge distribution

The charge distribution function $\varphi_p$, belonging to the charge density function $\varphi_p'$ is

$$\varphi_p(t) := \frac{t}{2\pi} + \sum_{j=1}^{p-1}\left(\frac{\alpha_j}{j}\sin jt - \frac{\beta_j}{j}\cos jt\right) + \frac{\alpha_p}{2p}\sin pt$$

An approximate solution $z_0, z_1, \ldots, z_{N+M}$ of the minimization problem (4) is obtained by solving the equations

$$\varphi_p(t_v) = \frac{v+1/2}{N+M+1},\quad v = 0,1,\ldots,N+M$$

for $t_v$ and letting $z_v := \zeta(t_v)$.

Numerical examples

We will give numerical examples for a few regions. To see how well the points $z_v := \zeta(t_v)$ are allocated, we introduce the function

$$\hat{S}(t) := c \cdot \frac{\prod_{v=0}^{N+M} |\zeta(t)-z_v|}{|W(\zeta(t))|},\quad z_v = \zeta(t_v)$$

The constant c is chosen to make $\max \hat{S}(t) = 1$. For optimally allocated $z_k$:s, all local maxima of $\hat{S}$ are also global maxima.

Below we show graphs of $\hat{S}(t)$, with $z_v = \zeta(t_v)$ placed with the distribution function $\varphi_p(t)$. Also $\varphi_p(t)$ is shown. Finally we show pictures of contours $\partial\Omega$ with the $z_k$:s marked with dots on the contour. The points $w_1^{(q)}, w_2^{(q)} \ldots w_N^{(q)}$ defining $W(z)$ are marked with crosses in $\Omega$.

Ellipse with semi-axes 2:1/2

$\zeta(t) = 2\cdot\cos t + i\frac{1}{2}\sin t$

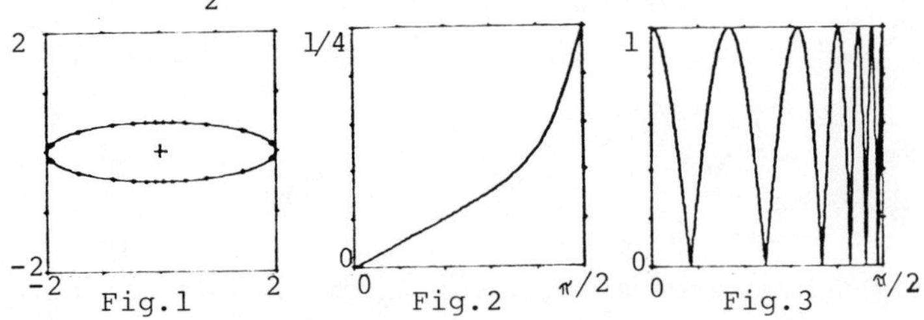


Fig.1 Fig.2 Fig.3

Fig.1. Contour $\partial\Omega$ with the computed nodes $z_1, z_2, \ldots, z_{24}$ for $N = 12$, $M = 11$. $W(z) = z^{12}$ i.e. $w_k^{(q)} = 0$, $k = 1(1)12$.

Fig.2. The distribution function $\varphi_{16}(t)$ used to allocate the nodes on fig.1.

Fig.3. $\hat{S}(t)$ for $N = 12$, $M = 11$ and the nodes $z_k$, $k = 1(1)24$ placed with $\varphi_{32}$.

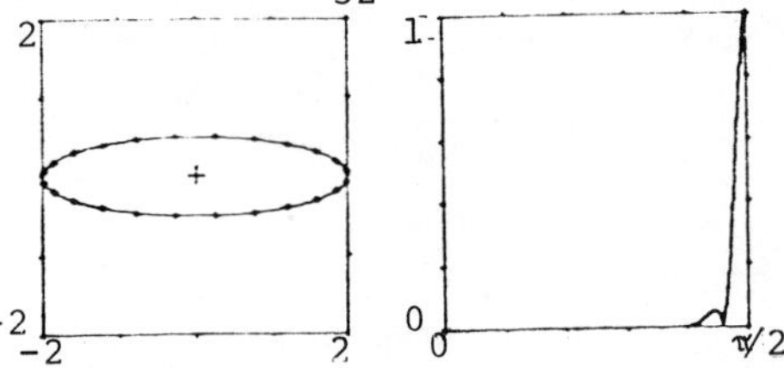


Fig.4 Fig.5

Fig.4. Contour $\partial\Omega$ with nodes $z_k$, $k = 1(1)24$, allocated equidistantly with respect to the parameter. This is the placement obtained for $N = 0$, $M = 23$, and is suitable for polynomial interpolation.

Fig.5. $\hat{S}(t)$ for $N = 12$, $M = 11$ and the points $z_k$ shown on fig.4. At the least local maximum $S(t)$ equals $6\cdot10^{-8}$, showing the necessity of a method to allocate the $z_k$.

Superellipse

The superellipse $x^4 + y^4 = 1$ is represented by $\zeta(t) :=$ $= \xi(t) + y(t)$

$$\begin{cases} \xi(t) := (1-2\sin^4(t))^{1/4} \\ y(t) := 2^{1/4}\sin(t) \end{cases} \qquad 0 \le t \le \pi/4$$

The rest of the contour is obtained by symmetry. This parametric representation is smoother than $\xi(t) := \cos^{1/2}(t)$, $\tilde{y}(t) := \sin^{1/2}(t$

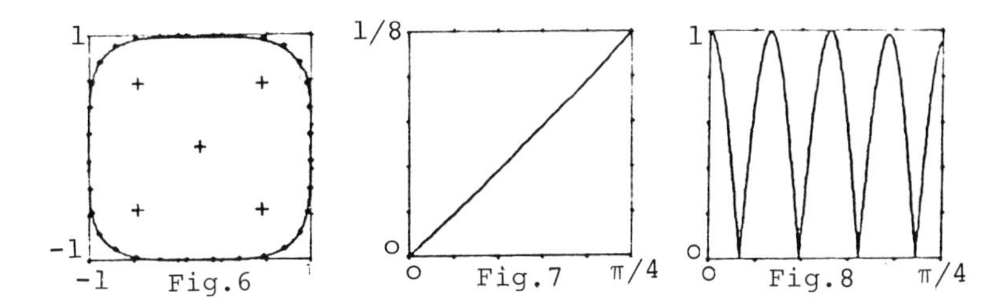


Fig.6. Contour $\partial\Omega$ and the computed points $z_k$, $k = 1(1)32$ for $N = 16$, $M = 15$ and $W(z) := z(z(z^2-w^2)(z^2-\bar{w}^2))^3$, where $w = 0.56(1+i)$.

Fig.7. The corresponding distribution function $\varphi_{16}(t)$.

Fig.8. $S(t)$ for $N = 16$, $M = 15$ and the points shown on fig.6.

Acknowledgement

I wish to state my gratitude to Professor Germund Dahlquist and Dr. Bengt Lindberg for many helpful and inspiring discussions. I also want to thank Professor Hans Wallin for valuable literature references, and Mr. Per Olof Risman for introducing me to problems in microwave technology.

References

1. Bates, R.H.T., Analytic constraints on electromagnetic field computations, IEEE Trans. Microwave Theory Tech., Vol. MTT-23, 605-623, August 1975.

2. Collatz, L., The numerical treatment of differential equations, Springer, Berlin 1966.

3. Reichel, L., On the determination of boundary collocation points for solving some problems for the Laplace operator, Dept. of Num. Anal. and Comp. Sci., Royal Institute of Technology, Stockholm, TRITA-NA-8006, 1980.

4. Walsh, J.L., Interpolation and approximation by rational functions in the complex domain, 4th ed., American Mathematical Society, Providence, R.I., 1965.

Department of Numerical Analysis and Computing Science
The Royal Institute of Technology
S-100 44 Stockholm 70, Sweden

# VALUE DISTRIBUTION OF QUASIREGULAR MAPPINGS

Seppo Rickman

1. Introduction. Quasiregular mappings form a natural generalization of the analytic functions in the plane to real n-dimensional space. The theory of these mappings was initiated by Rešetnjak around 1966 in a series of papers and a systematic study was continued mainly by Martio, Rickman, and Väisälä some years later.

Let $G$ be a domain in the Euclidean n-space $R^n$ and let $f: G \to R^n$ be continuous. We say that $f$ is quasiregular if (1) $f \in W^1_{n,loc}(G)$, i.e. $f$ has distributional partial derivatives which are locally $L^n$-integrable, and (2) there exists $K$, $1 \leq K < \infty$, such that

$$(1.1) \qquad |f'(x)|^n \leq KJ_f(x) \quad \text{a.e.}$$

Here $f'(x)$ is the formal derivative defined by means of the partial derivatives which by (1) exist a.e., $f'(x)$ is its supremum norm, and $J_f(x)$ is the Jacobian determinant. For the purpose of this article we call a quasiregular mapping K-quasiregular if (1.1) is satisfied although this is not common terminology. The definition of quasiregularity extends immediately to maps $f: M \to N$ where $M$ and $N$ are Riemannian n-manifolds, see for example [6]. The term quasimeromorphic is reserved for the case where $M$ is a domain in $R^n$ or $\bar{R}^n = R^n \cup \{\infty\}$ and $N = \bar{R}^n$. $\bar{R}^n$ is equipped with the spherical metric. A quasiregular homeomorphism is called a quasiconformal mapping.

If $n = 2$, 1-quasiregular mappings are exactly the analytic functions and any quasiregular mapping $f$ can be written as

$f = g \circ h$ where $h$ is quasiconformal and $g$ analytic. If $n \geq 3$, the situation is different. The 1-quasiregular mappings are very rigid, they are restrictions of Möbius transformations or constants. Hence a distortion factor $K$ is allowed in order to obtain a class of mappings which is interesting in the function theoretic sense.

One of the main early results by Rešetnjak is that a nonconstant quasiregular mapping is always discrete and open. Quasiregular mappings $f$ can have a nonempty branch set $B_f = \{x \in G \mid f$ is not locally homeomorphic at $x\}$. For $n \geq 3$ locally homeomorphic quasiregular mappings have a rigidity property which is missing in the plane case. To illustrate this we quote a theorem by Zorič [15] which states that a quasiregular mapping $f\colon R^n \to R^n$, $n \geq 3$, which is a local homeomorphism is in fact a homeomorphism. Hence, in the function theoretic sense the interesting quasiregular mappings are those with $B_f \neq \emptyset$. Zorič's theorem can for example be used to give light into the fact that holomorphic mappings of several complex variables form a generalization of the one complex variable theory which goes completely into a different direction from quasiregular mappings. Namely, let $f\colon C^p \to C^p$ be holomorphic, $p \geq 2$, and let the corresponding mapping $f\colon R^{2p} \to R^{2p}$ be quasiregular. Then $f$ is in fact affine [3].

The methods of proofs for quasiregular mappings in dimensions $n \geq 3$ differ essentially from the classical methods in the plane. The main tool in the theory developed so far is the method of moduli of path families.

2. Omitted values. The starting point of the classical value distribution theory of one complex variable was the fundamental result by Picard than a nonconstant entire analytic function cannot omit more than one finite point. The question of the existence of a corresponding result for quasiregular mappings for dimensions $n \geq 3$ was raised already by Zorič in [15]. In fact , such a theorem was proved a couple of years ago which shows that the number of omitted values is finite, more precisely as follows.

2.1. Theorem [11]. For each integer $n \geq 3$ and each $K \geq 1$ there exists a positive integer $q = q(n,K)$ such that if $f: R^n \to R^n \setminus \{a_1,\ldots,a_q\}$ is K-quasiregular and $a_1,\ldots,a_q$ are distinct points in $R^n$ , then $f$ is constant.

The best earlier result, which states that the set of omitted values for a nonconstant entire quasiregular mapping is of conformal capacity zero, was proved in [5] and [7]. This implies in particular that the omitted set has zero Hausdorff dimension.

The proof in [11] consists of the use of two tools, the method of moduli of path families and estimates for solutions of quasilinear partial differential equations. Here I will outline a different proof where quasilinear partial differential equations are avoided.

2.2. Outline of proof of Theorem 2.1. This proof is a simplification of the proof of the defect relation [12] which will be discussed in Section 3.

Let $f: R^n \to R^n \setminus \{a_1,\ldots,a_q\}$ be a nonconstant quasiregular mapping. We may assume that $a_j \in B(1/2) = \{x \in R^n \mid |x| < 1/2\}$ for all $j$ . For $y \in R^n$ and a Borel set $E \subset R^n$ we set

$$n(E,y) = \sum_{x \in f^{-1}(y) \cap E} i(x,f)$$

where $i(x,f)$ is the local topological index of $f$ at $x$ (see [4]). The integer $n(E,y)$ is the number of points in $f^{-1}(y) \cap E$ with multiplicity regarded. The average over the unit sphere $S$ is denoted by $\mu(E)$, i.e.,

$$\mu(E) = \frac{1}{\omega_{n-1}} \int_S n(E,y)\, dH^{n-1}(y)$$

where $\omega_{n-1}$ is the $(n-1)$-dimensional measure of $S$. The <u>counting function</u> is $n(r,y) = n(\bar{B}(r),y)$. We also write

$$\nu(r,s) = \frac{1}{\omega_{n-1}} \int_S n(r,sy)\, dH^{n-1}(y) ,$$

which is then the average of $n(r,y)$ over a sphere $S(s) = \{x \in R^n \,\|x| = s\}$, and abbreviate $\nu(r) = \nu(r,1) = \mu(\bar{B}(r))$. From [5,4.1] it follows that $\nu(r) \to \infty$ as $r \to \infty$.

The basic lemma is the following.

2.3. <u>Lemma</u> ([9,4.1]). The inequality

$$r\nu(\Theta r,t) \geq \nu(r,s) - \frac{K^{n-1} |\log \frac{t}{s}|^{n-1}}{(1-\frac{1}{c})(\log \Theta)^{n-1}}$$

holds for all $c,\Theta > 1$ and $r,s,t > 0$.

The proof of 2.3 is based on an important inequality for moduli of path families by Väisälä [13,3.1] and a special path lifting technique from [8]. For analytic functions a sharper form

$$\nu(\Theta r,t) \geq \nu(r,s) - |\log \tfrac{t}{s}|(\log \Theta)^{-1}$$

was proved in [2].

To start the proof of 2.1 we use an idea of Borel to obtain a set $E \subset [1,\infty[$ of finite logarithmic measure, i.e. $\int_E dr/r < \infty$, such that

$$(2.4) \qquad \nu(r') \leq 2\nu(r) , \quad r \in [1,\infty[ \setminus E ,$$

where

$$r' = r + \frac{r}{\nu(r)^{\alpha}}, \quad \alpha = \frac{1}{2(n-1)} .$$

To prove (2.4) let $r_o'' \geq 1$ be such that $\nu(r_o'') \geq 1$. Set $F = \{r \in ]r_o'',\infty[ \,|\, \nu(r') > 2\nu(r)\}$. We define inductively a sequence $r_o'' \leq r_1 < r_1'' \leq r_2 < r_2'' \leq \ldots$ by

$$r_k = \inf \{r \in F \mid r > r_{k-1}''\}, \; r_k'' = r_k + \frac{2r_k}{\nu(r_k)^{\alpha}} .$$

Set $E_1 = \cup_{k \geq 1} [r_k, r_k'']$. Then $F \subset E_1$ and it is enough to estimate the logarithmic measure of $E_1$. Since $\nu(r_{k+1}) \geq \nu(r_k'') \geq 2\nu(r_k)$, we get

$$\int_{E_1} \frac{dr}{r} \leq \sum_{k \geq 1} \frac{r_k'' - r_k}{r_k} = \sum_{k \geq 1} \frac{2}{\nu(r_k)^{\alpha}} < \infty .$$

We fix $r \in [\kappa,\infty[ \setminus E$ where $\kappa$ is taken sufficiently large later. We decompose the ball $\bar{B}(r)$ into disjoint Borel sets $U_1,\ldots,U_p$ with the following properties (we denote by $b_k$ positive constants which depend on $n$ only):

(a) $p \leq b_1 (r/(r' - r))^{n-1} = b_1 \nu(r)^{1/2}$,

(b) there exist $K_o$-quasiconformal mappings $\varphi_i \colon R^n \to R^n$, $i \in I = \{1,\ldots,p\}$, such that

(1) $\varphi_i(\mathrm{int}(U_i)) = B(1/2)$, $\varphi_i(\bar{U}_i) = \bar{B}(1/2)$,

(2) $W_i = \varphi_i^{-1}(\bar{B}(2)) \subset \bar{B}(r')$,

(3) each point belongs to at most $b_2$ sets $W_i$.

The sets $U_i$ can be chosen so that their hyperbolic diameters, measured in the hyperbolic metric in the ball $B(r')$, are approximately the same.

Let

$$I_o = \{i \in I \mid \mu(W_i) \geq 3b_2\mu(U_i)\} ,$$

$$I_1 = \{i \in I \mid \mu(U_i) \leq \nu(r)^{1/4}\} .$$

It follows from (2.4),(a),(3) in (b), and from the fact that $\nu(r) \to \infty$ as $r \to \infty$ that $I \setminus (I_o \cup I_1) \neq \emptyset$ if $\kappa$ is sufficiently large. Fix $i \in I \setminus (I_o \cup I_1)$ . Write

$$\sigma_o = \frac{1}{4} \min_{j \neq k} |a_j - a_k| .$$

Let $G_j : R^n \to R^n$ be a $K_1$-quasiconformal mapping, $K_1$ depending only on $n$, such that $G_j$ is the identity outside $B(1)$ and the translation $x \mapsto x - a_j$ in $B(a_j, \sigma_o)$ . We apply 2.3 to the map $h = h_j = G_j \circ f \circ \varphi_i^{-1}$ with $c = 2$, $\Theta = 2$, $s = 1$, $t = \sigma < \sigma_o$ . We obtain (by $c_k$ we denote positive constants which depend only on $n$ and $K$)

$$2\nu_h(1,\sigma) \geq \nu_h(1/2) - c_o(\log \frac{1}{\sigma})^{n-1} .$$

Since $i \notin I_1$, by increasing $\kappa$ if necessary, we may choose $\sigma$ so that

$$\nu_h(1/2) = 2c_o(\log \frac{1}{\sigma})^{n-1} \leq 4c_o(\log \frac{\sigma_o}{\sigma})^{n-1} . \tag{2.5}$$

It follows that $h_j^{-1}\bar{B}(\sigma) \cap \bar{B}(1) \neq \emptyset$ for every $j$ . From this one can show (see [12,Section 6] for details) that there exists a path family $\Gamma_j$ in the ring $B(3/2) \setminus \bar{B}(1)$ whose modulus satisfies the estimate

$$M(\Gamma_j) \geq b_3 q^{1/(n-1)}$$

and every path in $\Gamma_j$ connects $h_j^{-1}\bar{B}(\sigma)$ to some $h_k^{-1}\bar{B}(\sigma), k \neq j$ .

Since every path in the image family $h_j\Gamma_j$ connects the spheres $S(\sigma)$ and $S(\sigma_o)$, the ratio $M(\Gamma_j)/M(h_j\Gamma_j)$ has a lower bound of the form $b_4 q^{1/(n-1)}(\log(\sigma_o/\sigma))^{n-1}$. This feature and a somewhat tricky application of Lemma 2.3 leads to the following result.

2.6. Lemma ([12,6.7]). There exist constants $c_1, c_2 > 0$ such that $q \geq c_2$ implies

$$\mu(W_i) \geq c_1 q^{\frac{1}{n-1}} (\log \frac{\sigma_o}{\sigma})^{n-1}$$

provided $\kappa$ is sufficiently large.

To draw the final conclusion we estimate as follows. We assume $q \geq c_2$ where $c_2$ is the constant in 2.6. Then 2.6, (2.5), and $i \notin I_o$ imply

$$\mu(W_i) \geq c_1 q^{1/(n-1)}(\log \frac{\sigma_o}{\sigma})^{n-1} \geq c_3 q^{1/(n-1)}\mu(U_i) \geq c_4 q^{1/(n-1)}\mu(W_i).$$

This is a contradition for $q > c_4^{1-n}$. The theorem is proved and we can take for $q(n,K)$ an integer $> \max(c_2, c_4^{1-n})$.

Until very recently it has been conjectured that the Picard's theorem is true in the same form for $n \geq 3$ for quasiregular mappings as in the plane. However, it seems now that there is some hope to disprove this and in fact to show (at least for $n = 3$) that Theorem 2.1 is qualitatively sharp, more precisely, for every positive integer $k$ there exists a nonconstant entire quasiregular mapping which omits $k$ points. Although the basic idea exists for such a construction, there are still too many details to be filled so that one could at this point decide whether it really works. It is easy to construct an example of a nonconstant quasiregular mapping $f: R^n \to R^n \setminus \{a_1\}$. This was done already by Zorič in [15].

3. <u>A defect relation</u>. Classical value distribution for meromorphic functions was established in 1925 by R.Nevanlinna. His theory is a far reaching extension of Picard's theorem. The culminating part is the so called defect relation. Ten years later Ahlfors created in his famous article [1] a parallel theory which is very geometric in character.

Ahlfors's theory includes the following result. Let $f\colon R^2 \to \bar{R}^2$ be a nonconstant meromorphic function, let $n(r,y)$ be the counting function as in Section 2, and let $A(r)$ be the average of $n(r,y)$ over $\bar{R}^2$ with respect to the spherical metric. Then there exists a set $E \subset [1,\infty[$ of finite logarithmic measure such that

$$(3.1) \qquad \limsup_{\substack{r\to\infty \\ r\notin E}} \sum_{j=1}^{q} \left(1 - \frac{n(r,a_j)}{A(r)}\right)_+ \leq 2$$

whenever $a_1,\ldots,a_q$ are distinct points in $R$ . Here $\alpha_+ = \max(0,\alpha)$ for any real number.

The inequality (3.1) bounds for $r \notin E$ asymptotically the total contribution of the numbers $1 - n(r,a_j)/A(r)$ . Such a number measures how much less $a_j$ is covered by $f|\bar{B}(r)$ compared to the average covering $A(r)$ , and we can call it the defect of $a_j$ in the ball $\bar{B}(r)$ and (3.1) a defect relation.

For $n \geq 3$ there exists now a substitute for (3.1) in the following form.

3.2. <u>Theorem</u> [12]. Let $n \geq 3$ and let $f\colon R^n \to \bar{R}^n$ be a nonconstant K-quasimeromorphic mapping. Then there exists a set $E \subset [1,\infty[$ of finite logarithmic measure and a constant $C(n,K) < \infty$ depending only on $n$ and $K$ such that

$$\limsup_{\substack{r\to\infty \\ r\notin E}} q\left[\frac{1}{q}\sum_{j=1}\left(1-\frac{n(r,a_j)}{A(r)}\right)_+\right]^{n-1} \le C(n,K) \tag{3.3}$$

whenever $a_1,\ldots,a_q$ are distinct points in $\bar{R}^n$.

The proof of 3.2 is rather long. It consists of a careful analysis of lifts of paths together with a generalization of the method of moduli of path families. Theorem 2.1 is clearly a corollary of 3.2 and a part of the spirit of the proof of 3.2 can be seen from the outline of the proof of 2.1 in the preceding section.

For $n=2$ the left hand side of (3.3) reduces to that of (3.1). It is an open question whether the exponent $n-1$ is the right one. There is some indication that it could possibly be 1 as in (3.1). About $C(n,K)$ we refer to the discussion in the end of Section 2.

4. Averages of the counting function. In [1,p. 164,165] Ahlfors presented two covering theorems for meromorphic functions. The first says for functions of $R^2$ roughly the following. The average cover over any domain is arbitrarily close to the spherical average $A(r)$ (see Section 3) outside a set of finite logarithmic measure. The second is a similar statement for an arc with a regularity assumption. These results can be extended for quasiregular mappings and for averages with respect to any measure with a regularity assumption which prevents too strong singularities at points. In [6] we consider quasiregular mappings $f\colon M\to N$ where $M$ and $N$ are connected Riemannian n-manifolds, $M$ noncompact and $N$ compact. To avoid the discussion of exhaustions of $M$ we will here present the result for $M = R^n$.

4.1. Theorem [6, Theorem 5.11(1)]. Let $f\colon R^n \to N$ be a nonconstant quasiregular mapping. Then there exists a set $E \subset [1,\infty[$ of finite logarithmic measure such that the following holds. Let $\mu$ be a measure in $N$ such that Borel sets are measurable, $0 < \mu(N) < \infty$, and for any ball $B(x,\rho)$ in $N$

$$\mu(B(x,\rho)) \leq h(\rho)$$

holds, where $h$ is an increasing, continous, and positive function such that

$$\int_0^1 \frac{h(\rho)^{1/pn}}{\rho}\, d\rho < \infty \tag{4.2}$$

for some $p > 2$ . Then

$$\lim_{\substack{r \to \infty \\ r \notin E}} \frac{\nu_\mu(r)}{A(r)} = 1 \tag{4.3}$$

where $\nu_\mu(r)$ is the average of the counting function $n(r,y)$ with respect to $\mu$ .

The proof of 4.1 is based on an idea which originates from the proof of 2.3. A factor like $(\log \Theta)^{1-n}$ in the error term in 2.3 plays an essential role.

Recently Hinkkanen [2] sharpened the result of 4.1 for meromorphic functions to the form that (4.2) can be replaced by

$$\int_0^1 \frac{h(\rho)}{\rho}\, d\rho < \infty \quad .$$

4.4. Remark. The defect relation 3.2 gives of course a much stronger result than 4.1 in one direction, namely, it bounds the ratios $n(r,y)/A(r)$ from below very effectively in average. Hence the value of (4.3) is in the fact that the left hand side does not exceed 1. For examples which show that a regularity assumption is really needed we refer to [6,6.1].

REFERENCES:


[1] Ahlfors,L.: Zur Theorie der Überlagerungsflächen. Acta Math. 65 (1935), 157-194.

[2] Hinkkanen,A.: On the averages of the counting function of a meromorphic function. Ann.Acad.Sci.Fenn.A I, Dissertationes 26 (1980), 1-31.

[3] Marden,A. and Rickman,S.: Holomorphic mappings of bounded distortion. Proc.Amer.Math.Soc. 46 (1974), 226-228.

[4] Martio,O., Rickman,S. and Väisälä,J.: Definitions for quasiregular mappings. Ann.Acad.Sci.Fenn. A I 448 (1969), 1-40.

[5] Martio,O., Rickman,S. and Väisälä,J.: Distortion and singularities of quasiregular mappings. Ann.Acad.Sci.Fenn. A I 465 (1970), 1-13.

[6] Mattila,P. and Rickman,S.: Averages of the counting functions of a quasiregular mappings. Acta Math.143(1979),273-305.

[7] Rešetnjak,J.G.: Extremal properties of mappings with bounded distortion (Russian). Sibirsk.Mat.Ž.10(1969),1300-1310

[8] Rickman,S.: Path lifting for discrete open mappings. Duke Math. J. 40 (1973), 187-191.

[9] Rickman,S.: On the value distribution of quasimeromorphic maps. Ann.Acad.Sci.Fenn.A I 2 (1976), 447-466.

[10] Rickman,S.: On the theory of quasiregular mappings. Proceedings of the Rolf Nevanlinna Symposium on Complex Analysis, Silivri 1976.

[11] Rickman,S.: On the number of omitted values of entire quasiregular mappings. J.Analyse Math. 37 (1980), 100-117.

[12] Rickman,S.: A defect relation for quasimeromorphic mappings. Annals of Math., to appear.

[13] Väisälä,J.: Modulus and capacity inequalities for quasiregular mappings. Ann.Acad.Sci.Fenn.A I 509 (1972),1-14.

[14] Väisälä,J.: A survey of quasiregular maps in $R^n$. Proceedings of the International Congress of Mathematicians, Helsinki 1978.

[15] Zorič,V.A.: The theorem of M.A.Lavrentiev on quasiconformal mappings of space (Russian). Mat.Sb. 74 (1967),417-433.



University of Helsinki

Helsinki, Finland

# HOMOLOGY OF LOOP SPACES AND OF LOCAL RINGS

Jan-Erik ROOS

*"There are more things in heaven and earth, Horatio,*
*than are dreamt about in your philosophy."*

(Hamlet, Act I, scene 5.)

## § 0. Introduction

The aim of this paper is to show that Hamlet was right. More precisely, I will explain how seven conjectures, two of them very reasonable ones (these two have figured in mathematics for almost 30 years!), have recently found their counter-examples. Some of these conjectures are about how resolutions of modules over certain rings can behave or about how complicated the loop spaces of finite, simply-connected CW-complexes can be (more details about this are given in § 3 and § 4 below). Other conjectures are about the growth functions (or better: generating series) of graded algebras (cf. § 1 below), etc. The conjectures are due to Serre, Kaplansky, Moore, Kostrikin, Šafarevič, Govorov, Levin, Lemaire,.... In the search of proofs of these conjectures new mathematics has been developed, and unsuspected precise connections have been found between these conjectures themselves, and also e.g. between algebraic topology and the theory of local rings, combinatorics and invariant theory. The mathematics developed in the search of proofs of the conjectures is due to many mathematicians, e.g. Assmus, Avramov, Backelin, Fröberg, Golod, Govorov, Gulliksen, Herzog, Kostrikin-Šafarevič, Lech, Lemaire, Levin, Löfwall, Rahbar-Rochandel, Scheja, Schoeller, Serre, Shamash, Sjödin, Steurich, Tate, Ufnarovskij, Wiebe... and myself. The constructions of counter-examples are due to Shearer, Anick, Ufnarovskij, Lemaire, Serre, Löfwall, Babenko, Bøgvad and myself.

Essentially, it has been found that most of the conjectures above can be reduced to various generalizations and variants of a result that Hilbert found 90 years ago. We will begin with this theory of Hilbert.

§ 1. Trying to generalize a result of Hilbert. The Govorov conjecture. The following is very well-known and classical: Let k be a fixed field and let $k[X_1,\ldots,X_n]$ be the (commutative) polynomial ring in n variables over k. This is a graded ring if we give each homogeneous polynomial its ordinary degree. Let $\underline{\underline{a}} \subset k[X_1,\ldots,X_n]$ be a graded ideal, i.e. an ideal such that a polynomial $P \in \underline{\underline{a}}$, if and only if each homogeneous component of P is in $\underline{\underline{a}}$. Then the commutative ring $\Psi = k[X_1,\ldots,X_n]/\underline{\underline{a}}$ is a graded k-algebra, i.e. $\Psi = \bigoplus_{i\geq 0} \Psi_i$, $\Psi_i . \Psi_m \subset \Psi_{i+m}$, which is also connected, i.e. $\Psi_0 = k$.

Put $|\Psi_m| \overset{\text{def}}{=} \dim_k \Psi_m$, and consider the series

$$\Psi(Z) = \sum_{m\geq 0} |\Psi_m| Z^m$$

This is the so-called Hilbert series of $\Psi$, and it can also be defined in more general cases, e.g. for graded vector spaces. If $\underline{\underline{a}} = 0$, then clearly $\Psi(Z) = (1-Z)^{-n}$. More generally, if $\underline{\underline{a}} \neq 0$:

THEOREM (Hilbert-Serre).-*Let* $\Psi$ *be a graded algebra as above*. *Then* $\Psi(Z)$ *is a rational function*. *More precisely*:

$$(*) \qquad \Psi(Z) = P(Z)/(1-Z)^v$$

*where* P(Z) *is a polynomial with integer coefficients*, *and* v *is an integer*. *Furthermore*, *the order of the pole* Z=1 *is equal to the* (Krull) *dimension of* $\Psi$.

For the proof, we refer the reader to § 2 below or to [3], Chapter 11, where an even more general version is given.

Let us now pass to the non-commutative version of the preceding theory! Thus, let $k\langle T_1,\ldots,T_n\rangle$ be the free (non-commutative if n>1) associative algebra in n variables. Clearly $k\langle T_1,\ldots,T_n\rangle$ is graded if we give each variable $T_i$ the degree 1. Let $\underline{\underline{a}}$ be any homogeneous two-sided ideal of $k < T_1,\dots,T_n >$ and consider the graded algebra $\Gamma = k<T_1,\dots,T_n>/\underline{\underline{a}}$ and the corresponding "Hilbert series" $\Gamma(Z) = \sum_{i \geq 0} |\Gamma_i| Z^i$. If $a = 0$, we have $\Gamma(Z) = (1-nZ)^{-1}$, and from this it is easy to see that for $\underline{\underline{a}} \neq 0$ the convergence radius of $\Gamma(Z)$ is at least $1/n$.

In view of the preceding (commutative) results, it is natural to ask for conditions on $\underline{\underline{a}}$ that assure that $\Gamma(Z)$ is rational. The Hilbert-Serre theorem above can be formulated as saying that if $\underline{\underline{a}}$ is generated by at least all commutators $T_iT_k - T_kT_i$, and perhaps by some more elements, then $\Gamma(Z)$ is rational of the form (*). Furthermore in this case $\underline{\underline{a}}$ is automatically a finitely generated ideal (this last assertion is the Hilbert basis theorem). <u>However</u>, if $\underline{\underline{a}}$ does not contain all the $T_iT_k - T_kT_i$, then $\Gamma$ is not commutative and $\underline{\underline{a}}$ need not be finitely generated, even as a two-sided ideal, and it is also rather easy to give examples of non finitely generated $\underline{\underline{a}}$:s for which the corresponding $\Gamma(Z)$ is <u>not</u> rational (<u>cf</u>. [23], Example 1, where Govorov gives an example, where $\underline{\underline{a}}$ is generated by an infinite set of monomials. However, the example of Govorov has to be slightly modified, in order that his reasoning should work...).

However, Govorov proved rationality of $\Gamma(Z)$ if $\underline{\underline{a}}$ was generated by a <u>finite</u> set of monomials in the $T_i$:s, and in some other cases too [23-24]. This led him to conjecture:

*I (Question of Govorov 1972* [23-24]*): Let* $\Gamma$ *be a quotient of* $k< T_1,\dots,T_n>$ (deg $T_i = 1$) *by a homogeneous two-sided ideal* $\underline{\underline{a}}$ *generated by a finite number of elements. Is* $\Gamma(Z)$ *a rational function of* Z *?*

Backelin proved in [8], using an important theorem of Gerasimov [22] that the answer to *I* is positive if $\underline{\underline{a}}$ is generated by <u>one</u> element. Furthermore, Backelin (unpublished) and Ufnarovskij [52] have also shown that the answer to *I* is positive if $\underline{\underline{a}}$ is generated by <u>two</u> quadratic elements.

However, in the fall of 1978, by an ingenious purely combinatorial reasoning, James B. Shearer [47] was able to construct in $k\langle T_1,\ldots T_{11}\rangle$ (11 variables!) a two-sided ideal $\underline{\underline{a}}$, generated by 77 quadratic relations of the form $T_iT_j - T_kT_m$, for which the corresponding $\Gamma(Z)$ was non rational. In fact, Shearer's $\Gamma(Z)$ was of the form

$$R_1(Z) + R_2(Z).(1 - 4Z^2)^{-\frac{1}{2}}$$

where $R_1(Z)$ and $R_2(Z)$ are rational functions of Z (they can be calculated explicitly). Jörgen Backelin noted that if we divide out by one of the variables $T_i$ (the variable noted z in [47] we still get an algebra $\Gamma'$ having 10 generators of degree 1 and 57 relations of the form $T_iT_j - T_kT_m$ or $T_qT_r$, such that $\Gamma(Z) = \Gamma'(Z) + (1 - Z)^{-1}$, so that $\Gamma'(Z)$ is still non rational (but still algebraic!). By taking more generators (of degree 1) and more relations (of degree 2), Shearer also found examples where $\Gamma(Z)$ was transcendental. (Simpler such examples are given in § 5 below.). Both Shearer [47] and Ufnarovskij [51] have independently obtained even simpler counter-examples to the more general Govorov conjecture *I* [23-24] when the $T_i$:s are allowed to have degrees higher than 1. These examples may easily be turned into examples with deg $T_i$ =1, but some of the relations then still have degrees higher than 2. For the applications given below it is important to have relations of degree 2.

All this settles the Govorov conjectures in the negative. But this is only a beginning!

## § 2. Enters the homological algebra. A conjecture of Kostrikin and Šafarevič.

Let us first recall that an easy way of proving part of the Hilbert-Serre theorem of § 1 for $\Psi = k[X_1,\ldots,X_n]/\underline{\underline{a}}$ is to do the following:

Take a finite number of homogeneous generators $\{f_i\}_{i=1}^m$ of $\underline{\underline{a}}$, consider the map

$$(1) \qquad \bigoplus_1^m k[X_1,\ldots,X_n] \longrightarrow \underline{\underline{a}}$$

defined by $\{P_i\}_{i=1}^m \longmapsto \sum_{i=1}^m P_i f_i \in \underline{a}$ . Let q be any integer, and let $k[X_1,\ldots,X_n][-q]$ denote the new graded $k[X_1,\ldots,X_n]$-module structure on $k[X_1,\ldots X_n]$, obtained by means of the new grading:

$$k[X_1,\ldots,X_n][-q]^i \overset{\text{def}}{=} k[X_1,\ldots,X_n]^{i-q}$$

Now (1) can be considered as a homogeneous $k[X_1,\ldots,X_n]$-map of degree 0

$$(2) \qquad \bigoplus_{i=1}^m k[X_1,\ldots,X_n][-\deg f_i] \longrightarrow \underline{a}$$

and the kernel of (2) is a finitely generated graded $k[X_1,\ldots,X_n]$-module $M_1$. Now chosing homogeneous $k[X_1,\ldots X_n]$-generators of $M_1$, representing $M_1$ as in (2), and continuing in this manner, we finally obtain a homogeneous resolution of the $k[X_1,\ldots,X_n]$-module $\Psi$ of the following form

$$(3) \qquad \ldots \longrightarrow F_j \longrightarrow F_{j-1} \longrightarrow \ldots \longrightarrow F_1 \longrightarrow F_o \longrightarrow \Psi \longrightarrow 0$$

where $F_o = k[X_1,\ldots,X_n]$ and where each $F_j$ is a direct sum of a finite number of $k[X_1,\ldots,X_n][-q]$:s, for varying q:s ($q \geq 0$). Now, according to another well-known result of Hilbert, we can always make choices of generators so that (3) breaks off; more precisely, we may assume that $F_j = 0$ for $j > n$. Since the Hilbert series of the graded vector space $k[X_1,\ldots,X_n][-q]$ is $Z^q.(1-Z)^{-n}$ , and since the alternating sum of the Hilbert series of an exact sequence of graded modules is zero, it follows from (3) that

$$\Psi(Z) - F_o(Z) + F_1(Z) - \ldots + (-1)^{n+1} F_n(Z) = 0$$

which can be written

$$\Psi(Z) = \text{Pol}(Z).(1-Z)^{-n} ,$$

i.e. a rationality assertion is related to the existence of a certain kind of resolution. The proof of the Hilbert-Serre theorem, just given, suggests that in the non-commutative case, where there have been constructed with great efforts examples of finitely presented graded algebras, whose Hilbert series are non rational, we should try to go "backwards" and try to interpret these examples in terms of resolutions. This is indeed possible. However, the historical development was exactly the opposite, and it started with a paper by Kostrikin and Šafarevič [32], which we will discuss now. (The Kostrikin-Šafarevič paper was perhaps also influenced by the conjectures of § 3 and § 4 below).

Let k be a field, and let N be a k-algebra, which is also a finite-dimensional vector space over k, and which is further <u>nilpotent</u>, i.e., $N^n = 0$ for some integer n. Let $H^i(N,k)$ be the Hochschild cohomology groups of N , with values in the trivial N-module k [18] [32]. Under the hypotheses made, these groups are finite-dimensional vector spaces and Kostrikin-Šafarevič [32] considered the "Poincaré-Betti" series

$$P_N(Z) = \sum_{i \geq 0} |H^i(N,k)| Z^i \tag{4}$$

associated to N. In all cases, where they could calculate $P_N(Z)$, they found that it was a rational function of Z. They were thus led to the following conjecture:

*II (Question of Kostrikin and Šafarevič 1957* [32]*): Let* k *be a field,* N *a nilpotent, finite-dimensional, associative algebra over* k*,* $H^i(N,k)$ *the Hochschild cohomology groups of* N *with values in* k*, and* $P_N(Z)$ *the corresponding Poincaré-Betti series* (4) *of* N. *Is* $P_N(Z)$ *a rational function of* Z *?*

Consider the algebra $\tilde{N} = k \oplus N$, obtained by adding a unit to N in the well-known way [18]. This $\tilde{N}$ is a local algebra, with radical N, and k is in a natural way an $\tilde{N}$-module. Furthermore, we have isomorphisms [18]

$$H^i(N,k) \simeq \mathrm{Ext}^i_{\tilde{N}}(k,k) \simeq \mathrm{Hom}_k(\mathrm{Tor}_i^{\tilde{N}}(k,k),k) .$$

It follows that $|H^i(N,k)|$ is the rank of the $i^{th}$ step of a minimal free resolution of the left (say) $\tilde{N}$-module k. Therefore *Question II* asks whether there are any simple systematic phenomena that can be revealed in the minimal resolution of the residue field k of $\tilde{N}$. If $N^2 = 0$, then it is easy to see that $P_N(Z) = (1 - |N|Z)^{-1}$, so that we do have rationality in this case. If $N^3 = 0$, it turns out that if we consider $\mathrm{Ext}^*_{\tilde{N}}(k,k) = \bigoplus_{i \geq 0} \mathrm{Ext}^i_{\tilde{N}}(k,k)$ as a graded algebra under the Yoneda product $\mathrm{Ext}^i_{\tilde{N}}(k,k) \times \mathrm{Ext}^j_{\tilde{N}}(k,k) \longrightarrow \mathrm{Ext}^{i+j}_{\tilde{N}}(k,k)$, then the <u>subalgebra</u>, generated by $\mathrm{Ext}^1_{\tilde{N}}(k,k)$ is exactly of the form

$$\Gamma(N) = \Gamma = k\langle T_1,\dots,T_v\rangle \,/\, (\text{certain quadratic relations}) .$$

Furthermore, $P_N(Z)$ and $\Gamma(Z)$ are related by the explicit formula

$$(5) \quad P_N(Z)^{-1} = (1 + Z^{-1}).\Gamma(Z)^{-1} - Z^{-1}(1 - |N/N^2|Z + |N^2|Z^2) .$$

This was proved by Löfwall in [38], by specializing results from his thesis [37]. Furthermore, $N \longmapsto \Gamma(N)$ is a <u>one-one correspondence</u> between <u>all</u> the finite-dimensional nilpotent algebras N with $N^3 = 0$, and <u>all</u> the finitely presented $\Gamma$:s of the form

$$\Gamma = k\langle T_1,\dots,T_v\rangle/(\text{quadratic relations}) .$$

The inverse of this correspondence is defined as follows: Let $\Gamma$ be given as $\Gamma = T(V)/(F)$, where V is a finite-dimensional vector space (all elements are given the degree 1), T(V) its tensor algebra and $F \subset V \otimes V$ is the generating vector space for the relations. Let $V^*$ be the dual of V and put $N(\Gamma)$ = the elements of positive degree of

$$T(V^*)/(F^o + V^* \otimes V^* \otimes V^*)$$

where $F^o = \{ f \in V^* \otimes V^* \mid f(V) = 0 \}$. For more details we refer the reader to [38].

Using the preceding machinery we can now (as Löfwall did in [38] ) answer the *Question II* <u>negatively</u>. More precisely, using the Backelin variant $k\langle T_1,\dots,T_{10}\rangle/(57$ quadratic relations) of Shearer's example, mentioned above, with Hilbert series

$$R_1(Z) + R_2(Z).(1 - 4Z^2)^{-\frac{1}{2}}$$

where $R_1$ and $R_2 \neq 0$ are rational functions, that can be given explicitly, we obtain a nilpotent k-algebra N, with $N^3 = 0$, $\dim_k N =$ $= 10 + 57 = 67$, for which (use formula (5))

$$P_N(Z) = Q_1(Z) + Q_2(Z).(1 - 4Z^2)^{-\frac{1}{2}} \tag{6}$$

where $Q_1$ and $Q_2 \neq 0$ are some other rational functions that can also be calculated explicitly.

The function (6) is clearly an algebraic function of Z, which is non-rational. We will later construct smaller (even commutative) examples of nilpotent N:s for which $P_N(Z)$ is non-rational. However, in all these smaller examples $P_N(Z)$ is _non-algebraic_!

_Remark_ 1.- Formula (5) was proved in [38], using explicit resolutions. It might also be obtained from some more "superficial" observations, that we made in [43] in the commutative case. It might also be of interest to study the corresponding 2-variable series, that were studied in [43](_cf_. also § 6 below).

We will come back to the Kostrikin-Šafarevič question in § 4, but first we will take up another line of investigation (that will later be related to the previous theory):

## § 3. _Homology of loop spaces of_ CW-_complexes_. _Results of Bott and Serre_, _and conjectures of Serre and Moore_.

Recall that if X is any space with base point $x_o$, then $\Omega X$, the loop space of X at $x_o$, is defined to be the set of continuous maps $\varphi : [0,1] \longrightarrow X$, with $\varphi(0) = \varphi(1) = x_o$, with the compact-open topology. In the fifties Bott(and Samelson) had obtained some very precise results about the homology of loop spaces of compact, simply-connected homogeneous spaces H of Lie groups, showing in particular that in all cases in sight the series

$$\sum_{i \geq 0} |H_i(\Omega H, k)| \, Z^i \tag{7}$$

were rational (k = a field of coeficients) [13-16]. Serre, who in his thesis [44] had obtained fundamental results about the loop spaces of more general spaces, was thereby led to the following "conjecture":

*III (Question of Serre): Let* X *be a finite, simply-connected* CW-*complex, and let* $\Omega X$ *be the space of loops on* X. *Is the series*

$$\sum_{i \geq 0} |H_i(\Omega X, k)| . Z^i \tag{8}$$

*a rational function of* Z *?* (Here k is any field of coefficients.)

Although this question is much older, the first place in the literature where it occurs seems to be [45], p. IV-52. (Cf. also [41].) John C. Moore has even asked the more general question:

*IV (Question of Moore): Let* X *be a finite,* n-*connected (* $n \geq 1$*)* CW-*complex, and let* $\Omega^n X$ *be the* $n^{th}$ *iterated loop space of* X. *Is it true that* $H_*(\Omega^n X, k)(Z)$ *is a rational function of* Z *?*

This question was rather quickly disposed of by Serre (the results were published much later in [46]), who proved directly, using p-adic analysis, that $H(\Omega^2(S^3 v S^3), \mathbb{Q})(Z)$ was not rational. Here $S^3 v S^3$ means the wedge of two 3-spheres (i.e. they are identified along a common base point). Recently Babenko [7] announced the following general theorem, whose proof is based on a detailed study of the analytic properties of the series $H_*(\Omega Y, \mathbb{Q})(Z)$:

THEOREM [7].- Let X be a finite n-connected CW-complex, having an infinite number of non-zero rational homotopy groups. Then among the series $H_*(\Omega^j X, \mathbb{Q})(Z)$ $(1 \leq j \leq n)$, at most one is rational.

In the Serre case, $S^3 v S^3$ has an infinite number of non-zero rational homotopy groups and $H_*(\Omega(S^3 v S^3), \mathbb{Q})(Z) = (1 - 2Z^2)^{-1}$, so that the Babenko theorem applies, showing again that the double loop space series is not rational.

But the *Question III* still remained. In his thesis [34] (cf. also

[33]), J.-M. Lemaire showed <u>in particular</u>, that for simply-connected CW-complexes X, that were obtained as the mapping cones of maps f between suspensions of finite, connected complexes; i.e. as pushouts of diagrams (here CY means the cone over Y , and SY is the suspension of Y):

$$\begin{array}{ccc} CSX_2 & \dashrightarrow & X \\ \uparrow & & \uparrow \\ SX_2 & \xrightarrow{f} & SX_1 \end{array}$$

the series $H_*(\Omega X, \mathbb{Q})(Z)$ depends in a rational way (similar to formula (5) in § 2 ) in the Hilbert series of the cokernel H of the map (induced by f) between Hopf algebras:

$$H_*(\Omega SX_2, \mathbb{Q}) \xrightarrow{H_*(\Omega f)} H_*(\Omega SX_1, \mathbb{Q}).$$

We have indeed the formula

$$(9)\quad H_*(\Omega X, \mathbb{Q})(Z)^{-1} = (1 + Z).H(Z)^{-1} - Z\{1 - \overline{H}_*(X_1, \mathbb{Q})(Z) + \overline{H}_*(X_2, \mathbb{Q})(Z)\}$$

where e.g.

$$\overline{H}_*(X_1, \mathbb{Q})(Z) = \sum_{i \geq 0} |H_i(X_1, \mathbb{Q})| Z^i.$$

The question therefore arises as to which graded cocommutative Hopf algebras H, that we obtain in this manner. Actually, <u>all</u> finitely presented, cocommutative Hopf algebras (over $\mathbb{Q}$) occur (<u>cf</u>. p. 69 of Lemaire's thesis [34]).

An interesting special case is obtained if we take

$$SX_1 = \vee S^2 \qquad , \quad SX_2 = \vee S^3$$

(wedges of a finite number of spheres). The associated X:s form a subclass of the simply-connected, 4-dimensional, finite CW-complexes, and the corresponding H:s are <u>all</u> the Hopf algebras of the form ($k = \mathbb{Q}$ ):

$$k\langle T_1, \ldots, T_v\rangle/(\text{quadratic Hopf relations}) .$$

Here the $T_i$:s are given the degree 1, and a quadratic Hopf relation is a k-linear combination of graded commutators

$$[T_i,T_k] = T_iT_k + T_kT_i \quad (i < k) \quad \text{and } T_i^2 .$$

Thus we see that *Question III* is intimately related to a special case of *Question I*. More details about this will be given in § 4 and § 5 below.

§ 4. Homology of local rings and conjectures of Serre and Kaplansky.

A useful way of calculating $H_*(\Omega X)$, first used by Serre [44], is to study the spectral sequence of a certain fibration $\Omega X \longrightarrow LX \xrightarrow{p} X$:

(10) $$E^2_{p,q} = H_p(X,H_q(\Omega X)) \Rightarrow H_n(LX) .$$

Here LX is the space of paths, starting at $x_o$, and p is the natural map, associating to each path its end point. Furthermore, LX is contractible, so that $H_n(LX) = 0$ for $n > 0$, and the spectral sequence is obtained by considering the filtration of the chain complex of LX, by means of $p^{-1}$(skeleton filtration of X).

Now, if $(R,\underline{m})$ is a commutative, noetherian local ring with maximal ideal $\underline{m}$ and residue field $k = R/\underline{m}$, and if $P_*(k)$ is an R-projective resolution of the residue field k, we can filter the complex $P_*(k)$ as follows:

(11) $$P_*(k) = R \otimes_R P_*(k) \supset \underline{m} \otimes_R P_*(k) \supset \underline{m}^2 \otimes_R P_*(k) \supset \dots$$

Associated to the filtered complex (11) there is a spectral sequence

(12) $$E^1_{p,q} = gr_p(R) \otimes_k Tor^R_q(k,k) \Rightarrow Tor^R_n(R,k), \text{ where } gr_p(R)=\underline{m}^p/\underline{m}^{p+1} .$$

The analogy between (10) and (12) led Serre to make the following conjecture (cf. e.g. [45] IV-52), which was also made independently by I. Kaplansky (cf. [25] ):

*V (Question of Kaplansky and Serre): Let* $(R,\underline{m})$ *be a commutative local noetherian ring and let*

$$P_R(Z) = \sum_{i>0} |Tor^R_i(k,k)| Z^i$$

*be the Poincaré-Betti-Series of* $(R,\underline{m})$. *Is* $P_R(Z)$ *a rational function of* Z *?*

Positive results about *V* were obtained early by e.g. Tate, Assmus,

Golod, Scheja, Shamash, Gulliksen, Levin, Wiebe.... More details can be found in Gulliksen's and Levin's book [25].

In 1974 Levin applied the Artin-Rees lemma in an ingenious way and obtained the following beautiful result [36]:

For each local ring $(R,\underline{m})$, there exists an integer n such that for all $v \geq n$

(13) $$P_R(Z)^{-1} - P_{R/\underline{m}^v}(Z)^{-1} = (-1)^v . Z^{-(v-2)} . H_R(-Z)|_{\geq v}$$

where $H_R(Z)$ denotes the Hilbert series $\sum_{i\geq 0} |\underline{m}^i/\underline{m}^{i+1}| Z^i$ of R, and $H_R(-Z)|_{\geq v}$ denotes that part of the series for $H_R(-Z)$, that contains the powers of Z of degree $\geq v$.

Since the right hand side of (13) is always rational, this reduces *Question V* to artinian local rings, i.e. local rings $(R,\underline{m})$ for which $\underline{m}^v = 0$ for some v. In particular, if $(R,\underline{m})$ is equicharacteristic, we have reduced *Question V* to the commutative version of *Question II*, in § 2. As we mentioned in § 2, the first non-trivial case is $\underline{m}^3 = 0$, and in this commutative case there is of course also a formula ,due to Löfwall [37], and true for all local rings $(R,\underline{m})$ with $\underline{m}^3 = 0$:

(14) $$P_R(Z)^{-1} = (1 + Z^{-1}).A(Z)^{-1} - Z^{-1}\{1 - |\underline{m}/\underline{m}^2|Z + |\underline{m}^2| Z^2\}$$

where A is the subalgebra of $\mathrm{Ext}_R^*(k,k)$, generated by $\mathrm{Ext}_R^1(k,k)$, and (14) was indeed found before (5) of § 2. It turns out that the A:s that occur here are exactly those (Hopf) algebras H that occurred in connexion with the loop space problem at the end of § 3 (generators of degree 1 and relations of the Hopf type of degree 2), and all such A:s do occur.

In [43], I gave an alternative ("resolution-free") proof of (14) and also a two-variable version of (14) (this last version worked at least for equicharacteristic local rings). Combining Lemaire's and Löfwall's results, I obtained in particular precise relations between

$P_R(Z)$ and $H_*(\Omega X, \mathbb{Q})(Z)$, showing in particular that the *Questions III* and *V* were equivalent (base field $\mathbb{Q}$) if X was of dimension 4 and $(R, \underline{m})$ satisfied $\underline{m}^3 = 0$. The two-variable version of (14) gave in particular a negative answer to a problem of Levin (cf. § 7 below and [43] for this and other more precise results).

It should also be remarked that meanwhile Fröberg [19-20] had proved that the answer to *V* was positive if e.g. $R = k[[X_1, \ldots, X_n]]/$(some quadratic monomials in the $X_i$). Here $k[[X_1, \ldots X_n]]$ is the formal power series ring. Also Gulliksen-Ghione [28] and Gulliksen [27] had proved rationality of $\mathrm{Tor}_*^R(k,M)(Z)$ if M was a finitely generated R-module and R a Golod ring or a complete intersection, respectively.

Furthermore, Avramov had obtained several interesting results by means of the so-called Avramov spectral sequence, in particular e.g. the rationality of $k[[X_1, \ldots, X_n]]$ /(3 relations) [4].

One of the most spectacular later results was found by Gulliksen [29], who essentially proved that *Question V* could be reduced to the case of "quadratic relations". But now the time was ripe for deciding the question:

## § 5. Constructing finitely presented Hopf algebras with "bad" properties.

Some of the results of the last two sections can be summed up by saying that the *Question V* (for local rings $(R, \underline{m})$ with $\underline{m}^3 = 0$) and the *Question III* for finite, simply-connected CW-complexes X, with dim $X \leq 4$, would both be answered positively, if and only if we could answer yes to:

*VI (Question asked in Stockholm in the mid-1970:s): Let* H *be a finitely presented, primitively generated, Hopf algebra of the form:*

$$H = k\langle T_1, \ldots, T_n\rangle/(\varphi_1, \ldots, \varphi_t) \qquad (\deg T_i = 1)$$

*where each* $\varphi_s$ *is a linear combination of graded commutators*

$[T_i, T_k] = T_iT_k + T_kT_i$ $(i<k)$ *and* $T_i^2$ . *Is* H(Z) *a rational function of* Z ?

The answer to *VI* is positive if $n \leq 3$ [9], and probably also so if $n = 4$.

When Shearer came up with his counter-examples to *I* in the fall of 1978, many mathematicians began to doubt that the answer to *VI* would be yes, especially since earlier Lemaire [33-34] , inspired by examples of Stallings, had constructed H:s that behaved badly in other ways (more about that in § 7). It was however not until the fall of 1979 that David Anick [1-2] finally managed to construct a counter-example to *VI*. The idea of Anick is to study in detail certain quotients of semi-tensor products of free algebras. We refer the reader to [1] and [2] for more details about Anick's ingenious study, in particular his nice use of a certain map $\phi$ to a Lie algebra from its enveloping algebra. Here I will present a completely different approach to these questions, due to Löfwall and myself [39-40] , which is based on homological methods (cohomology and extensions of graded Lie algebras; the Hochschild-Serre spectral sequence and the properties of the differentials in this spectral sequence...) and which seems to be more suited for studying the two-variable series (a still open question of Lemaire should be answered negatively in this way, cf. § 7 below):

The idea of the construction of [39] : We want a finitely presented k-Hopf algebra H with a minimal set of generators, all of degree 1 (i.e., $\mathrm{Tor}_1^H(k,k)$ concentrated in degree 1 [34] ), and with a minimal set of relations, all in degree 2 (i.e. $\mathrm{Tor}_2^H(k,k)$ concentrated in degree 2 [34] ), such that H(Z) is transcendental. We will construct H as the enveloping algebra $U(\underline{g})$ of a graded Lie algebra $\underline{g} = \bigoplus_{i \geq 1} \underline{g}_i$ .

We first start with the "simplest" graded Lie algebra $\underline{h}$ with $U(\underline{h})(Z)$ transcendental, namely the abelian Lie algebra, having one generator

$e_i$ in each degree i:

$$\underline{h} = \bigoplus_{i\geq 1} k.e_i$$

Here

(15) $$U(\underline{h}) = E(x_1,x_3,x_5,\dots) \otimes_k P(x_2,x_4,x_6,\dots)$$

where E is an exterior algebra and P is a polynomial ring and the degrees of the variables are indicated as indices. Furthermore

(16) $$U(\underline{h}) = \prod_{i=1}^{\infty} (1 + Z^{2i-1})/(1 - Z^{2i})$$

and the function (16) is transcendental. However, the algebra $U(\underline{h})$, given by (15) is far from finitely presented. We will now remedy this fact! If we could construct an extension of graded Lie algebras ($\underline{h}$ as above)

(17) $$0 \longrightarrow \underline{h} \longrightarrow \underline{g} \longrightarrow \Phi \longrightarrow 0$$

where $U(\underline{g})$ is finitely presented, with generators of degree 1 and relations of degree 2, and where $U(\Phi)(Z)$ is a rational function, then, since $U(\underline{g})(Z) = U(\underline{h})(Z).U(\Phi)(Z)$, we would have that $U(\underline{g})(Z)$ is transcendental, and our problem would be solved.

This construction is indeed possible! As $\Phi$ we can take e.g. the product of two free graded Lie algebras, each on two generators of degree 1 (denoted $T_1,T_2$ and $T_3,T_5$ respectively). We define an $U(\Phi)$-module structure on $\underline{h}$ by $T_1.e_i = T_3.e_i = 0$, $T_2.e_i = e_{i+1}$ and $T_5.e_i = (-1)^{i+1}e_{i+1}$. Then the graded Lie algebra extensions (17), inducing the given $U(\Phi)$-module structure on $\underline{h}$ are classified by the degree 0 part of the second cohomology group $H^2(\Phi,\underline{h})$ of the graded Lie algebra $\Phi$. One shows that this vector space $H^2(\Phi,\underline{h})_o$ is one-dimensional and that a generating cocycle $\xi$ can be defined by $\xi(T_1,T_3) = \xi(T_3,T_1) = e_2$, $\xi(T_i,T_k) = 0$ for $(i,k) \neq (1,3),(3,1)$ ( $\xi$ on higher degree elements of $\Phi \times \Phi$ is given by the cocycle condition).

One proves, using the Hochschild-Serre spectral sequence for the extension (17) corresponding to this $\xi$ , and an explicit formula for the differential $d^2$ in terms of this $\xi$ , that $\mathrm{Tor}_2^{U(\underline{g})}(k,k)$ and $\mathrm{Tor}_1^{U(\underline{g})}(k,k)$ are concentrated in degrees 2 and 1 respectively [39-40]. It follows that

$$U(\underline{g}) = \frac{k\langle T_1,T_2,T_3,T_4,T_5\rangle}{(T_4^2,[T_4,T_2]-[T_4,T_5],[T_4,T_2]-[T_1,T_3],[T_1,T_4],[T_1,T_5],[T_2,T_3],[T_2,T_5],[T_3,T_4])}$$

where the generator corresponding to $e_1 \in \underline{h}$ is denoted by $T_4$. Clearly $U(\underline{g})(Z) = U(\Phi)(Z).U(\underline{h})(Z) = (1-2Z)^{-2}.U(\underline{h})(Z)$, where $U(\underline{h})(Z)$ is given by (16).

The corresponding local ring is

$$(18)\qquad R = \frac{k[X_1,X_2,X_3,X_4,X_5]}{(X_1^2,X_2^2,X_3^2,X_5^2,X_1X_2,X_3X_5,X_1X_3+X_2X_4+X_4X_5,\ \underline{n}^3)}$$

where $\underline{n}$ is the maximal ideal $(X_1,\ldots,X_5)$ of $k[X_1,\ldots,X_5]$ . The corresponding Poincaré-Betti series $P_R(Z)$ is given by (cf. (14))

$$(19)\quad P_R(Z)^{-1} = (1+Z^{-1})(1-2Z)^2\prod_{i=1}^{\infty}(1-Z^{2i})/(1+Z^{2i-1}) - Z^{-1}(1-5Z+8Z^2)$$

Clearly, the convergence radius of the Taylor series for $P_R(Z)^{-1}$ is 1. For more details about the preceding results, we refer the reader to [39] and [40]. More complicated extensions (17) are also studied in [39] and [40]. By means of these extensions it is e.g. rather easy to construct a local ring $(S,\underline{m})$, having $\underline{m}^3 = 0$ and $|\underline{m}/\underline{m}^2| = 11$, for which the corresponding $P_S(Z)^{-1}$ has convergence radius 1/2, and is transcendental, etc.

## § 6. Theta functions and double Poincaré-Betti series of local rings.

Parts of this section are highly conjectural. Let $(R,\underline{m})$ be a local ring, $B = \mathrm{Ext}_R^*(k,k)$ the corresponding Yoneda Ext-algebra. Then each $\mathrm{Tor}_p^B(k,k)$ is an (upper)graded vector space, finitedimensional in each degree, so that the double series

(20) $$P_B(X,Y) = \sum_{p,q\geq 0} |\mathrm{Tor}_p^B(k,k)^q| X^p Y^q$$

is well-defined. Furthermore, $P_B(-1,Z)^{-1} = P_R(Z)^{-1}$ , [34] , and the formulae (9),(13),(14),... are all special cases (put X = -1, Y = Z) of(simpler!) two variable formulae, the one corresponding to (14) being e.g.

(21) $$P_B(X,Y) = P_A(X,Y).(1+X^{-2}Y^{-1}) - X^{-2}Y^{-1}H_R(XY) ,$$

where $H_R(Z)$ is the Hilbert series of $(R,\underline{m})$ with $\underline{m}^3 = 0$, and A is the subalgebra of $\mathrm{Ext}_R^*(k,k)$, generated by $\mathrm{Ext}_R^1(k,k)$ (cf.[43] , p. 305 ; we suppose that we are in the equicharacteristic case).

The double series (20) is of general interest (it gives e.g. the generators of $\mathrm{Ext}_R^*(k,k)$, their relations, etc.) and in view of (21) it is sufficient to calculate $P_A(X,Y)$ (when $\underline{m}^3 = 0$...).

We will present some questions about $P_A(X,Y)$, at least for the case when $(R,\underline{m})$ is the local ring (18) of § 5.

We start by rewriting (recall that $P_A(-1,Z)^{-1} = A(Z) = U(\underline{g})(Z)$):

$$A(Z)^{-1} = (1-2Z)^2 \prod_{i=1}^{\infty} (1-Z^{2i})/(1+Z^{2i-1}) .$$

Put $\varphi(Z) = \prod_{j=1}^{\infty} (1-Z^j)$. It is clear that

$$\prod_{i=1}^{\infty} (1-Z^{2i})/(1+Z^{2i-1}) = \varphi(Z^2)^2 / \varphi(-Z).$$

Gauss proved 1809 [21] (for a modern version of this, see[31] ) that

(22) $$\varphi(Z^2)^2/ \varphi(Z) = \sum_{n=0}^{\infty} Z^{n(n+1)/2} .$$

From this it follows that

(23) $$A(Z)^{-1} = (1-2Z)^2 \sum_{n=0}^{\infty} (-Z)^{n(n+1)/2}$$

This gives a useful expansion for $P_R(Z)^{-1}$ and $A(Z)^{-1}$.

But the infinite sum in (23) is clearly related to theta functions, and these functions usually occur as functions of <u>two</u> variables, and identities such as (22) usually come up upon specialization of one of the variables [49]. The same phenomenom occurs for A(Z) which is equal to $P_A(-1,Z)^{-1}$. This suggests that there should be an explicit "theta-like" formula for $P_A(X,Y)$ for the local ring (18), and perhaps in some other cases too. Recall that for the local ring (18) we have $A = U(\underline{g})$, where $\underline{g}$ is obtained from the extension (17) (corresponding to $\xi$ , and the given $\Phi$-module structure on $\underline{h}$). If $\xi = 0$ and if $\Phi$ operates <u>trivially</u> on $\underline{h}$, then we <u>do</u> have a formula of the type we wish. The general case should be considered as some sort of perturbation of this simpler case (<u>cf</u>. also § 7).

§ 7. <u>Conjectures of Levin and Lemaire.</u>

In [25] and [35] Levin announced the following conjecture:

*VII (Question of Levin 1969* [25],[35]): *Let* $(R,\underline{m})$ *be a local (commutative, noetherian) ring with residue field* k, *and let* $\mathrm{Ext}^*_R(k,k)$ *be the Yoneda* Ext-*algebra of* $(R,\underline{m})$. *Is* $\mathrm{Ext}^*_R(k,k)$ *finitely generated as an algebra over* k *?*

The corresponding problem for the homology of loop spaces was disposed of by Lemaire in [33-34]. Inspired by group-theoretical constructions of Stallings, Lemaire had indeed constructed finite, simply-connected CW-complexes X that were mapping cones of maps between wedges of spheres, such that $H_*(\Omega X,\mathbb{Q})$ was <u>not</u> finitely generated as an algebra, the essential point being the construction of algebras $A = k\langle T_1,\ldots,T_n\rangle$ /(quadratic Hopf relations) for which the graded vector space $\mathrm{Tor}_3^A(k,k)$ occurred in an infinite number of degrees. If we take the $T_i$:s of degree 1 in one of Lemaire's A:s, and if $(R,\underline{m})$ is the corresponding local ring with $\underline{m}^3 = 0$, then,using our formula (21), we found in [43] that a minimal set of generators for $B = \mathrm{Ext}^*_R(k,k)$, i.e. a basis for $\mathrm{Tor}_1^B(k,k)$, needs at least a basis for $\mathrm{Tor}_3^A(k,k)$, so that $\mathrm{Ext}_R^*(k,k)$ needs an infinite number of generators. In particular it turns out that both the ring

(24) $S = k[X_1,X_2,X_3,X_4,X_5]/(X_1^2,X_2^2,X_3^2,X_4^2,X_5^2,X_1(X_2+\ldots+X_5),X_2X_3,X_4X_5)$

and $R = S/(X_1X_2X_4)$ (we assume char $k \neq 2$) have Ext-algebras that are not finitely generated and the Levin conjecture was thereby negatively answered. However $P_R(Z)$ and $P_S(Z)$ are both rational, although not of the form $(1+Z)^n$/polynomial(cf.[43] p. 314-315).

In connexion with these studies, Lemaire had asked the following question:

*VIII (Question of Lemaire* [33], p. 120*): Let* A *be a finitely presented Hopf algebra over* k . *Is e.g. the series* $\mathrm{Tor}_{3,*}^A(k,k)(Z)$ always *a rational function of* Z *?*

It is quite possible that the answer to this question is no. For more details we refer the reader to a forthcoming paper by C. Jacobsson [30], where e.g. the $\mathrm{Tor}_3$:s of the graded Lie algebra extensions of § 5 and the more general extensions of [39] are studied in detail. He has found that the homological properties of U(g) as an Ad-g-module play an important role here.... This theory is clearly related to invariant theory....

## § 8. Gorenstein rings.

Gorenstein rings were introduced a long time ago as a natural generalization of complete intersections (cf. e.g. 11-12 , in particular the introduction of 12 for the early theory and early history of these rings). A local ring (S,m) is a Gorenstein ring if and only if S has a finite injective resolution. Tate proved that local complete intersections have rational Poincaré-Betti series $P_S(Z)$. However, there are local Gorenstein rings having transcendental $P_S(Z)$! This result is due to Bøgvad [17] , who also proved some more general results:

Let us first recall that Gulliksen proved among other things in [26], that if R is any local artinian ring, I(k) the injective envelope of the residue field k of R, and $\tilde{R} = R \ltimes I(k)$ the "trivial" extension of R by the R-module I(k), then $\tilde{R}$ is a Gorenstein ring. Furthermore, Gulliksen proved also in [26] that for all trivial extensions $R \ltimes M$ (M any finitely generated R-module)

$$(25) \qquad P_{R \ltimes M}(Z) = P_R(Z)/(1-Z.P_R^M(Z))$$

where

$$P_R^M(Z) = \sum_{i \geq 0} |\operatorname{Tor}_i^R(k,M)| Z^i .$$

I suggested to Richard Bøgvad to use the preceding information and to calculate $P_{\tilde{R}}(Z)$ for the ring $\tilde{R} = R \ltimes I(k)$, when e.g. $(R,\underline{m})$ was the local ring (18). It is well-known [26] that the dual of $\operatorname{Tor}_i^R(k,I(k))$ is isomorphic to $\operatorname{Ext}_R^i(k,R)$ for any artinian $(R,\underline{m})$. Therefore it is sufficient to study the series $\operatorname{Ext}_R^*(k,R)(Z)$.

THEOREM (R. Bøgvad [17] ).- a) _For any local ring_ $(R,\underline{m})$, $\underline{m}^3 = 0$, _such that the subalgebra of_ $\operatorname{Ext}_R^*(k,k)$, _generated by_ $\operatorname{Ext}_R^1(k,k)$ _is_ not _graded commutative, we have an exact sequence of right_ $\operatorname{Ext}_R^*(k,k)$-_modules_ (the module structure is given by the Yoneda product):

$$(26) \quad 0 \longrightarrow \operatorname{Ext}_R^{*-1}(k,k) \longrightarrow \operatorname{Ext}_R^0(k,R/\underline{m}^2) \otimes_k \operatorname{Ext}_R^*(k,k) \longrightarrow \operatorname{Ext}_R^*(k,R/\underline{m}^2) \longrightarrow 0 .$$

b) _Under some extra conditions, which are e.g. verified for the local ring_ $(R,\underline{m})$ _of_ (18) _we also have an exact sequence_

$$(27) \quad 0 \longrightarrow \operatorname{Ext}_R^{*-1}(k,R/\underline{m}^2) \longrightarrow \operatorname{Ext}_R^0(k,R) \otimes_k \operatorname{Ext}_R^*(k,k) \longrightarrow \operatorname{Ext}_R^*(k,R) \longrightarrow 0$$

_which is analogous to_ (26).

From this beautiful result, Bøgvad deduces the following

COROLLARY.- _Let_ $(R,\underline{m})$ _be e.g. the local ring of_ (18) _and_ $\tilde{R} = R \ltimes I(k)$ _the associated Gorenstein ring. Then_

$$(28) \quad P_{\tilde{R}}(Z)^{-1} = P_R(Z)^{-1} - Z^3 H_R(-Z^{-1}), \text{ where } H_R(Z) = 1 + |\underline{m}/\underline{m}^2| Z + |\underline{m}^2| Z^2 .$$

In particular $(\widetilde{R},\widetilde{\underline{m}})$ is a Gorenstein ring with $\widetilde{\underline{m}}^4 = 0$ and $|\widetilde{\underline{m}}/\widetilde{\underline{m}}^2| = 13$, having a transcendental Poincaré-Betti series.

PROOF OF THE COROLLARY: From (26) and (27) we obtain

$$\text{(29)} \qquad \operatorname{Ext}^*_R(k,R)(Z) = P_R(Z).Z^2.H_R(-Z^{-1})$$

where we have used that for the R of (18) $\operatorname{Ext}^0_R(k,R) = \operatorname{Ext}^0_R(k,\underline{m}^2)$.

Now combine (29) with (25) for $M = I(k)$.

Remark.- It is well-known that if $(S,\underline{n})$ is a Gorenstein ring with $\underline{n}^3 = 0$, then $P_S(Z)$, and even $P^N_S(Z)$ ( N finitely generated S-module) are rational. In this sense the Corollary is the best possible one.

§ 9. Coherence.

We say that a graded algebra is (left) coherent if every finitely generated homogeneous left ideal is finitely presented. We proved in [42] that the algebra $\operatorname{Ext}^*_R(k,k)$ is coherent when R is a Golod ring (or more generally when R comes form a complete intersection by a Golod map). This gives in particular that $\operatorname{Ext}^*_R(k,M)(Z)$ and $\operatorname{Ext}^*_R(M,k)(Z)$ are rational for all finitely generated M over such R:s (cf.[42] , p. 12 ).

In joint work with Löfwall, we found that the ring

$$R = k[X_1,X_2,X_3,X_4]/\,(X_1^2,X_1X_2,X_2^2,X_3^2,X_3X_4,X_4^2)$$

had the property that

$$A = \operatorname{Ext}^*_R(k,k) = k\langle T_1,T_2\rangle \otimes_k k\langle T_3,T_4\rangle$$

is not coherent. (This might be the simplest example and it is inspired by a corresponding example in the theory of groups [50].). Indeed, the left ideal of A, generated by $T_1+T_3$, $T_2$, $T_4$ is not finitely presented, as is easily seen. This non-coherence property of A is in fact responsible for the local ring (18) of § 5 having a transcendental Poincaré-Betti series. It seems not unprobable that all algebras $k\langle T_1,T_2,T_3\rangle$ /(quadratic Hopf relations) are coherent and this indicates that there should not exist a counterexample to V of the type (18) with fewer that 5 variables.

§ 10. Open problems.

i) Let $(R,\underline{m})$ be a local commutative noetherian ring, and assume that $\mathrm{Ext}^*_R(k,k)$ is a noetherian algebra. Does it follow that $(R,\underline{m})$ is a complete intersection ? For a partial result, cf. [48] .

ii) Assume that $\mathrm{Ext}^*_R(k,k)$ is not noetherian. Does it contain as a subalgebra a free associative algebra in 2 (primitive) variables ? Cf. [6] , [5] for partial results.

iii) What can be said about the class of local rings S that are such that there exists a finite sequence of Golod maps:

$$R_{regular} \longrightarrow R_1 \longrightarrow R_2 \longrightarrow \dots \longrightarrow R_t \tag{30}$$

with $R_t = S$ ? Is $P_S(Z)$ rational for these S ? In [43] , p. 320-321 we thought that the answer was yes. Now we have doubts, even for monomial rings (cf. however [10] ). One can show that the Hopf algebra kernel of

$$\mathrm{Ext}^*_S(k,k) \longrightarrow \mathrm{Ext}^*_{R_{reg}}(k,k)$$

has global dimension $\leq t$, if S is an $R_t$ of a sequence of Golod maps (30). Is the converse true ?

iv) For all local rings $(R,\underline{m})$ with $|\underline{m}/\underline{m}^2| \leq 5$, we know of, the convergence radius of the series $P_R(Z)^{-1}$ is 1, unless $P_R(Z)^{-1}$ is a polynomial. This is no longer true if e.g. $|\underline{m}/\underline{m}^2| = 11$ (cf. § 5). How are the analytic properties of the series $P_R(Z)^{-1}$ (or of the double series $P_{\mathrm{Ext}^*_R(k,k)}(X,Y)$) related to properties of R, in general ?

v) The Löfwall formula (14) can be restated by saying that for any local ring $(R,\underline{m})$, with $\underline{m}^3 = 0$, the map $R \xrightarrow{\varphi} R/\underline{m}^2$ is almost Golod, in the sense that (put $S = R/\underline{m}^2$):

$$P_S(Z) = P_R(Z) \Big/ \left\{ (1+Z) \frac{P_R(Z)}{(\mathrm{Im}\varphi^*)(Z)} - Z.P^S_R(Z) \right\} \tag{31}$$

where $\mathrm{Ext}^*_S(k,k) \xrightarrow{\varphi^*} \mathrm{Ext}^*_R(k,k)$ is the Hopf algebra map induced by $\varphi$ , and $A(Z) = (\mathrm{Im}\varphi^*)(Z)$ is the Hilbert series of the image of $\varphi^*$.

If $\varphi^*$ is onto, then $P_R(Z) = (\mathrm{Im}\varphi^*)(Z)$, and (31) is exactly Golod's formula [36].

More generally we have proved, inspired by joint work with John C.Moore about the homology of loop spaces, that (at least in the graded case) any local ring map onto

$$R \xrightarrow{\varphi} S$$

where $\mathrm{Ker}\,\varphi \subseteq \mathrm{socle}(R) \subseteq \underline{m}^2$, for which the Hopf algebra kernel $\mathrm{Ker}\,\varphi^*$ of $\mathrm{Ext}_S^*(k,k) \longrightarrow \mathrm{Ext}_R^*(k,k)$ has global dimension $\leq 2$, is a almost Golod map in the sense that (31) holds. It is easy to see that gldim $\mathrm{Ker}\,\varphi^* \leq 2$, if S is an $R_2$ of iii). It follows that the map

$$k[X,Y,Z]/(X^2,YZ,XZ+Y^2) \longrightarrow k[X,Y,Z]/(X^2,YZ,XZ,Y^2)$$

is almost Golod ( but not Golod ! ). For this example, cf.[9] . I have also deduced an explicit, but slightly more complicated formula than (31), when gldim $\mathrm{Ker}\,\varphi^* = 3$. How generalize this ? A complete answer to this question would give $P_R(Z)$ for all $(R,\underline{m})$ (it is of course also of interest to try to deduce the corresponding double series results). Indeed, consider the following sequences of morphisms

$(32)_n$ $\qquad R \longrightarrow R/\underline{m}^n$

$(32)_{n-1}$ $\qquad R/\underline{m}^n \longrightarrow R/\underline{m}^{n-1}$

............

$(32)_3$ $\qquad R/\underline{m}^4 \xrightarrow{\varphi_3} R/\underline{m}^3$

$(32)_2$ $\qquad R/\underline{m}^3 \xrightarrow{\varphi_2} R/\underline{m}^2$

If we choose n sufficiently big, then $(32)_n$ is always Golod by Levin's theorem [36] ( it even has slightly better properties...). The map $(32)_2$ is always <u>almost Golod</u>. If gldim $\mathrm{Ker}\varphi^*_3 \leq 2$, then $(32)_3$ is <u>also</u> almost Golod (at least in the graded case), and if only gldim $\mathrm{Ker}\varphi^*_3 \leq 3$ , we have an explicit, slightly more complicated formula than (31). It is clear that lots of hard problems still remain here.

BIBLIOGRAPHY

[1] D. ANICK, A counterexample to a conjecture of Serre, Thesis, Mass. Inst. Techn., 1980. Has also appeared in: Reports, Department of Math., Univ. of Stockholm, Sweden, n° 8, 1980.

[2] D. ANICK, Construction d'espaces de lacets et d'anneaux locaux à séries de Poincaré-Betti non rationnelles, Comptes rendus Acad. Sc. Paris, 290, série A, 1980, p. 729-732.

[3] M.F. ATIYAH - I.G. MACDONALD, Introduction to Commutative Algebra, Addison-Wesley, Reading, Mass. 1969.

[4] L. AVRAMOV, Small homomorphisms of local rings, J. Algebra, 50, 1978, p. 400-453.

[5] L. AVRAMOV, On the convergence radius of Poincaré series of local rings, Reports, Department of Math., Univ. of Stockholm, Sweden, n° 4, 1979.

[6] L. AVRAMOV, Sur la croissance des nombres de Betti d'un anneau local, Comptes rendus Acad. Sc. Paris, 289, série A, 1979, p. 369-372.

[7] I.K. BABENKO, O rjadach Poincaré kratnych prostranstv' petel', Uspechi Matem. Nauk, 34, vyp. 3, 1979, p. 191-192.

[8] J. BACKELIN, La série de Poincaré-Betti d'une algèbre de type fini à une relation est rationnelle, Comptes rendus Acad. Sc. Paris, 287, série A, 1978, p. 843-846.

[9] J. BACKELIN - R. FRÖBERG, Studies on some k-algebras giving the Poincaré series of graded k-algebras of length $\leq 7$ and local rings of embedding dimension 3 with $\underline{m}^3 = 0$, Reports, Department of Math., Univ. of Stockholm, Sweden, n° 9, 1978.

[10] J. BACKELIN, Monomial rings of embedding dimension four have rational Poincaré series, Reprts, Department of Math., Univ. of Stockholm, Sweden, n° 14, 1980.

[11] H. BASS, Injective dimension in noetherian rings, Trans.Amer.Math.Soc., 102, 1962, p. 18-29.

[12] H. BASS, On the ubiquity of Gorenstein rings, Math.Z., 82, 1963, p.8-28.

[13] R. BOTT, On torsion in Lie groups, Proc. Nat. Acad. Sc., 40, 1954, p. 586-588.

[14] R. BOTT, An application of the Morse theory to the topology of Lie groups, Bull. Soc. Math. France, 84, 1956, p. 251-281.

[15] R. BOTT, The space of loops on a Lie group, Mich. Math. J., 5, 1958, p. 35-61.

[16] R. BOTT - H. SAMELSON, Application of the theory of Morse to symmetric spaces, Amer. J. Math., 80, 1958, p. 964-1029.

[17] R. BØGVAD, to appear.

[18] H. CARTAN - S. EILENBERG, Homological Algebra, Princeton Univ. Press, Princeton, 1956.

[19] R. FRÖBERG, Determination of a class of Poincaré series, Math. Scand., 37, 1975, p. 29-39.

[20] R. FRÖBERG, Some complex constructions with applications to Poincaré series, Lecture Notes in Mathematics, 740, p. 272-284, Springer-Verlag, Berlin, 1979.

[21] C.F. GAUSS, Summatio quarumdam serierum singularum, Comm. Soc. Gotting., 1, 1809 = Werke, Band 2, p. 20, Göttingen, 1876.

[22] V.N. GERASIMOV, Distributivnye resjetki podprostvanstv i problema ravenstva dlja algebr' s odnim sootnosjeniem, Algebra i Logika, 15, 1976, p. 384-435. English translation: Distributive lattices of subspaces and the equality problem for algebras with a single relation, Algebra and Logic, 15, 1976, p. 238-274.

[23] V. E. GOVOROV, O graduirovannych algebr, Matem. Zametki, 12, 1972, p. 197-204. English translation: Graded algebras, Math. Notes of the Acad. of the Sci. of the USSR, 12, 1972, p. 552-556.

[24] V.E. GOVOROV, O razmernosti graduirovannych algebr, Matem. Zametki, 14, 1973, p. 209-216. English translation: On the dimension of graded algebras, Math. Notes of the Acad. of the Sci. of the USSR, 14, 1973, p. 678-682.

[25] T.H. GULLIKSEN - G. LEVIN, Homology of local rings, Queen's Papers in Pure Appl. Math.,n° 20, Queen's Univ., Kingston, Ontario, 1969.

[26] T.H. GULLIKSEN, Massey operations and the Poincaré series of certain local rings, J. Algebra, 22, 1972, p, 223-232.

[27] T.H. GULLIKSEN, A change of ring theorem with applications to Poincaré series and intersection multiplicity, Math. Scand., 34,1974, p.167-183.

[28] T.H. GULLIKSEN - F. GHIONE, Some reduction formulas for the Poincaré series of modules, Atti Accad. naz. Lincei VIII. Ser., Rend.,Cl. Sci. fis. mat. natur., 58, 1975, p. 82-91.

[29] T.H. GULLIKSEN, Reducing the Poincaré series of local rings to the case of quadratic relations, Preprint series, Institute of Math., Univ. of Oslo, Norway, n° 10, 1979 (Submitted to Math. Scand.).

[30] C. JACOBSSON, to appear.

[31] V.G. KAC, Infinite-dimensional algebras, Dedekind's η-function, classical Möbius function and the very strange formula, Adv. Math., 30, 1978, p. 85-136.

[32] A.I. KOSTRIKIN - I.R. ŠAFAREVIČ, Gruppy gomologij nil'potentnych algebr, Doklady Akad. Nauk SSSR, 115, 1957, p. 1066-1069.

[33] J.M. LEMAIRE, A finite complex, whose rational homotopy is not finitely generated, Lecture Notes in Mathematics, 196, p. 114-120, Springer-Verlag, Berlin, 1971.

[34] J.M. LEMAIRE, Algèbres connexes et homologie des espaces de lacets, Lecture Notes in Mathematics, 422, Springer-Verlag, Berlin, 1974.

[35] G. LEVIN, Two conjectures in the homology of local rings, J. Algebra, 30, 1974, p. 56-74.

[36] G. LEVIN, Local rings and Golod homomorphisms, J. Algebra, 37, 1975, p. 266-289.

[37] C. LÖFWALL, On the subalgebra generated by the one-dimensional elements of the Yoneda Ext-algebra, Reports, Department of Math., Univ. of Stockholm, Sweden, n° 5 , 1976.

[38] C. LÖFWALL, Une algèbre nilpotente dont la série de Poincaré-Betti est non rationelle, Comptes rendus Acad. Sc. Paris, 288, série A, 1979, p. 327-330.

[39] C. LÖFWALL - J.-E. ROOS, Cohomologie des algèbres de Lie graduées et séries de Poincaré-Betti non rationnelles, Comptes rendus Acad. Sc. Paris, 290, série A, 1980, p. 733-736.

[40] C. LÖFWALL - J.-E. ROOS, Detailed version of [39] (to appear).

[41] Proceedings of Symposia in Pure Mathematics, vol. 32, Algebraic and Geometric Topology, Part 2, p. 252, Problem 3, Amer. Math. Soc., Providence, 1978.

[42] J.-E. ROOS, Sur l'algèbre Ext de Yoneda d'un anneau local de Golod, Comptes rendus Acad. Sc. Paris, 286, série A, 1978, p. 9-12.

[43] J.-E. ROOS, Relations between the Poincaré-Betti series of loop spaces and of local rings, Lecture Notes in Mathematics, 740, p. 285-322, Springer-Verlag, Berlin, 1979.

[44] J.-P. SERRE, Homologie singulière des espaces fibrés, Ann. Math., 54, 1951, p. 425-505.

[45] J.-P. SERRE, Algèbre locale. Multiplicités, Lecture Notes in Mathematics, 11, Seconde édition, Springer-Verlag, Berlin, 1965.

[46] J.-P. SERRE, Un exemple de série de Poincaré non rationnelle, Proc. Kon. Nederl. Akad., ser. A, 82, 1979, p. 469-471 [= Indag. Math., 41, 1979, p. 469-471] .

[47] J.B. SHEARER, A graded algebra with a non-rational Hilbert series, J. Algebra, 62, 1980, p. 228-231.

[48] G. SJÖDIN, A characterization of local complete intersections in terms of the Ext-algebra, J. Algebra, 64, 1980, p. 214-217.

[49] N.J.A. SLOANE, Binary codes, lattices and sphere packings, p. 117--164 of Combinatorial Surveys (ed. P.J. Cameron), Academic Press, London, 1977.

[50] J. STALLINGS, Coherence of 3-manifold groups, Lecture Notes in Mathematics, 567, p. 167-173, Springer-Verlag, Berlin, 1977.

[51] V.A. UFNAROVSKIJ, O rjadach Poincaré graduirovannych algebr, Matem. Zametki, 27, 1980, p. 21-32.

[52] V.A. UFNAROVSKIJ, Thesis abstract, Moscow, 1980.

Jan-Erik ROOS
Department of Mathematics
University of Stockholm
Box 6701
S-113 85 STOCKHOLM
SWEDEN

# ANALYTIC SINGULARITIES OF SOLUTIONS OF BOUNDARY VALUE PROBLEM.

I. If $u$ is a distribution on an open set $X$ in $R^n$, we first recall that the analytic wavefront set (or analytic singular spectrum) of $u$ is a closed conic set $WF_a(u) \subset T^*X \setminus 0$ which projects down to the analytic singular support of $u$. This notion was introduced around 1970 by Bros and Iagolnitzer, Sato, Hörmander.

II. If $P$ is a differential operator with analytic coefficients on $X$ and $Pu$ is a real-analytic function ($Pu \in a(X)$) then it is well-known that $WF_a(u) \subset p^{-1}(0)$, where $p$ denotes the principal symbol of $P$. This was proved by Hörmander and Sato.

III. The general question is now what subsets in $p^{-1}(0)$ can be analytic wavefrontsets of solutions to $Pu \in a(X)$ ? This has been studied quite a lot during the last ten years, but perhaps more for ordinary (mod. $C^\infty$) wavefrontsets than for analytic ones (despite the fact that analytic wavefrontsets are technically at least as simple and that no complicated hyperfunction machinery is needed). We have the following general result of N.Hanges:

<u>Theorem 1</u>. Let $\gamma : [a,b] \to p^{-1}(0)$ be a real bicharacteristic strip of $p$ and suppose that $Pu \in a(X)$. Then either the image of $\gamma$ is contained in $WF_a(u)$ or disjoint from $WF_a(u)$.

This result, which is recent but not very difficult to prove, generalizes earlier results of Hörmander [3] and Sato-Kashiwara [6] In [2] Hanges formulated a general conjecture, of which Theorem 1 is a special case.

IV. Theorem 1 applies in particular to the waveoperator $\Box = D_t^2 - D_x^2$, $t \in R$, $x$ $R^{n-1}$. We are now interested in the boundary value problem

$$\Box u \in a(R_t \times X) \quad , \quad u|_{R\times\partial X} \in a(R_t \times \partial X) \;, \tag{1}$$

where $X \subset R^{n-1}$ is the complement of an open set $\Omega$ with boundary given by an analytic hyper-surface. Locally, (1) can be reduced to

$$Pu = ( D_{x_n}^2 + R(x,D_{x'}) )u \in a(M) \quad , \quad u|_{\partial M} \in a(\partial M) \;, \tag{2}$$

where $M$ is the halfspace $x_n \geq 0$ in $R^n$. Here $R(x,D_{x'})$ is of real principal type, and $R(x',0,D_{x'})$ can be thought of as a "wave operator" on $R_t \times \partial X$.

V. As in the $C^\infty$-theory (Taylor, Melrose, Andersson, Ivrii,..) we introduce $r_0(x',\xi') = r(x',0,\xi')$ and decompose $T^*\partial M \setminus 0$ into $H \cup E \cup G$, where $H$ is given by $r_0 < 0$, $E$ is given by $r_0 > 0$, and $G$ is given by $r_0 = 0$. An analytic ray is a certain continuou curve in $\Sigma_b = H \cup G \cup p^{-1}(0)|_{\overset{\circ}{M}}$ (equipped with a very natural topology) which was defined in [9] as a slight modification of a corresponding notion of $C^\infty$ ray, which was defined by Melrose-Sjöstrand (see [9] for references). The following picture shows the projection in x-space for the wave-equation of various rays :

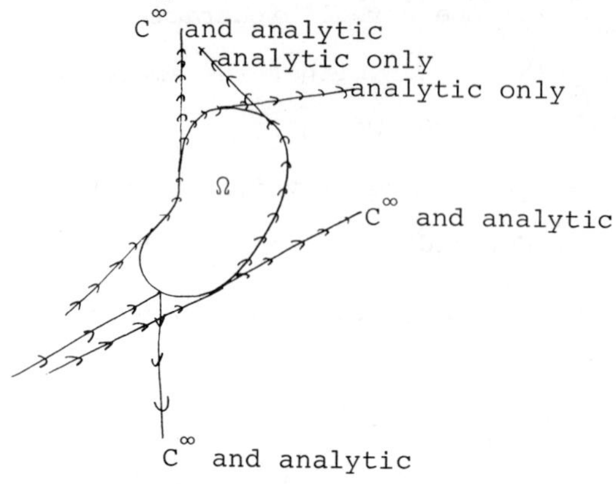

VI. If $Pu \in a(M)$, there is a natural way of defining the reduced analytic wavefrontset $WF_{ba}(u) \subset (T^*\partial M \setminus 0) \subset (T^*\overset{\circ}{M} \setminus 0)$ and general theorems of Schapira [7] show that $WF_{ba}(u) \subset \Sigma_b$ is a closed set when $u$ solves (2) .

VII. The basic result for the problem (2) is

Theorem 2. ( [9]). If $u$ is a solution of (2) then $WF_{ba}(u)$ is a union of maximally extended analytic rays.

The case of transversal reflection was treated by Schapira [7] . Theorem 2 is a special case of a general result ([8]) concerning boundary value problems satisfying a microlocal weak Lopatinsky condition.

VIII. Since analytic rays are not always uniquely extendible, many further problems arise naturally. Consider first the problem with a point source:

$$(3) \qquad \begin{cases} \Box u = 0 & , \quad u_{|R \times \partial X} = 0 \\ u|_{t=0} = 0 & , \quad \partial_t u_{|t=0} = \delta(x-x_0) \end{cases} ,$$

where $x_0 \in \overset{\circ}{X}$ .

Theorem 3. (Rauch-Sjöstrand [5]). If $\Omega$ is convex and $n-1 = 2$ , then $WF_{ba}(u)$ is the union of all analytic rays passing over $(0,x_0)$ .

The proof of this theorem uses Theorem 2 and the Holmgren uniqueness theorem as in Rauch [4], see also Friedlander - Melrose [1]. The fact that analytic singularities may penetrate into the $C^\infty$-shadow was proved implicitly in [4] for the wave-equation and in [1] for a special operator of the form (2).

IX. We now return to the problem (2). Define the diffractive region $G_+ \subset G$ by $\partial r/\partial x_n < 0$ and let $\rho_0 \in G_+$. In a neighbourhood of $\rho_0$ we then have a unique analytic ray in $G_+$ that passes through $\rho_0$ and also a $C^\infty$-ray passing through $\rho_0$ such that this is the only point over the boundary of $M$.

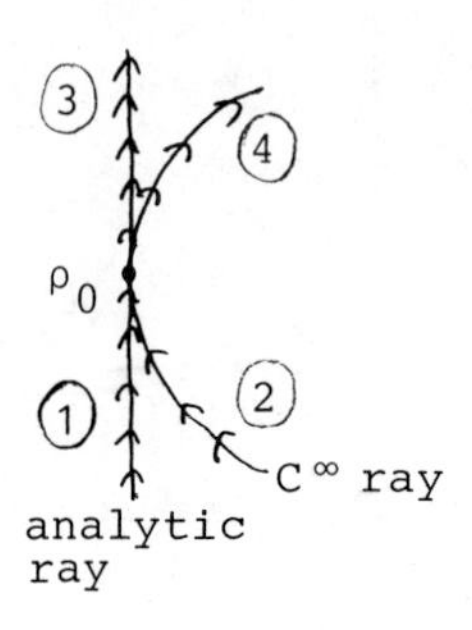


The picture shows the two rays ① $\cup \{\rho_0\} \cup$ ③ and ② $\cup \{\rho_0\} \cup$ ④, where the unions are disjoint. Let $u$ solve (2). Theorem 2 implies that (① $\cup$ ②) $\cap\ WF_{ba}(u) = \emptyset \Rightarrow \rho_0 \notin WF_{ba}(u)$. We also have

<u>Theorem 4 ( [10])</u>. (② $\cup$ ④) $\cap\ WF_{ba}(u) = \emptyset \Rightarrow \rho_0 \notin WF_{ba}(u)$.

Kataoka [12] proved independently a stronger result, namely that the conclusion still holds if we drop the boundary condition in (2).

<u>Theorem 5 ( [11])</u>. (② $\cup$ ③) $\cap\ WF_{ba}(u) = \emptyset \Rightarrow \rho_0 \notin WF_{ba}(u)$.

However,

<u>Theorem 6</u>. There exists a solution $u$ of (2) such that $WF_{ba}(u)$ is the cone generated by the $C^\infty$-ray ② $\cup \{\rho_0\} \cup$ ④.

This result was proved in a special case in [11], but the general case is not very much harder. The asymptotic solutions constructed in the proof can also be used to prove Kataoka's result, mentioned above.

The arguments of [4], [5] and the result above give quite easily:

Theorem 7 ([11]). Let $u$ solve (3), where $n \geq 3$ and $\Omega$ is strictly convex. Then we have the same conclusion as in Theorem 3.

Using the closedness of $WF_{ba}(u)$ it is actually possible to weaken the assumption about strict convexity ([11]).

Johannes Sjöstrand
Université de Paris-Sud,
F-91405 ORSAY, France.

References:

[ 1 ] Friedlander,F.G., Melrose,R.B., Math.Proc.Camb.Phil.Soc. (1977), 81, 97-120.

[ 2 ] Hanges,N, Duke Math.Journal (47) 1980, 17-25.

[ 3 ] Hörmander,L., C.P.A.M.,24 (1971), 671-704.

[ 4 ] Rauch,J., Bull.Soc.Roy.Sci.Liége, 46(1977), 156-161.

[ 5 ] Rauch,J., Sjöstrand,J., Indiana Journal of Math., to appear.

[ 6 ] Sato,M, Kawai,T., Kashiwara,M., Springer L.N.Math. No. 287.

[ 7 ] Shapira,P., Publ.RIMS, Kyoto Univ., 12 Suppl.(1977), 441-453.

[ 8 ] Sjöstrand,J., To appear.

[ 9 ] Sjöstrand,J., Comm.P.D.E., 5(1)(1980), 41-94.

[10] Sjöstrand,J., Comm.P.D.E., 5(2)(1980),187-207.

[11] Sjöstrand,J., To appear.

[12] Kataoka,K., Microlocal Theory of boundary value problems II, to appear.

# Compactification of a family of vector bundles on $\mathbb{P}^3$.

Jose Meseguer
Ignatio Sols
Stein Arild Strømme

<u>Introduction</u> Denote by $M(c_1,c_2)$ the moduli space of stable rank-2 vector bundles on $P = \mathbb{P}^3$ (projective 3-space over an algebraically closed field $k$) with Chern classes $c_1$ and $c_2$. Maruyama has shown in [4] that there exists a projective scheme $M(c_1,c_2,0)$ acting as a coarse moduli space for equivalence classes of semistable rank-2 torsionfree sheaves on $P$ with the given first two Chern classes and third Chern class equal to zero, in which $M(c_1,c_2)$ lies as an open subspace. The equivalence relation mentioned reduces to isomorphism for stable sheaves (see [4] for the details). The purpose of this paper is to study the closure $\overline{M}(c_1,c_2)$ of $M(c_1,c_2)$ in $M(c_1,c_2,0)$, in particular, to provide an answer to the natural question: Which torsionfree sheaves are limits of vector bundles?

Posed in this generality, the question is, of course, too hard. As far as we know, the only known case is $c_1 = 0$, $c_2 = 1$, where Wever [5] showed that $M(0,1,0) \cong \mathbb{P}^5$, and that $M(0,1)$ is the complement of the Grassmannian $G(1,3) \subseteq \mathbb{P}^5$ via the Plücker embedding. He also gives an explicit description of the sheaves on the boundary.

In this paper we answer the question for $c_1 = -1$, $c_2 = 2$. Hartshorne and Sols have shown [3,thm.3.1] that $M(-1,2)$ is an irreducible, nonsingular and rational variety of dimension 11. Applying essentially the method from that paper, we are able to strengthen their result as follows:

Let $G$ be the Grassmannian of lines in $P$. Then there exists a certain rank-6 vector bundle $A$ on $G$, a rank-3 vector bundle $B$ on $H := \mathbb{P}_G(A)$, and a family $\mathcal{E}$ of stable, torsionfree sheaves on $P$ parametrized by $M := \mathbb{P}_H(B)$, such that the induced morphism

$$i : M \to M(-1,2,0)$$

is a closed embedding onto $\overline{M}(-1,2)$ (4.8). In particular this is also a nonsingular variety (4.9).

The complement of $M(-1,2)$ in its closure is the union of two irreducible divisors (4.10). We compute their classes in $\operatorname{Pic} M \cong \mathbb{Z}^{\oplus 3}$ and deduce that $\operatorname{Pic} M(-1,2) \cong \mathbb{Z} \oplus \mathbb{Z}/2\mathbb{Z}$ (4.11).

Any sheaf $E$ corresponding to a closed point of $M$ can be obtained as an extension of $I_Z$ by $\mathcal{O}_P(-1)$, where $Z \subseteq P$ is a Z-curve (definition 2.1). We give an explicit description of which extension can occur (4.6). The Chern classes of $E^{\vee\vee}$ are computed in (5.4).

Finally, we give some examples of stable, torsionfree sheaves that are not limits of vector bundles (5.5).

We would like to express our thanks to Robin Hartshorne for suggesting the problem, and for many inspiring conversations.

## § 1. Preliminaries

We work over an algebraically closed field $k$ of any characteristic. Our notion of stability is due to Mumford and Takemoto:

A torsionfree rank-2 sheaf $E$ on $\mathbb{P}^n$ is stable (resp. semistable) if for any torsionfree rank-1 quotient $R$ of $E$, the following inequality holds:

$$2c_1(R) > c_1(E) \quad (\text{resp.} \geq).$$

Note that if $c_1(E)$ is odd, stability and semistability are equivalent.

(This definition, in general, not equivalent to the one used by Maruyama in [4]. For $c_1$ odd, however, they are equivalent).

For a coherent sheaf $E$ on a projective $k$-scheme $X$, we will often write $H^i(E)$ for $H^i(X,E)$ and $h^i(E)$ for $\dim_k H^i(E)$. The Euler characteristic of $E$ is written $\chi(E) = \sum_{i \geq 0} (-1)^i h^i(E)$.

We shall need the following result of Hartshorne-Sols:

(1.1) <u>Proposition</u> [3,prop.2.1]. Let $E$ be a stable rank-2 vector bundle on $P$ with $c_1(E) = -1$, $c_2(E) = 2$. Then there is an exact sequence

$$0 \to \mathcal{O}_P(-1) \to E \to I_Z \to 0$$

where $I_Z$ is the ideal sheaf of a locally complete intersection curve $Z$ of degree 2 with $\omega_Z \cong \mathcal{O}_Z(-3)$. In particular, $h^0(E(1)) = 1$.

## § 2. Classification of Z-curves.

(2.1) <u>Definition</u>: A <u>generic</u> Z-<u>curve</u> is a locally complete intersection curve in $P$ of degree 2 with $\omega_Z \cong \mathcal{O}_Z(-3)$. A Z-<u>curve</u> is any curve that can be obtained as a flat specialization of generic Z-curves. A Z-curve that is not generic is said to be <u>special</u>.

By (1.1), any stable bundle with $c_1 = -1$ and $c_2 = 2$ corresponds to a unique generic Z-curve. To compactify $M(-1,2)$ we are thus forced to consider special Z-curves also, which is the motivation for this §.

(2.2) Let Z be a generic Z-curve. It is easily seen that Z must be some multiplicity 2 structure on a line $L = Z_{red}$ in P. If $J = I_L/I_Z$ denotes the ideal sheaf of L in Z, then $J^2 = 0$, so J carries the structure of an $\mathcal{O}_L$-module, which must be invertible since Z is Cohen-Macaulay. Consideration of the dualizing sheaf shows that $J \cong \mathcal{O}_L(1)$. On the other hand, a theorem of Ferrand [1,1.5] implies conversely that given any surjection $u : I_L \to \mathcal{O}_L(1)$, the kernel of u is the ideal of a generic Z-curve The map u factors through the conormal bundle $I_L/I_L^2 \cong 2\mathcal{O}_L(-1)$, so we get the following basic commutative diagram:

$$
\begin{array}{ccccccccc}
 & & 0 & & 0 & & & & \\
 & & \downarrow & & \downarrow & & & & \\
 & & I_L^2 & = & I_L^2 & & & & \\
 & & \downarrow & & \downarrow & & & & \\
0 & \to & I_Z & \longrightarrow & I_L & \xrightarrow{u} & J & \to & 0 \\
 & & \downarrow & & \downarrow & & \| & & \\
0 & \to & \mathcal{O}_L(-3) & \xrightarrow{v} & 2\mathcal{O}_L(-1) & \to & J & \to & 0 \\
 & & \downarrow & & \downarrow & & & & \\
 & & 0 & & 0 & & & &
\end{array} \qquad (2.3)
$$

(2.4) <u>Proposition</u>: The diagram (2.3) sets up a 1-1 correspondence between the set of injective maps $v : \mathcal{O}_L(-3) \to 2\mathcal{O}_L(-1)$, modulo scalars, and the set of Z-curves supported on L. Under this correspondence the generic Z-curves correspond to maps v without zeros (i.e. the two quadratic forms defining v have no common factor).

The proof is straightforward and will be omitted.

From (2.3) it is clear that $J \cong \mathcal{O}_L(1-x) \oplus \delta_x$ for $x = 0,1$ or $2$, where $\delta_x$ is a (monic) $\mathcal{O}_L$-module of length $x$. The corresponding Z-curve will be said to be of the x-th kind. It consists of a double line $Y$ with $\omega_Y = \mathcal{O}_Y(x-3)$ plus $x$ embedded points (possibly infinitely near, if $x = 2$). Note that $I_L^2 \subseteq I_Z$.

(2.5) Proposition: Let $Z$ be a Z-curve supported on $L$, and let $L' \subset P$ be any line. Then

$$h^1(I_Z \otimes \mathcal{O}_{L'}(1)) = \begin{cases} 1 & \text{if } L' = L \\ 0 & \text{if } L' \neq L \end{cases}$$

Proof: Use the left column of (2.3) to get an exact sequence

$$I_L^2 \otimes \mathcal{O}_{L'}(1) \to I_Z \otimes \mathcal{O}_{L'}(1) \to \mathcal{O}_L \otimes \mathcal{O}_{L'}(-2) \to 0$$

If $L' = L$, the first term is $3\mathcal{O}_L(-1)$ and the last $\mathcal{O}_L(-2)$, showing our assertion in this case. If $L' \neq L$, denote by $L_2$ the scheme defined by $I_L^2$. Then $\mathcal{O}_{L_2} \otimes \mathcal{O}_{L'}$ is of finite length $y \leq 2$. Hence we get an exact sequence

$$0 \to \underline{\mathrm{Tor}}_1^P(\mathcal{O}_{L'}(1), \mathcal{O}_{L_2}) \to I_L^2 \otimes \mathcal{O}_{L'}(1) \to \mathcal{O}_{L'}(1) \to \mathcal{O}_{L'} \otimes \mathcal{O}_{L_2} \to 0$$

showing that $I_L^2 \otimes \mathcal{O}_{L'}(1) \cong \mathcal{O}_{L'}(1-y) \oplus$ torsion. In particular, it has no $H^1$. Q.E.D.

(2.6) Proposition: If $Z$ is generic, then $h^0(N_{Z/P}) = 9$.

Proof: It suffices to show that $h^1(N_{Z/P}) = 1$, since $\chi(N_{Z/P}) = 4\deg Z = 8$. Note that $N_{Z/P} = I_Z \otimes \mathcal{O}_Z(1)$ since $\omega_Z = \mathcal{O}_Z(-3)$. From the exact sequence

$$0 \to \mathcal{O}_L(1) \to \mathcal{O}_Z \to \mathcal{O}_L \to 0$$

we obtain

$$I_Z \otimes \mathcal{O}_L(2) \to I_Z \otimes \mathcal{O}_Z(1) \to I_Z \otimes \mathcal{O}_L(1) \to 0.$$

Now the conclusion follows from (2.5). Q.E.D.

Remark: A different proof can be based on the fact that $Z$ is a divisor on a smooth cubic surface.

(2.7) It is clear that $Z$-curves correspond to an irreducible component of the Hilbert scheme of $P$. The diagram (2.3) indicates how to give a complete description of this component, to be called $H$ below. $H$ will be a bundle over the Grassmannian $G$ of lines in $P$ with fiber $\mathbb{P}^5 = \mathbb{P}(\mathrm{Hom}(\mathcal{O}_L(-3), 2\mathcal{O}_L(-1))^\vee)$, by (2.4). We shall need the following preliminaries on $G$:

Let $0 \to K \to 4\mathcal{O}_G \to Q \to 0$ be the universal exact sequence on $G$, and let $\tilde{L} \subseteq P \times G$ be the universal line. Then $\tilde{L} = \mathbb{P}_G(Q)$; let $\pi : \tilde{L} \to G$ be the natural map, and denote by $\mathcal{O}_{\tilde{L}}(\tau)$ the associated $\pi$-tautological linebundle. Note that $\mathcal{O}_{\tilde{L}}(\tau) = pr_P^* \mathcal{O}_P(1)|\tilde{L}$, where $pr_P : P \times G \to P$ is the first projection, and $\pi_* \mathcal{O}_{\tilde{L}}(\tau) = Q$. Denote by $\sigma \in \mathrm{Pic}(G)$ the first Chern class of $Q$. Then $\sigma$ is the positive generator of $\mathrm{Pic}(G) \cong \mathbb{Z}$. As a matter of notation, if $f : X \to G$ is any morphism, $\mathcal{O}_X(\sigma)$ will denote $f^* \mathcal{O}_G(\sigma) = \overset{2}{\wedge} Q_X$, and similarly for other divisor classes.

Now $\tilde{L}$ has the following Koszul resolution in $P \times G$

$$0 \to \mathcal{O}_{P\times G}(-2\tau-\sigma) \to \mathcal{O}_P(-\tau) \boxtimes K \to \mathcal{O}_{P\times G} \to \mathcal{O}_{\tilde{L}} \to 0$$

where $\tau$ denotes the hyperplane class in $P$. In particular, the conormal bundle of $\tilde{L}$ in $P \times G$ is $I_{\tilde{L}}/I_{\tilde{L}}^2 = K_{\tilde{L}}(-\tau)$. We are now in the position to construct a "universal" version of the diagram (2.3). More precisely, the set of maps $v$ is parametrized by the $\mathcal{O}_G$-module

$$\pi_* \underline{\mathrm{Hom}}(\mathcal{O}_{\tilde{L}}(-3\tau), K_{\tilde{L}}(-\tau)) = \pi_* K_{\tilde{L}}(2\tau) = K \otimes S^2 Q.$$

Define $H = \mathbb{P}_G((K \otimes S^2 Q)^\vee) \xrightarrow{\lambda} G$ and denote the associated $\lambda$-tautological linebundle by $\mathcal{O}_H(\rho)$. On $H$ there is a canonical section of $K_H \otimes S^2 Q_H(\rho)$, giving a map of sheaves on $\tilde{L}_H = L \times_G H$

$$\tilde{v} : \mathcal{O}_{\tilde{L}_H}(-3\tau-\rho) \to K_{\tilde{L}_H}(-\tau).$$

Let $\tilde{J} = \mathrm{coker}(\tilde{v})$ and construct the following diagram of sheaves on $P \times H$, defining $I_{\tilde{Z}}$:

$$
\begin{array}{ccccccccc}
 & & 0 & & 0 & & & & \\
 & & \downarrow & & \downarrow & & & & \\
 & & I_{\tilde{L}_H}^2 & = & I_{\tilde{L}_H}^2 & & & & \\
 & & \downarrow & & \downarrow & & & & \\
0 & \longrightarrow & I_{\tilde{Z}} & \longrightarrow & I_{\tilde{L}_H} & \longrightarrow & \tilde{J} & \to & 0 \\
 & & \downarrow & & \downarrow & & \| & & \\
0 & \to & \mathcal{O}_{\tilde{L}_H}(-3\tau-\rho) & \xrightarrow{\tilde{v}} & K_{\tilde{L}_H}(-\tau) & \to & \tilde{J} & \to & 0 \\
 & & \downarrow & & \downarrow & & & & \\
 & & 0 & & 0 & & & &
\end{array}
\tag{2.8}
$$

It is straightforward that the formation of the diagram (2.8) commutes with base change on $H$, inducing diagrams of the type (2.3) on the closed fibers. Let $\tilde{A} \subseteq \tilde{L}_H$ be the zero-scheme of $\tilde{v}$ and $A \subseteq H$ the image of $\tilde{A}$ under the natural map $\tilde{L}_H \to H$, also denoted by $\pi$. We want to prove the following

(2.9) <u>Proposition</u>: The map $\mathbf{i}$ from $H$ to the Hilbert scheme of $P$ induced by $\tilde{Z}$ is a closed embedding, by which we may identify $H$

with the component of the Hilbert scheme parametrizing Z-curves. Over the open subset $H^o = H-A$ corresponding to generic Z-curves the map is an isomorphism.

<u>Proof</u>: Let Hilb be the Hilbert scheme and $\text{Hilb}^o$ the open subscheme corresponding to generic Z-curves. Let $F$ be the closure of $\text{Hilb}^o$ in Hilb. The morphism $i: H \to F$ is clearly bijective on closed points. It follows that $\dim F \geq 9$. By (2.6), it follows that $\text{Hilb}^o$ is nonsingular. In particular, $F$ is reduced. We proceed to construct an inverse $j: F \to H$ to $i$. Let $\mathcal{Z} \subseteq P \times F$ be the universal Z-curve. Consider the diagram

$$\begin{array}{ccc} \tilde{L} \times F & \xrightarrow{\varphi} & P \times F \\ \downarrow{\scriptstyle \pi_F} & & \\ G \times F & & \end{array}$$

By (2.5), the sheaf $R^1\pi_{F*}\varphi^* I_{\mathcal{Z}}(\tau)$ is a locally free $\mathcal{O}_F$-module of rank one, hence gives a morphism $k: F \to G$ with the property that a closed point of $F$ corresponding to a Z-curve maps to the line supporting it. Let $L_F \subseteq P \times F$ be the induced family of lines, and let $J_F$ be the ideal of $L_F$ in $\mathcal{O}_{\mathcal{Z}}$. Now one can construct a diagram of type (2.8) for the family $\mathcal{Z}$. Since (2.8) is universal, we obtain a map $j: F \to H$ which gives the inverse of $i$. Q.E.D.

(2.10) Let us conclude this § by computing the Picard group of $H^o$. The intersection ring $A^\bullet(H)$ is of course determined by $A^\bullet(G)$ and the Chern classes of $K \otimes S^2Q$, which are well known. In particular, $\text{Pic}(H) = \mathbb{Z}\sigma \oplus \mathbb{Z}\rho$. We need to compute the class of the complement $A = H - H^o$. Since $\tilde{A}$ is the zero-scheme of $\tilde{v}$

from (2.6), its class is given by $[\widetilde{A}] = c_2 K_{\widetilde{L}_H}(2\tau + \rho) =$ $c_2(K) - \sigma(2\tau + \rho) + (2\tau + \rho)^2$, since $c_1(K) = -\sigma$.

(2.11) Lemma. $A$ is reduced and irreducible, and $\widetilde{A} \to A$ is birational.

Proof: It is sufficient to consider a fiber of $H \to G$. In such a fiber, $A$ is given by the resultant of two generic quadratic forms, which is clearly a reduced and irreducible divisor (of degree 4) in $\mathbb{P}^5$ = fiber. If a pair of quadratic forms corresponds to a general point in this divisor, they have only one common zero, which shows that $\widetilde{A} \to A$ has degree 1. Q.E.D.

(2.12) Proposition: The class of $A$ in $A^\bullet(H)$ is $[A] = 2\sigma + 4\rho$.

Proof: We know that $\pi_*(\tau) = 1$ and $\pi_*(\tau^2) = \sigma$ from the relation $\tau^2 - \tau\cdot\sigma + c_2(Q)$ in $A^\bullet(\widetilde{L})$. Using the projection formula, we get

$$[A] = \pi_*(c_2 K - \sigma(2\tau+\rho) + (2\tau+\rho)^2)$$

$$= 0 - 2\sigma + 0 + 4\sigma + 4\rho + 0 = 2\sigma + 4\rho. \qquad \text{Q.E.D.}$$

(2.13) Corollary $\mathrm{Pic}(H^o) = (\mathbb{Z}\sigma \oplus \mathbb{Z}\rho)/(2\sigma + 4\rho) \cong \mathbb{Z} \oplus \mathbb{Z}/2\mathbb{Z}$.

Proof: This follows from the irreducibility of $A$ and the resulting exact sequence

$$\begin{array}{ccccccc} \mathbb{Z} & \to & \mathrm{Pic}(H) & \to & \mathrm{Pic}(H^o) & \to & 0 \\ \cup & & \cup & & & & \\ 1 & \to & [A] & & & & \end{array}$$

Q.E.D.

## § 3. Construction of a universal family.

Let us return to vector bundles again. In § 1 we saw that any bundle $E \in M(-1,2)$ can be obtained as an extension

(3.1) $$0 \to \mathcal{O}_P(-1) \to E \to I_Z \to 0$$

for some generic Z-curve Z. A converse is easily proved:

(3.2) Proposition: Let Z be any curve, and $\xi$ a non-zero element of $Ext^1_P(I_Z, \mathcal{O}_P(-1))$ giving rise to a sequence (3.1). Then E is a stable torsion-free sheaf with $c_1(E) = -1$.

Proof: The only non-trivial statement is that E is stable. Assume the contrary, and take some rank-1 torsion-free quotient R of E with $c_1(R) = x \leq -1$. Then $R^{\vee\vee} = \mathcal{O}_P(x)$. Consider the composed map $\mathcal{O}_P(-1) \to E \to R \to R^{\vee\vee} = \mathcal{O}_P(x)$. Since $x \leq -1$, this is either zero or an isomorphism. In the first case there is induced a non-zero map $I_Z \to \mathcal{O}_P(x)$, in the second case the isomorphism gives a splitting of (3.1). Both are impossible. Q.E.D.

(3.3) If, in (3.2), Z is a Z-curve, the Chern classes of E are $c_1(E) = -1$, $c_2(E) = 2$ and $c_3(E) = 0$. Even if Z is generic, however, E is not necessarily locally free. In fact, necessary and sufficient conditions for E to be locally free are that Z be generic, and that under the isomorphism $\underline{Ext}^1_P(I_Z, \mathcal{O}(-1)) \to \underline{Ext}^2(\mathcal{O}_Z, \mathcal{O}(-1)) \cong \omega_Z(3) \cong \mathcal{O}_Z$, the section $\xi \in H^0(\mathcal{O}_Z)$ generate $\mathcal{O}_Z$, or equivalently, that the image of $\xi$ in $H^0(\mathcal{O}_{Z_{red}}) = k$ be non-zero. Let us summarize all this in the following.

(3.4) <u>Proposition:</u> Let $Z$ be a generic $Z$-curve. Then there is a natural 1-1 correspondence between the set of isomorphism classes of bundles belonging to $Z$ and the set of generating sections of $\underline{Ext}^2_{\mathbb{P}^3}(I_Z, \mathcal{O}_{\mathbb{P}^3}(-1))$ modulo scalars.

<u>Proof:</u> There remains to show that if $E_i$ corresponds to $\xi_i$ for $i = 1,2$, then $E_1 \cong E_2$ if and only if $\xi_1 = \alpha\xi_2$ for some scalar $\alpha \in k^*$. If $\xi_1 = \alpha\xi_2$ it is clear that $E_1 \cong E_2$. Conversely, if $E_1 \cong E_2$, this isomorphism must leave the subsheaf $\mathcal{O}_P(-1)$ invariant, since $h^0(E_i(1)) = 1$ by (1.1). Since $Aut(\mathcal{O}_P(-1)) = k^*$, we are done. Q.E.D.

(3.5) It is now clear how to construct a universal family of torsionfree sheaves. For convenience we introduce the following notation: If $X$ is any $k$-scheme and $F$ is a coherent sheaf on $P \times X$, let $E_X^i(F) = \underline{Ext}^i_{pr_X}(F, \mathcal{O}_{P\times X}(-\tau))$ denote the sheaf on $X$ associated to the presheaf $U \to Ext^i_{P\times U}(F_U, \mathcal{O}_{P\times U}(-\tau))$. If $F$ is flat over $X$ and $Y \to X$ is any base change, there are natural base-change maps $E_X^i(F)_Y \to E_Y^i(F_Y)$, which are isomorphisms if $Y \to X$ is flat.

Recall the varieties $H^0 \subseteq H$ from § 2, and the ideal $I_{\tilde{Z}} \subseteq \mathcal{O}_{\mathbb{P}^3\times H}$ defined by the diagram (2.8).

(3.6) <u>Proposition:</u> The $\mathcal{O}_H$-module $E_H^1(I_{\tilde{Z}})$ is locally free of rank three. Furthermore, its formation commutes with base change on $H^0$. In particular, its fiber over a closed point in $H^0$ corresponding to the generic $Z$-curve $Z$ is $Ext^1_{\mathbb{P}^3}(I_Z, \mathcal{O}_{\mathbb{P}^3}(-1))$.

<u>Proof:</u> The assertions concerning $H^0$ all follow from standard base-change theorems via the fact that $E_H^i(I_{\tilde{Z}})$ is supported

in $H-H^o$ for $i \geq 2$. There remains to show that $E^1_H(I_{\tilde{Z}})$ is locally free.

Let $\underline{E}^i(F) = \underline{\mathrm{Ext}}^i_{P\times H}(F, {}_{P\times H}(-\tau))$ for any coherent sheaf $F$ on $P\times H$. There is a standard spectral sequence $R^i pr_{H*}\underline{E}^j(F) \Rightarrow E^{\bullet}_H(F)$. Recall the diagram (2.6). The middle row gives an exact sequence

$$(3.7) \qquad 0 \to E^1_H(I_{\tilde{L}_H}) \to E^1_H(I_{\tilde{Z}}) \to E^2_H(\tilde{J}) \to 0.$$

The first term is clearly locally free of rank 2, allowing us to concentrate on $E^2_H(\tilde{J}) = pr_{H*}\underline{E}^2(\tilde{J})$. The bottom row of (2.8) gives an exact sequence

$$(3.8) \quad 0 \to \underline{E}^2(\tilde{J}) \to \underline{E}^2(K_{\tilde{L}_H}(-\tau)) \xrightarrow{\tilde{v}_1} \underline{E}^2(\mathcal{O}_{\tilde{L}_H}(-3\tau-\rho))$$

where we recognize $\tilde{v}_1$ as the transpose of $\tilde{v}$, suitably twisted. In particular, the cokernel of $\tilde{v}_1$ is a twist of the structure sheaf of $\tilde{A} \subseteq \tilde{L}_H$, and $\underline{E}^2(\tilde{J})$ can be identified as a linebundle on $\tilde{L}_H$, inducing the trivial linebundle on the fibers of $\pi_H\colon \tilde{L}_H \to H$. Hence $E^2_H(\tilde{J}) = \pi_{H*}(\underline{E}^2(\tilde{J}))$ is a linebundle on $H$. Q.E.D.

(3.9) <u>Remark</u> Explicit computations show that

$$\underline{E}^1(I_{\tilde{L}_H}) = \mathcal{O}_{\tilde{L}}(\tau+\sigma) \text{ and } E^1_H(I_{\tilde{L}_H}) = Q_H(\sigma)$$

$$\underline{E}^2(\mathcal{O}_{\tilde{L}_H}(-3\tau-\rho)) = \mathcal{O}_{\tilde{L}_H}(4\tau+\rho+\sigma)$$

$$\underline{E}^2(K_{\tilde{L}_H}(-\tau)) = K^{\vee}_{\tilde{L}_H}(2\tau+\sigma)$$

$$\underline{E}^2(\tilde{J}) = \mathcal{O}_{\tilde{L}_H}(2\sigma-\rho) \text{ and } E^2_H(\tilde{J}) = \mathcal{O}_H(2\sigma-\rho).$$

So (3.7) can be rewritten as

$$(3.10) \quad 0 \to Q_H(\sigma) \to E^1_H(I_{\tilde{Z}}) \to \mathcal{O}_H(2\sigma-\rho) \to 0$$

(3.11) Define $M = \mathbb{P}_H(E^1_H(I_{\tilde{Z}})^\vee) \xrightarrow{\mu} H$ and denote the $\mu$-tautological linebundle by $\mathcal{O}_M(\omega)$. On $M$, we get a canonical section of $E^1_H(I_{\tilde{Z}})_M(\omega) \tilde{\to} E^1_M(I_{\tilde{Z}_M}(-\omega))$. This gives a short exact sequence

$$(3.12) \qquad 0 \to \mathcal{O}_{P\times M}(-\tau) \to \mathcal{E} \to I_{\tilde{Z}_M}(-\omega) \to 0$$

For any closed point $m$ of $M$, let $\mathcal{E}_m$ denote the restriction of $\mathcal{E}$ to $P \cong P\times\{m\} \subseteq P\times M$. In the next §, we prove that $\mathcal{E}$ is a universal family of stable torsion-free sheaves, showing that $M$ can be identified with the closure of $M(-1,2)$ in $M(-1,2,0)$.

## § 4. Properties of the family $\mathcal{E}$.

(4.1) Let $h \in H$ be a closed point corresponding to a $Z$-curve $Z \subseteq P$. If $m \in M$ is a closed point lying over $h$, restriction of (3.12) to $P\times\{m\} \cong P$ gives an exact sequence

$$(4.2) \qquad 0 \to \mathcal{O}_P(-1) \to \mathcal{E}_m \to I_Z \to 0$$

and hence an element $\xi_m \in \mathrm{Ext}^1_P(I_Z, \mathcal{O}_P(-1))$. By the very fact that his comes from restriction of (3.12), $\xi_m$ must lie in the image of the base-change map

$$\beta_m : E^1_H(I_{\tilde{Z}}) \otimes k(h) \to \mathrm{Ext}^1_P(I_Z, \mathcal{O}_P(-1)).$$

If $Z$ is a generic $Z$-curve, $\beta_m$ is an isomorphism, but if $Z$ is special, we need to study $\beta_m$ in more detail.

Assume that $Z$ is special of the x'th kind, $x = 1$ or $2$ ($x = 0$ corresponds to generic $Z$). Consider the diagram (2.3) for $Z$. The dual of the middle row gives

$$0 \to \mathrm{Ext}^1_P(I_L, \mathcal{O}(-1)) \to \mathrm{Ext}^1_P(I_Z, \mathcal{O}(-1)) \xrightarrow{\alpha} \mathrm{Ext}^2_P(J, \mathcal{O}(-1)) \to 0$$

The dual of the bottom row gives

$$\begin{array}{ccccccc} 0 \to & \underline{\mathrm{Ext}}^2(J, \mathcal{O}(-1)) \to & \underline{\mathrm{Ext}}^2(2\mathcal{O}_L(-1), \mathcal{O}(-1)) & \to & \underline{\mathrm{Ext}}^2(\mathcal{O}_L(-3), \mathcal{O}(-1)) \\ & & \wr\wr & & \wr\wr \\ & & 2\mathcal{O}_L(2) & \xrightarrow{v^t} & \mathcal{O}_L(4) \end{array}$$

Let the map $v$ of (2.2) be given by two quadratic forms $(f_o, g_o)$. If $f_o = ab$, $g_o = ac$ where $b$ and $c$ have no common factor, then $a$ has degree $x$ since $Z$ is special of the $x$'th kind. This shows that the kernel of $v^t$ is $\underline{\mathrm{Ext}}^2(J, \mathcal{O}(-1)) = \mathcal{O}_L(x)$, and hence $\mathrm{Ext}^2(J, \mathcal{O}(-1)) = H^o(\mathcal{O}_L(x))$. Let $V \subseteq H^o(\mathcal{O}_L(x)$ be the one-dimensional subspace generated by $a$.

(4.3) Proposition. The base-change map $\beta_m$ is injective, and its image is the three-dimensional subspace $\alpha^{-1}V \subseteq \mathrm{Ext}^1_P(I_Z, \mathcal{O}_P(-1))$.

Proof: The question is local on $H$, so we may pass to an open subset $U \subseteq H$ trivializing $\mathcal{O}_H(\rho)$ and $K_H$. The restriction of (3.8) to $U$ can be written

$$\begin{array}{ccccc} 0 \to & \underline{E}^2(\tilde{J})_U \to & 2\mathcal{O}_{\tilde{L}_U}(2\tau) & \xrightarrow{\tilde{v}_1} & \mathcal{O}_{\tilde{L}_U}(4\tau) \\ & \wr\wr & & & \\ & \mathcal{O}_{\tilde{L}_U} & & & \end{array}$$

where $\tilde{v}_1$ is given by two quadratic forms $f$ and $g$ with no common factors, and reducing to $f_o$ and $g_o$ on the closed fiber over $h \in U$. We get the following base-change diagram (note that the middle row is not exact)

$$\begin{array}{ccccccc}
0 \to \mathcal{O}_{\tilde{L}_U} & \xrightarrow{\binom{g}{-f}} & 2\mathcal{O}_{\tilde{L}_U}(2\tau) & \xrightarrow{(f,g)} & \mathcal{O}_{\tilde{L}_U}(4\tau) \\
\downarrow & & \downarrow & & \downarrow \\
0 \to \mathcal{O}_{\tilde{L}} & \xrightarrow{\binom{g_o}{-f_o}} & 2\mathcal{O}_L(2) & \xrightarrow{(f_o,g_o)} & \mathcal{O}_L(4) \\
a\downarrow & & \| & & \| \\
0 \to \mathcal{O}_L(x) & \xrightarrow{\binom{c}{-b}} & 2\mathcal{O}_L(2) & \xrightarrow{a(b,c)} & \mathcal{O}_L(4)
\end{array}$$

showing that the base-change map $\underline{E}^2(\tilde{J}) \otimes k(h) \to \mathrm{Ext}^2_{\mathbb{P}^3}(J, \mathcal{O}_{\mathbb{P}^3}(-1))$ is nothing but the inclusion $\mathcal{O}_L \xrightarrow{a} \mathcal{O}_L(x)$. From this (4.3) follows easily. Q.E.D.

(4.5) <u>Corollary</u>: All the fibers $\mathcal{E}_m$ of $\mathcal{E}$ are stable and mutually non-isomorphic.

<u>Proof</u>: This follows from the injectivity of $\beta_m$ and (3.2).

(4.6) <u>Corollary</u>: For any Z-curve Z, and any extension $0 \neq \xi \in \mathrm{Ext}^1_P(I_Z, \mathcal{O}_P(-1))$, the induced stable torsion-free sheaf is a limit of vector bundles iff $\alpha(\xi) \in V$.

<u>Proof</u>: This is precisely the second half of (4.3). Q.E.D.

(4.7) By (4.5) and the universal mapping property of $M(-1,2,0)$, there is induced a map $i : M \to M(-1,2,0)$. It is clear that the image $N = i(M)$ is precisely the closure $\overline{M}(-1,2)$ of $M(-1,2)$.

(4.8) <u>Proposition</u>: The map $i : M \to N$ is an isomorphism.

<u>Proof</u>: We proceed to construct an inverse. By (4.5) $i$ is bijective on k-rational points. Let $\mathcal{F}$ be a universal family on $N \times P$, which exists by [4,6.11]. By (4.2) and (2.4),

$h^o(\mathcal{F}_n(1)) = 1$ for any closed point $n \in N$. Since $N$ is reduced, by [3, thm.3.1], it follows that $\mathcal{L} = pr_{N*}\mathcal{F}(\tau)$ is a linebundle on $N$. The induced section of $\mathcal{F}(\tau) \otimes pr_N^* \mathcal{L}^{-1}$ must vanish in a subscheme $\mathcal{Z} \subseteq \mathbb{P}^3 \times N$, easily checked to be a flat family of Z-curves. Hence there is induced a map $k : N \to H$ such that $\mathcal{Z} = k^* \tilde{Z}$. There is an exact sequence

$$0 \to pr_N^* \mathcal{L}(-\tau) \to \mathcal{F} \to I_{\mathcal{Z}} \otimes pr_N^* \mathcal{M} \to 0$$

for some linebundle $\mathcal{M}$ on $N$. This gives a nowhere vanishing section of $E_N^1(I_{\mathcal{Z}}) \otimes \mathcal{L} \otimes \mathcal{M}^{-1}$, and hence a surjection

$$E_N^1(I_{\mathcal{Z}})^\vee \cong E_H^1(I_{\tilde{Z}})_N^\vee \to \mathcal{L} \otimes \mathcal{M}^{-1}.$$

By the defining property of $M = \mathbb{P}_H(E_H^1(I_{\tilde{Z}})^\vee)$, this gives a map $j : N \to M$ of schemes over $H$. Clearly, $j$ is an inverse of $i$.

Q.E.D.

Our main theorem is an immediate corollary:

(4.9) <u>Theorem</u>: The closure $\overline{M}(-1,2)$ of $M(-1,2)$ in $M(-1,2,0)$ is a nonsingular, rational and irreducible projective variety of dimension 11.

<u>Remark</u>: It is not true that $M(-1,2,0)$ itself is nonsingular at all points of $\overline{M}(-1,2)$. In fact, it is singular at any point lying over a special Z-curve. (See 5.5(i) below.)

(4.10) <u>Proposition</u>. Let $M^o = i^{-1}M(-1,2) \subseteq M$. Then the complement of $M^o$ in $M$ is the union of the following two irreducible divisors:

$$C_1 = \mu^{-1}(\tilde{A})$$

$C_2 = \mathbb{P}_H(Q_H(\sigma))$ included in $M$ via (3.10).

Proof: It is clear that $C_1$ is contained in $M - M^o$. For $C_2$, use (3.3) and the fact that, over a closed point in $H^o$ corresponding to a generic Z-curve Z, (3.10) reduces to

$$0 \to H^o(\mathcal{O}_L(1)) \to H^o(\mathcal{O}_Z) \to H^o(\mathcal{O}_L) \to 0$$

where $L = Z_{red}$ is the reduced line. Q.E.D.

(4.11) Corollary: The map $\mu^* : \mathbb{Z} \oplus \mathbb{Z}/2\mathbb{Z} = \operatorname{Pic} H^o \to \operatorname{Pic} M^o$ is an isomorphism.

Proof: $C_2$ induces hyperplanes on the fibers of $M \to H$, so $M^o \to H^o$ is an $\mathbb{A}^2$-bundle. Then use (2.13). Q.E.D.

## § 5. The sheaves on the boundary.

(5.1) Let Z be a Z-curve of the x'th kind, $x = 0,1$ or $2$, where $x = 0$ corresponds to a generic Z-curve. Let L be the reduced line, and recall J from diagram (2.2). Note that $J \cong \mathcal{O}_L(1-x) \oplus \delta_x$, where $\delta_x$ is an $\mathcal{O}_L$-module of length x. We define a curve $Y \subseteq Z$ by the following diagram

$$\begin{array}{ccccccccc}
 & & 0 & & 0 & & & & \\
 & & \downarrow & & \downarrow & & & & \\
 & & \delta_x & = & \delta_x & & & & \\
 & & \downarrow & & \downarrow & & & & \\
0 & \to & J & \longrightarrow & \mathcal{O}_Z & \to & \mathcal{O}_L & \to & 0 \\
 & & \downarrow & & \downarrow & & \| & & \\
0 & \to & \mathcal{O}_L(1-x) & \longrightarrow & \mathcal{O}_Y & \to & \mathcal{O}_L & \to & 0 \\
 & & \downarrow & & \downarrow & & & & \\
 & & 0 & & 0 & & & &
\end{array}$$

By Ferrand's theorem [1,thm.1.5] $Y$ is a double line with $\omega_Y \cong \mathcal{O}_Y(x-3)$. Furthermore, the exact sequence

$$0 \to I_Z \to I_Y \to \delta_x \to 0$$

shows that $\underline{\mathrm{Ext}}^1_P(I_Z, \mathcal{O}_P(-1)) \cong \underline{\mathrm{Ext}}^1_P(I_Y, \mathcal{O}_P(-1)) \simeq \mathcal{O}_Y(x)$.

(5.2) Let $0 \neq \xi \in \mathrm{Ext}^1_P(I_Z, \mathcal{O}_P(-1)) = H^0(\mathcal{O}_Y(x))$ correspond to an extension

$$0 \to \mathcal{O}_{\mathbb{P}3}(-1) \to E \to I_Z \to 0 .$$

Dualizing, we get an exact sequence

$$0 \to \mathcal{O}_P(-1) \to E^\vee(-1) \to \mathcal{O}_P \xrightarrow{\varphi} \underline{\mathrm{Ext}}^1(I_Z, \mathcal{O}_P(-1)) \to \underline{\mathrm{Ext}}^1(E, \mathcal{O}_P(-1)).$$

Almost per definition of the correspondence between extensions and $\mathrm{Ext}^1$-groups, the map $\varphi$ is given by the section $\xi$. This enables us to compute the Chern classes of $E^{\vee\vee} = E^\vee(-1)$.

Consider the following diagram, where $\bar{\varphi} \circ p = \varphi$ and $p$ is the canonical map:

$$\begin{array}{ccccc}
 & & & & 0 \\
 & & & & \downarrow \\
 & & & & \mathcal{O}_L(1) \\
 & & & & \downarrow \\
\mathcal{O}_{\mathbb{P}3} & \xrightarrow{p} & \mathcal{O}_Y & \xrightarrow{\bar{\varphi}} & \mathcal{O}_Y(x) = \underline{\mathrm{Ext}}^1(I_Z, \mathcal{O}_P(-1)) \\
 & & \downarrow & & \downarrow \\
 & & \mathcal{O}_L & \xrightarrow{\bar{\bar{\varphi}}} & \mathcal{O}_L(x) \\
 & & \downarrow & & \downarrow \\
 & & 0 & & 0
\end{array}$$

(5.3) <u>Lemma</u>: If $\bar{\bar{\varphi}} \neq 0$, then $\bar{\varphi}$ is injective.

Proof: Since $Y$ is Cohen-Macaulay, it suffices to consider the generic point of $L$. But there, $\bar{\bar{\varphi}}(1)$ is invertible, and since $\mathcal{O}_Y \to \mathcal{O}_L$ has nilpotent kernel, also $\bar{\varphi}(1)$ is invertible.

Q.E.D.

(5.4) Proposition: Let $E$ be a torsion-free sheaf corresponding to a closed point of the boundary $M - M^o = C_1 \cup C_2$.

(i) If $E \in C_2$, then $c_1(E^{\vee\vee}) = -1$, $c_2(E^{\vee\vee}) = 1$ and $c_3(E^{\vee\vee}) = 1$.

(ii) If $E \in C_1 - C_2$, then $c_1(E^{\vee\vee}) = -1$ and $c_2(E^{\vee\vee}) = 2$.
If $E$ corresponds to a Z-curve of the x-th kind, then $c_3(E^{\vee\vee}) = 2x$.

(iii) In any case, $E^{\vee\vee}$ is stable.

Proof: First note that $E \in C_2 \Leftrightarrow \bar{\bar{\varphi}} = 0$. In this case, the map $\bar{\varphi}$ factors through $\mathcal{O}_L(1)$, so the image of $\varphi$ is isomorphic to $\mathcal{O}_L$. From the resulting sequence

$$0 \to \mathcal{O}_P(-1) \to E^{\vee}(-1) \to \mathcal{O}_P \xrightarrow{\varphi} \mathcal{O}_L \to 0$$

we can easily compute the Chern classes. This proves (i). For (ii), (5.3) shows that the image of $\varphi$ is isomorphic to $\mathcal{O}_Y$. From there we compute the Chern classes the same way as above. (iii) We can see it directly from the proof of (i) and (ii), or note that the definition of stability easily implies $E$ stable $\Rightarrow E^{\vee\vee}$ stable, since they have the same $c_1$.

Q.E.D.

Remark: For any stable reflexive rank-2 sheaf with $c_1 = -1$, one can show [2,8.2] the inequality $0 \leq c_3 \leq c_2^2$. This, together with the congruence $c_1c_2 \equiv c_3 \pmod 2$ shows a priori that the Chern classes obtained by (5.4) are the only possible ones.

(5.5) Let us conclude by giving some examples of stable, torsion-free sheaves which are not limits of vector bundles. Some examples are already provided by (4.6). Here are some others:

(i) Let $X$ be any closed subscheme of $P$ with Hilbert polynomial $\chi(\mathcal{O}_X(l)) = 2l+3$, for example two disjoint lines plus one isolated or embedded point, or a conic plus two isolated or embedded points. Let $E$ be any non-trivial extension of $I_X$ by $\mathcal{O}_P(-1)$. Then $E$ is a stable, torsion-free sheaf with $c_1 = -1$, $c_2 = 2$ and $c_3 = 0$. If $X$ is two disjoint lines plus a point, we get a family of dimension 15. If $X$ is a conic plus two points, we get a family of dimension 19. The corresponding components of $M(-1,2,0)$ intersect $M^o$ in $C_1$.

(ii) Let $Y$ be the union of two disjoint lines, and construct a stable, reflexive sheaf $F$ as an extension $I_Y$ by $\mathcal{O}_P(-1)$. Then $c_1(F) = -1$, $c_2(F) = c_3(F) = 2$. Since $Y$ is not plane, $F(1)$ has a unique section, vanishing precisely along $Y$. Take a point $p \in P - Y$ and a surjection $F \to k(p) \to 0$, and let $E$ be the kernel. Then $E$ is stable, torsion-free with $c_1(E) = -1$, $c_2(E) = 2$ and $c_3(E) = 0$. Since the only section of $F$ is nonzero at $p$, we see that $H^o(E(1)) = 0$. This gives a family of dimension 14.

References.

[1] Hartshorne, R.: Stable vector bundles of rank 2 on $\mathbb{P}^3$, Math. Ann. 238 (1978) 229-280.

[2] Hartshorne, R.: Stable reflexive sheaves, Preprint (1980).

[3] Hartshorne, R. and Sols, I.: Stable rank 2 vector bundles on $\mathbb{P}^3$ with $c_1 = -1$, $c_2 = 2$. Preprint (1980).

[4] Maruyama, M: Moduli of stable sheaves II, J. Math. Kyoto Univ. 18 (1978) 557-614.

[5] Wever, G.P.: The moduli of a class of rank 2 vector bundles on projective 3-space, thesis, Univ. Calif. Berkeley (1977).

Jose Meseguer and
Ignacio Sols
University of California, Berkeley
Dept. of Mathematics
Berkeley, CA 94720
USA

Stein Arild Strømme
Matematisk Institutt
P.O. Box 1053 - Blindern
Oslo 3
NORWAY

# THE DIRECTIONAL DERIVATIVE PROBLEM

Bengt Winzell

## INTRODUCTION.

A classical problem in mathematics and mathematical physics has been to solve Laplace´s equation with a boundary condition in the form $\frac{\partial u}{\partial \ell} = f$ where $\ell$ is a non-vanishing field. The Neumann problem, for which $\ell$ has to be along the normal, is of this kind, and it is well known how an integral equation for a single layer density leads to an existence theory with estimates for the solution in terms of the data.

In his great works on celestial mechanics Poincaré raised questions about the origin of tidal waves. Then he faced a boundary value problem of the form above and with an oblique direction for the derivative. Since $\ell$ was no longer normal, the single layer approach did not lead to

an equation with a compact integral operator. It took some 30 years until Giraud was able to give the solution in case $\ell$ had a non-vanishing normal component. Oblique boundary conditions are also present in certain problems of geodesy (the Molodensky problem solved by Hörmander) and in engineering applications as for cooling systems etc. (See Amundsen-Ramkrishna [2].)
As is well known from a priori estimates of Agmon, Douglis and Nirenberg [1], as long as $\ell$ never becomes tangential to the boundary, the problem is elliptic: There are uniqueness, existence and regularity similar to what one gets for the Neumann problem.

To the author's knowledge there are very few examples of physical situations where an oblique derivative along a field with a some-times tangential direction occurs in a natural way. However, in ground water flow theory, the model due to Baiocchi [3], in which the pressure in the liquid is obtained in terms of the derivative $\frac{\partial w}{\partial z}$ of a potential function $w$, is of that kind. In fact, one has to find a solution of $\Delta w \in \mathcal{H}(w)$ with mixed boundary conditions, among which $\frac{\partial w}{\partial z}$ has a given value at those parts of the boundary where the pressure is known beforehand. Thus at all points where the limiting surfaces for the flow has a vertical segment we have a problem which degenerates into a situation in which the elliptic estimates are no longer true. We will presently, mainly by observing some examples, see why this is so.

The two-dimensional case.

The oblique derivative in plane domains never looses its elliptic nature. This can easily be made plausible since locally every harmonic function $u$ has a conjugate harmonic $v$ such that $u + iv$ is holomorphic. There is a certain isomorphism between $u$ and $v$, and because of the Cauchy-Riemann equations a tangential derivative $\frac{\partial u}{\partial \ell}$ corresponds to a normal differentiation $\frac{\partial v}{\partial \ell^*}$.
However, we can be very precise, since in the case of a simply connected domain there is an almost explicit representation of the solutions (cf. the Lienard solution in [13]): The classical formulas have been derived in many textbooks. See for instance the very direct solution in Hörmander [10], p 266, or the presentation in Bitsadze [4]. Since we want to generalize, we will use a few lines here to do the calculations.
By means of a conformal mapping we can always obtain a situation in which the domain is the unit disc, and if the original domain is smooth the orientation of the field on the boundary, such as the occurrence of a tangential derivative, is the same after the transformation. We thus consider the problem

$$(P)\begin{cases} \Delta u = 0 & \text{for} \quad |z| < 1 \\ a\dfrac{\partial u}{\partial x} + b\dfrac{\partial u}{\partial y} = f & \text{when} \quad |z| = 1 \end{cases}$$

where $a^2 + b^2 = 1$ and $z = x + iy$. The variation of the argument function $\arg(a + ib)$ in one loop around the circle is an integer multiple $p$ of $2\pi$, and hence $\varphi(\theta) = i\cdot\log(a(\theta) + ib(\theta)) + p\cdot\theta$ is a smooth function on $|z| = 1$. This represents the boundary values of the real part of a holcmorphic function $\Phi(z)$ in $|z| < 1$ which is smooth up to the boundary. In fact, we can take

$$\Phi(z) = \frac{1}{2\pi}\cdot\int_0^{2\pi} \frac{e^{i\theta} + z}{e^{i\theta} - z}\,\varphi(\theta)\,d\theta.$$

Furthermore, $w(z) = \frac{\partial u}{\partial x} - i\frac{\partial u}{\partial y}$ is holomorphic in $|z| < 1$ if u is harmonic, and the boundary condition in (P) now means that

$$f = a\frac{\partial u}{\partial x} + b\frac{\partial u}{\partial y} = \operatorname{Re}\,((a + ib)\cdot w) =$$

$$= \operatorname{Re}\,[e^{-\Phi(z)}\cdot z^p\cdot w]\cdot\exp\,(\operatorname{Im}\Phi(z))$$

at $|z| = 1$. Hence we have reduced (P) to the equivalent problem of finding a holomorphic function $w$ for which

$$(P^*) \qquad \operatorname{Re}\,[z^p\cdot w] = f.$$

When $p > 0$, $z^p\cdot w$ is holomorphic. Thus it must be of the form

$$\frac{1}{2\pi}\cdot\int_0^{2} \frac{e^{i\theta} + z}{e^{i\theta} - z}\,f(\theta)\,d\theta \;+\; i\,Ç$$

with a real constant $\varsigma$. However, if we want a holomorphic $w$ it must be possible to divide by $z^p$ in this expression for some choice of $\varsigma$. This requires that $f$ satisfies the $2p - 1$ conditions

$$\int_0^{2\pi} f(\theta)\, d\theta = \dots\dots = \int_0^{2\pi} e^{i(p-1)\theta} f(\theta)\, d\theta = 0.$$

Hence the co-dimension of the range of boundary values $\frac{\partial u}{\partial \ell}$ is $2p - 1$. But $u$ is uniquely determined from $w$, and thus from $f$, as for additional constants, so the index of the problem is $2(p - 1)$.

When on the other hand $p \leq 0$ we get

$$z^p \cdot w = \frac{1}{2\pi} \cdot \int_0^{2\pi} \frac{e^{i\theta} + z}{e^{i\theta} - z} f(\theta)\, d\theta + z^p \cdot P(z)$$

where $P(z)$ is any polynomial of degree $2|p|$ for which $\text{Re}\,(z^p P(z)) = 0$ when $|z| = 1$. The only possibilities are

$$\begin{aligned} (1) \qquad z^p \cdot P(z) = i \cdot [\varsigma_0 + \varsigma_1 (z + z^{-1}) + \dots + \\ + \varsigma_{|p|} (z^{-p} + z^p)] + \varsigma_1' (z - z^{-1}) + \dots + \\ + \varsigma_{|p|}' (z^{-p} - z^p) \end{aligned}$$

with real coefficients $\varsigma_0, \dots, \varsigma_{|p|}, \varsigma_1', \dots \varsigma_{|p|}'$. Hence we have a kernel with dimension $2|p| + 1$ for $(P^*)$ and $2|p| + 2$ for $(P)$. Thus we again get the index $2(p - 1)$.

The reason for us to give this exposition was partially to point out that in case $p > 1$, the typical picture

of the behaviour of $\ell$ is

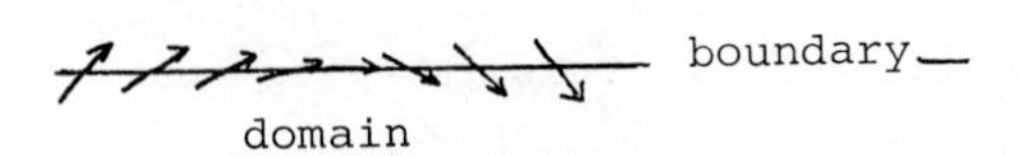


and then we have a minor lack of existence: f has to fulfil a finite number of compatibility conditions, while on the other hand, when $p \leq 0$, the picture is

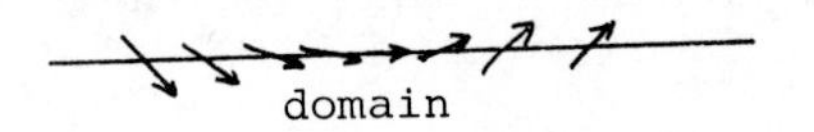


and we have a certain lack of uniqueness: There is a finite-dimensional kernel.

It is this very mild degree of deficiency which in the higher dimensional case grows worse into the non-solvability (with infinite-dimensional co-kernel) or non-uniqueness (with infinite-dimensional kernel) which was first reported by Hörmander [11]. To anticipate the reformulation of (P) due to Maljutov [14] and Egorov-Kondrat'ev [6] which resulted in a well posed problem, we make a generalization of the solution given above:

PROPOSITION 1: *Assume that* $p \leq 0$. *Then the solution* u *of* (P) *is uniquely determined by the data* f *and its prescribed values at the* $2|p| + 2$ *points on the unit circle where* $z^p$ *is tangential*.

The proof of this can be given along elementary lines and will not be reproduced here. It is enough to point out that the integration of $w$ obtained from (1) results in an expression with $2|p| + 2$ real parameters to be determined from the values of $u$, prescribed at the points of tangency.

We also have the follwing result which follows from the explicit construction of singular solutions $u_\upsilon$, proportional to $\arg (z - z_\upsilon)$ where $z_\upsilon$ are the solutions of $z^{2(p-1)} = -1$ for which $\frac{\partial u_\upsilon}{\partial \ell}$ are smooth on the circle.

PROPOSITION 2: *Assume that $p > 0$. Then, unless $f$ satisfies the $2p - 1$ conditions*

$$\int_0^2 e^{ik\theta} f(\theta)\, d\theta = 0, \quad k = 0,\ldots,p-1,$$

*one has to accept that a solution of (P) has jumps at the $2p - 2$ points where the field $z^p$ is tangential to the boundary. If one does allow for these jump singularities, then there is a solution of (P), unique as for additional constants, for every $f$ satisfying one single compatibility condition.*

The two propsitions say that if one modifies the regularity demands of the solutions or prescribes values at the points of tangency, then the oblique derivative problem is of *index zero*.

Two three-dimensional examples.

Consider the problem

$$\begin{cases} \Delta u = 0 & \text{in the unit ball of } \mathbb{R}^3 \\ \frac{\partial u}{\partial z} = f & \text{on the boundary}^{1)} \end{cases}$$

Obviously $w = \frac{\partial u}{\partial z}$ is harmonic in the ball and has boundary values $f$. Hence any solution $u$ must have the representation

$$u(x,y,z) = u(x,y,0) + \int_0^z w(x,y,\zeta)\, d\zeta$$

in which $v(x,y)$, defined to be $u(x,y,0)$, has to be a solution of

$$\Delta v = -\frac{\partial w}{\partial z}(\cdot,\cdot,0)$$

in the unit disc $x^2 + y^2 < 1$. Since the equation for $v$ can be solved for any boundary values on the equator $x^2 + y^2 = 1$, $z = 0$, we see that the kernel of the problem above consists of as many elements as there are (smooth) functions on the circle. At the same time, however, we have proved that the modified problem

$$\begin{cases} \Delta u = 0 & \text{in the ball} \\ \frac{\partial u}{\partial z} = f & \text{on the boundary} \\ u = h & \text{on the equator} \end{cases}$$

---

1) Now $x$, $y$ and $z$ are real coordinates in $\mathbb{R}^3$.

is uniquely solvable for all (smooth) data $f$ and $h$.
It is easy to check that the equator is precisely the manifold at which the field $\hat{\partial}$ of differentiation is tangential to the boundary.
Thus when the direction field changes its orientation according to the picture

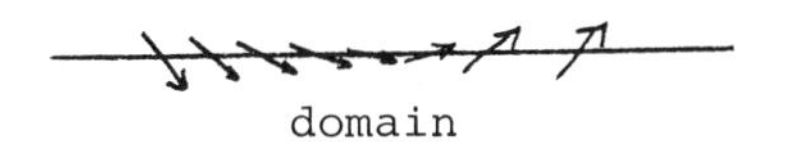


then a well posed oblique derivative problem requires information on the the values of $u$ along the manifold of tangency.
To get a situation in which the dual picture

is relevant we can solve the same problem in the outer domain, requiring that the solutions have a limit at infinity. (In fact, if the condition is that the limit is zero, then one can prove uniqueness for smooth solutions.)
Again $w = \frac{\partial u}{\partial z}$ is harmonic and has boundary values $f$ on the sphere. However, the integration of $w$ along lines, parallel to the $z$-axis and contained in the outer domain, does not always give the net result zero which is necessary not only for the condition of a limit at infinity, but

also for the very possibility to reconstruct u from w. This puts restrictions on the kind of data f which can be given. It is known from a general theorem of Hörmander [11] that these restrictions are not only in the shape of a finite number of conditions. On the other hand, as Maljutov proved ([14] and [15]), if u can have jumps along the equator, then there are unique solutions for all smooth f.

Thus, when the field $\ell$ has a behaviour according to the second of our pictures in this section, _one must accept jump singularities_ along the manifold of tangency _to have existence_.

## A generalized maximum principle and the uniqueness theory.

For the problem with non-tangential derivative boundary condition the basic tool in uniqueness theory is the maximum principle in a form like the following (see Protter-Weinberger [21]):

PROPOSITION 3: _Let u be harmonic and $C^1$ up to the $C^2$-boundary of a domain $\Omega$. Then u takes on its maximum at a point $p \in \partial\Omega$ and $\frac{\partial u}{\partial \ell}(p) > 0$ for any outward direction $\ell$ unless u is constant._

To discuss a generalized version, or rather a generalized _indirect_ version in case $\ell$ may be tangential we introduce some notation. Let $\ell = \alpha\hat{n} + X$ be the decomposition into a tangential component $X$ and a component measured along the outer normal. Reserve the name $Z$ for the subset of the boundary in which $\ell$ is tangential, i.e. $Z = \{\alpha = 0\}$. Then through every point $p \in Z$ there passes a non-trivial integral curve of $X$, which can be represented as $s \rightsquigarrow x_p(s)$ where $\dot{x}_p = X \circ x_p$ and $x_p(0) = p$. Now we can formulate

LEMMA 4: *Assume that the harmonic function $u$ is not constant, has a maximum at $p$, and is $C^1$ up to the boundary where it satisfies $\frac{\partial u}{\partial \ell} = 0$. Then $p \in Z$ and neither (i) nor (ii) below can be true for $p$:*

(i) *The set $\{s > 0: x_p(s) \notin Z\}$ is non-empty with infimum $s^+$, and for some $\delta > 0$, $\alpha(x_p(s)) \leq 0$ when $s^+ < s < s^+ + \delta$.*

(ii) *The set $\{s < 0: x_p(s) \notin Z\}$ is non-empty with supremum $s^-$, and for some $\delta > 0$, $\alpha(x_p(s)) \geq 0$ when $s^- - \delta < s < s^-$.*

A detailed proof of this is given in [24], Section 2, where also some examples can be found as well as certain extensions. However, the lemma is readily demonstrated: Since $u$ is constant along the arc consisting of $x_p(s)$ for $s \in [0,s^+]$ or $[s^-,0]$, Proposition 3 implies that $\frac{\partial u}{\partial n} > 0$ in a neighbourhood of that arc. It then suffices to write $\frac{\partial u}{\partial \ell} = 0$ as $\frac{\partial u}{\partial X} = -\alpha\frac{\partial u}{\partial n}$ and draw the conclusions from the sign of $\frac{\partial u}{\partial X}$.

As special cases we obtain

1) If $\alpha \geq 0$ everywhere and $Z$ does not contain any closed integral curve of $X$, or any such curve of infinite length, then the only harmonic functions which satisfy $\frac{\partial u}{\partial \ell} = 0$ are constant.

2) For smooth solutions Lemma 4 implies uniqueness (as for constants) in case $\alpha$ changes sign from plus to minus along $X$ but not in case $\alpha$ has the opposite behaviour. In fact, one of the previously given examples shows that there can be no uniqueness in the latter case.

We add two more points, which are not corollaries of the lemma, but are relevant to the matter of uniqueness:

3) As we have already mentioned, it is in general necessary to modify the statement of the problem. This calls for an extension of the lemma. The case in which $\alpha$ varies in such a way that one should prescribe the values of $u$ presents no difficulties. In the other case we have developed a uniqueness theory for solutions which have jumps along a submanifold in the boundary, contained in $Z$. See [26], Section 4, where the appropriate barrier functions are constructed.

4) Our uniqueness results in higher dimensions are of local nature. It is still an open question if there is a global characterization in analogy with the two-dimensional situation: There only the total winding number $p$ decides at how many points the value of the solution should be prescribed.

One way of proving existence.

We have already solved a typical oblique derivative problem by the decomposition

$$u = v + \int_0^z w$$

for the case of the three-dimensional ball and $\ell = \hat{z}$. It is not difficult to see that this method carries over to the case when $\ell$ has an extension $L$ to all of the domain such that

(i) $\Delta \frac{\partial u}{\partial L} = \frac{\partial}{\partial L} \Delta u$ for all smooth functions $u$

(ii) The domain $\Omega$ is a union of connected integral curves of $L$, all intersecting a smooth manifold $\Lambda$ with boundary $\partial\Lambda$ once and under a uniformly positive angle.

In fact, then we let $w$ be the harmonic function in $\Omega$ which equals $f$ on $\partial\Omega$ and put

$$(2) \qquad u(p) = v(p') + \int_{p'}^{p} w\, ds$$

where $p$ and $p'$ are on the same integral curve of $L$,

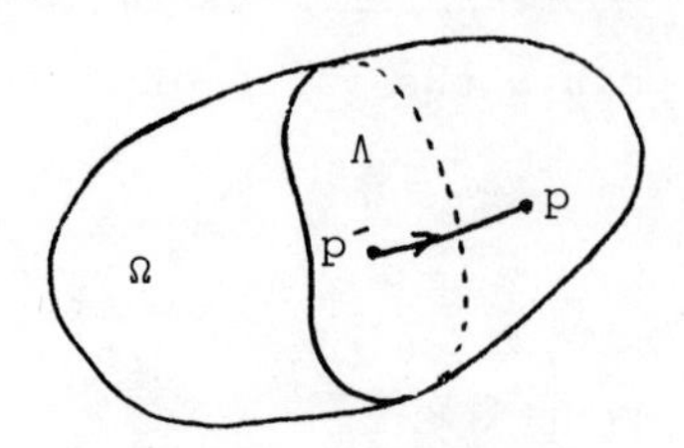


$p' \in \Lambda$ and the line integral is taken along the curve. If $v$ is the solution of a certain elliptic equation in $\Lambda$ with boundary values $h$ on $\partial\Lambda$ the formula gives a solution of

$$(3) \quad \begin{cases} \Delta u = 0 & \text{in } \Omega \\ \dfrac{\partial u}{\partial \ell} = f & \text{on } \partial\Omega \\ u = h & \text{on } \partial\Lambda. \end{cases}$$

In general, however, (i) is not satisfied, but the commutator $\frac{\partial}{\partial L}\Delta - \Delta\frac{\partial}{\partial L}$ is a second order operator. We now will describe a way of handling this more general case. First we use (2) to reduce the problem (3) to the form

$$\begin{cases} \Delta u = g & \text{in } \Omega \\ \dfrac{\partial u}{\partial \ell} = 0 & \text{on } \partial\Omega \\ u = 0 & \text{on } \partial\Lambda. \end{cases}$$

This is achieved by taking $w$ with boundary values $f$, and $v$ in $\Omega$ with boundary values $h$. At last we subtract in (3) to get the new formulation.

Now we observe that the application of the Laplacian to (2) gives a new identity in which the commutator $\mathcal{P} = \frac{\partial}{\partial L}\Delta - \Delta\frac{\partial}{\partial L}$ plays an important role and $\Delta'$ denotes the elliptic operator in $\Lambda$ which is induced by the action of the Laplacian on functions in $\Lambda$ which are constant along integral curves of $L$:

$$\Delta u\ (p) = \Delta' v\ (p') + \int_{p'}^{p} \Delta w\ ds + \int_{p'}^{p} \mathcal{P} u\ ds + \text{lower order terms in } w \text{ and } v.$$

Thus, if $w$ is defined as the solution of Poisson's equation $\Delta w = \frac{\partial g}{\partial L}$ in $\Omega$ with zero Dirichlet data, and $v$ satisfies $\Delta' v = g$ in $\Lambda$ with zero voundary values on $\partial\Lambda$, then the map $\mathcal{A}$: $g \sim \Delta u$ (via the representation (2)) is of the form

$$\mathcal{A} = \text{identity} + \mathcal{T} + \mathcal{K}$$

where $\mathcal{T}: g \sim \int \mathcal{P} u\ dL$ has norm less than one in some appropriate function space, provided the integration along the integral curves of $L$ are sufficiently short. The remainder term $\mathcal{K}$ can be considered as a compact operator as it is built from lower order terms only. Thus existence follows from standard application of the

Riesz-Schauder theory for compact operators.

For details and extensions we refer to [26] and [28].

In fact, it is possible to prove the

EXISTENCE THEOREM:

*Assumptions:* Let the boundary be the union of two closed sets $\partial\Omega^+$ and $\partial\Omega^-$ such that $\alpha \geq 0$ in $\partial\Omega^+$, $\alpha \leq 0$ in $\partial\Omega^-$ and $\partial\Omega^+ \cap \partial\Omega^-$ is a smooth manifold $Z_o$ of co-dimension one in $\partial\Omega$.

Assume that $X$ makes a uniformly positive angle with $Z_o$ and let $Z_o = Z_o^+ \cup Z_o^-$ such that $X$ points from $\partial\Omega^-$ to $\partial\Omega^+$ in $Z_o^+$ and in the other direction in $Z_o^-$.

Finally, assume that no integral curve of $X$ in $Z$ is closed or has infinite length. Then,

*Conclusions:* for every $f \in C^{1+\lambda}(\partial\Omega)$, $g \in C^{\lambda}(\bar{\Omega})$ and $h \in C^{2+\lambda}(Z_o^+)$, such that $f$ is $C^{2+\lambda}$ and $g$ is $C^{1+\lambda}$ near $Z$, there exists a unique, bounded solution $u \in$ $\in C^{2+\lambda}(\bar{\Omega} \setminus Z_o^-)$ of $\Delta u = g$ such that $\frac{\partial u}{\partial \ell} = f$ on $\partial\Omega \setminus Z_o^-$ and $u = h$ on $Z_o^+$.

Remark: The theorem is true also if $Z_o^-$ is empty. If, on the other hand, $Z_o^+ = \emptyset$, then of course the data $h$ are not introduced, and the solutions are unique as for constants, but exist only when one condition on the data is satisfied.

The question of best regularity: Sub-elliptic estimates.

The existence result above indicates that in order to have $u \in C^{2+\lambda}$ one has to give $\frac{\partial u}{\partial \ell} \in C^{2+\lambda}$, i.e. there could be a loss of one derivative compared with elliptic regularity. In fact, there are examples showing that this is generally the case. On the other hand, it was stated by Egorov [5] that in case the normal component $\alpha$ of $\ell$ vanishes only to a finite order along integral curves of X, some higher regularity can be gained. In fact, the result is essentially that if $|\alpha(x_p(s))| \geq c_o \cdot |s|^k$ when $p \in Z$, then

$$\frac{\partial u}{\partial \ell} \in C^{1+\lambda}(\partial\Omega) \Rightarrow u \in C^{1+\lambda+(k+1)^{-1}}(\overline{\Omega}).$$

(Here $s \sim x_p(s)$ is the representation of integral curves of X introduced above.)
We have proved this in [25] and [27], using a representation (2) for u which then requires estimates of $\overset{?}{u}$ only in the interior (where we can apply interior Schauder estimates, accounting for the distance to the boundary) and the integral of such interior estimates of $\frac{\partial u}{\partial L}$ (satisfying an identity $\Delta\frac{\partial u}{\partial L} = \mathcal{P}u$) in intermediate regions.

Concluding remarks.

The method of treating oblique derivative problems has here been based on curve geometric constructions. Before we end it should be pointed out that the problem can be reduced to one which is naturally embedded in a larger class of problems for degenerating pseudo-differential operators. In this setting the matter has been examined by Hörmander, Egorov, Eskin, Maz´ja-Panejah, Sjöstrand, Melin-Sjöstrand, Taira and others. We have collected a few references at the end of this note.
Another way of attacking the problem is to formulate a weak identity with carefully chosen weighting functions. This was done by Maz´ja [17], who also was able to let $\ell$ be tangential not only to $\partial\Omega$, but also to $Z$ (in this case equal to the thin set $Z_0$).

Finally we remark that though we have everywhere referred to the Laplacian, and to not very precise smoothness of the domain and the field, almost everything goes through for a uniformly elliptic second order operator, and we only have to impose regularity to a degree which is natural for estimates of the Agmon-Douglis-Nirenberg type to be true. It turns out that the index of the problem is zero under all these circumstances.

There are still open questions, for instance concerning

global characterizations. Work in that direction has been performed by Bitsadze, and along his lines Janusjauskas has obtained some results which, however, with a few exceptions are more of the kind of examples.

## REFERENCES


[1] S.Agmon, A.Douglis, L.Nirenberg: Estimates near the boundary for solutions of elliptic partial differential equations satisfying general boundary conditions. Comm Pure Appl Math 12(1959),623-727.

[2] N.R.Amundson, D.Ramkrishna: Boundary value problems in transport with mixed or oblique derivative boundary conditions. Chem Eng Sci 34(1979),301-308.

[3] C.Baiocchi: Su un problema di frontiera libera connesso a questioni di idraulica. Ann Mat Pura Appl(IV) 92(1972),107-127.

[4] A.V.Bitsadze: Boundary value problems for second order elliptic equations. North-Holland, Amsterdam 1968

[5] Ju.V.Egorov: Sub-elliptic pseudo-differential operators. Sov Math Dokl 10(1969),1056-1059.

[6] Ju.V.Egorov, V.A.Kondrat'ev: A problem with an oblique derivative. Sov Math Dokl 7(1966),1271-1273.

[7] -"- : The oblique derivative problem. Math USSR Sbornik 7(1969),148-176.

[8] G.I.Eskin: Elliptic pseudo-differential operators with a degeneracy of the first order in the space variables. Trans Moscow Math Soc 25(1971),91-130.

[9] G.Giraud: Équations à intégrales principales d'ordre quelconque. Ann Sci Ecole Norm Sup (3) 53(1936),1-40.

[10] L.Hörmander: Linear partial differential operators. Springer-Verlag 1964

[11] -"- : Pseudo-differential operators and non-elliptic boundary problems. Ann Math 83(1966),129-209.

[12] A.I.Janusjauskas: Global methods of studying the oblique derivative problem for elliptic equations. Diff Uravn 12(1976),159-171.

[13] A.Liénard: Problème plan de la dérivée oblique dans la théorie du potentiel. J École Polytech 144(1938), 35-96,97-158,177-226; 145(1939),55-84,88-137.

[14] M.B.Maljutov: Problem of inclined derivative in three dimensions. Sov Math Dokl 8(1967),87-90.

[15] -"- : On the Poincaré boundary value problem. Trans Moscow Math Soc 20(1969),173-204.

[16] V.G.Maz´ja, B.P.Panejah: Degenerate elliptic pseudo-differential operators on a smooth manifold without boundary. Funct Anal Appl 3(1969),159-160

[17] V.G.Maz´ja: On a degenerating problem with directional derivative. Math USSR Sbornik 16(1972),429--469.

[18] A.Melin, J.Sjöstrand: Fourier integral operators with complex phase functions and parametrix for an interior boundary problem. Comm Part Diff Equ 1(1976) 313-400.

[19] B.P.Panejah: On a problem with oblique derivative. Sov Math Dokl 19(1978),1568-1572.

[20] H.Poincaré: Leçons de mécanique céleste: Théorie de marées. Gauthiers-Villars, Paris 1910.

[21] M.H.Protter, H.F.Weinberger: Maximum principles in differential equations. Prentice Hall, New Jersey 1967.

[22] J.Sjöstrand: Operators of principle type with interior boundary conditions. Acta Math 130(1973),1--51.

[23] K.Taira: Sur le problème de la dérivée oblique. Ark f Matematik 17(1979),177-191.

[24] B.Winzell: The oblique derivative problem I. Math Ann 229(1977),267-278.

[25] -"- : Sub-elliptic estimates for the oblique derivative problem. Math Scand 43(1978),169-176.

[26] -"- : The oblique derivative problem II. Ark f Matematik 17(1979),107-122.

[27] -"- : The sub-ellipticity with best exponent for the oblique derivative problem. Report LiTH-MAT--R-79-11, Linköping University, 1979.

[28] -"- : A boundary value problem with an oblique derivative. To appear.

Department of Mathematics
Linköping University
S-581 83 Linköping, Sweden